A·N·N·U·A·L E·D·I·T·I·O·N·S

Developing World *04/05*

Fourteenth Edition

EDITOR
Robert J. Griffiths
University of North Carolina at Greensboro

Robert J. Griffiths is associate professor of political science and director of the International Studies Program at the University of North Carolina at Greensboro. His teaching and research interests are in the fields of comparative and international politics, and he teaches courses on the politics of development, African politics, international law and organization, and international political economy. His publications include articles on South African civil/military relations, democratic consolidation in South Africa, and developing countries and global commons negotiations.

McGraw-Hill/Dushkin
530 Old Whitfield Street, Guilford, Connecticut 06437

Visit us on the Internet
http://www.dushkin.com

Credits

1. **Understanding the Developing World**
   Unit photo—© 2004 by PhotoDisc, Inc.
2. **Political Economy and the Developing World**
   Unit photo—© 2004 by Sweet By & By/Cindy Brown.
3. **Conflict and Instability**
   Unit photo—© United Nations photo.
4. **Political Change in the Developing World**
   Unit photo—United Nations photo.
5. **Population, Development, Environment, & Health**
   Unit photo—United Nations photo.
6. **Women and Development**
   Unit photo—World Bank photo.

Copyright

Cataloging in Publication Data
Main entry under title: Annual Editions: Developing World. 2004/2005.
1. Developing World—Periodicals. I. Griffiths, Robert J., *comp*. II. Title: Developing World.
ISBN 0–07–286062–6   658'.05   ISSN 1096–4215

Fourteenth Edition

Cover image © 2004 PhotoDisc, Inc.
Printed in the United States of America   1234567890BAHBAH54   Printed on Recycled Paper

Editors/Advisory Board

Members of the Advisory Board are instrumental in the final selection of articles for each edition of ANNUAL EDITIONS. Their review of articles for content, level, currentness, and appropriateness provides critical direction to the editor and staff. We think that you will find their careful consideration well reflected in this volume.

To the Reader

In publishing ANNUAL EDITIONS we recognize the enormous role played by the magazines, newspapers, and journals of the public press in providing current, first-rate educational information in a broad spectrum of interest areas. Many of these articles are appropriate for students, researchers, and professionals seeking accurate, current material to help bridge the gap between principles and theories and the real world. These articles, however, become more useful for study when those of lasting value are carefully collected, organized, indexed, and reproduced in a low-cost format, which provides easy and permanent access when the material is needed. That is the role played by ANNUAL EDITIONS.

The developing world is home to the vast majority of the world's population. Because of its large population, increasing role in the international economy, frequent conflicts and humanitarian crises, and importance to environmental preservation, the developing world continues to be a growing focus of international concern. The developing world has also figured prominently in the recent protests over globalization and the September 11, 2001, terrorist attacks against the United States are tied to certain circumstances in developing countries.

Developing countries demonstrate considerable ethnic, cultural, political, and economic diversity, making generalizations about them difficult. Increasing differentiation further complicates our ability to understand developing countries and to comprehend the challenges of modernization, development, and globalization that they face. Confronting these challenges must take into account the combination of internal and external factors that contribute to the current circumstances throughout the developing world. Issues of peace and security, international trade and finance, debt, poverty, and the environment illustrate the effects of growing interdependence and the need for a cooperative approach to dealing with these issues. There is significant debate regarding the best way to address the developing world's problems. Moreover, the developing world's needs compete for attention on an international agenda that is often dominated by relations between the industrialized nations and more recently by the war on terrorism. Domestic concerns within the industrial nations also continue to overshadow the plight of the developing world.

This fourteenth edition of *Annual Editions: Developing World* seeks to provide students with an understanding of the diversity and complexity of the developing world and to acquaint them with the challenges that these nations confront. I am convinced that there is a need for greater awareness of the problems that confront the developing world and that the international community must make a commitment to address those problems, especially in the post-September 11, 2001, era. I hope that this volume contributes to students' knowledge and understanding and serves as a catalyst for further discussion.

Approximately two-thirds of the articles in this edition are new. I have chosen articles that I hope are both interesting and informative and that can serve as a basis for further student research and discussion. The units deal with what I regard as the major issues facing the developing world. In addition, I have attempted to suggest similarities and differences between developing countries, the nature of their relationships with the industrialized nations, and the differences in perspective regarding the causes of and approaches to the issues.

I would again like to thank McGraw-Hill/Dushkin for the opportunity to put together a reader on a subject that is the focus of my teaching and research. I would also like to thank those who have sent in the response forms with their comments and suggestions. I have tried to take these into account in preparing the current volume.

No book on a topic as broad as the developing world can be comprehensive. There are certainly additional and alternative readings that might be included. Any suggestions for improvement are welcome. Please complete and return the postage-paid article rating form at the end of the book with your comments.

Robert J. Griffiths
Editor

Contents

UNIT 1
Understanding the Developing World

Six selections examine how the developing world's problems are interrelated with world order.

Unit Overview   xvi

The concepts in bold italics are developed in the article. For further expansion, please refer to the Topic Guide and the Index.

UNIT 2
Political Economy and the Developing World

Ten articles discuss the impact that debt has on the developing world's politics and economy.

The concepts in bold italics are developed in the article. For further expansion, please refer to the Topic Guide and the Index.

UNIT 3
Conflict and Instability

The concepts in bold italics are developed in the article. For further expansion, please refer to the Topic Guide and the Index.

The concepts in bold italics are developed in the article. For further expansion, please refer to the Topic Guide and the Index.

UNIT 4
Political Change in the Developing World

Nine selections examine the innate problems faced by developing countries as they experience political change.

The concepts in bold italics are developed in the article. For further expansion, please refer to the Topic Guide and the Index.

UNIT 5
Population, Development, Environment, & Health

Five articles examine some of the effects that the developing world's growth has on
Earth's sustainability.

Unit Overview   166

The concepts in bold italics are developed in the article. For further expansion, please refer to the Topic Guide and the Index.

UNIT 6
Women and Development

Five articles discuss the role of women in the developing world.

The concepts in bold italics are developed in the article. For further expansion, please refer to the Topic Guide and the Index.

Topic Guide

This topic guide suggests how the selections in this book relate to the subjects covered in your course. You may want to use the topics listed on these pages to search the Web more easily.

On the following pages a number of Web sites have been gathered specifically for this book. They are arranged to reflect the units of this *Annual Edition.* You can link to these sites by going to the DUSHKIN ONLINE support site at *http://www.dushkin.com/online/.*

ALL THE ARTICLES THAT RELATE TO EACH TOPIC ARE LISTED BELOW THE BOLD-FACED TERM.

Afghanistan

13. Eyes Wide Open: On the Targeted Use of Foreign Aid

Africa

13. Eyes Wide Open: On the Targeted Use of Foreign Aid
33. The Many Faces of Africa: Democracy Across a Varied Continent

Agriculture

5. Why People Still Starve
9. Rich Nations' Tariffs and Poor Nations' Growth
15. The WTO Under Fire

AIDS

35. The Population Implosion
39. Withholding the Cure

Alternative visions of world politics

35. The Population Implosion
42. Women Waging Peace

Civil War

17. Market for Civil War

Communal conflict

17. Market for Civil War
18. Engaging Failing States
22. Mugabe's End-Game
25. Blaming the Victim: Refugees and Global Security
33. The Many Faces of Africa: Democracy Across a Varied Continent

Conditional lending

10. Unelected Government
11. The IMF Strikes Back
13. Eyes Wide Open: On the Targeted Use of Foreign Aid

Cultural values

1. The Great Divide in the Global Village
42. Women Waging Peace

Culture

4. Development as Poison: Rethinking the Western Model of Modernity

Debt

11. The IMF Strikes Back
31. New Hope for Brazil?

Democracy

26. Democracies: Emerging or Submerging?
27. Two Theories
30. Iran's Crumbling Revolution
32. Latin America's New Political Leaders: Walking on a Wire
33. The Many Faces of Africa: Democracy Across a Varied Continent

43. The True Clash of Civilizations

Developing world

33. The Many Faces of Africa: Democracy Across a Varied Continent

Development

1. The Great Divide in the Global Village
3. Institutions Matter, but Not for Everything
4. Development as Poison: Rethinking the Western Model of Modernity
6. Putting a Human Face on Development
8. Trading for Development: The Poor's Best Hope
12. Ranking the Rich
34. NGOs and the New Democracy: The False Saviors of International Development

Economic growth

6. Putting a Human Face on Development
7. The Free-Trade Fix

Economics and politics

33. The Many Faces of Africa: Democracy Across a Varied Continent

Education

40. Empowering Women

Environment

12. Ranking the Rich
36. Local Difficulties
37. A Dirty Dilemma: The Hazardous Waste Trade
38. 'Undoing the Damage We Have Caused'

Ethnic conflict

17. Market for Civil War
18. Engaging Failing States
22. Mugabe's End-Game
25. Blaming the Victim: Refugees and Global Security
33. The Many Faces of Africa: Democracy Across a Varied Continent

Failing states

18. Engaging Failing States
28. Not a Dress Rehearsal

Foreign aid

9. Rich Nations' Tariffs and Poor Nations' Growth
12. Ranking the Rich
13. Eyes Wide Open: On the Targeted Use of Foreign Aid
14. The Cartel of Good Intentions
41. Women & Development Aid

Foreign intervention

28. Not a Dress Rehearsal
29. One Country, Two Plans

World Wide Web Sites

The following World Wide Web sites have been carefully researched and selected to support the articles found in this reader. The easiest way to access these selected sites is to go to our DUSHKIN ONLINE support site at *http://www.dushkin.com/online/*.

AE: Developing World 04/05

The following sites were available at the time of publication. Visit our Web site—we update DUSHKIN ONLINE regularly to reflect any changes.

General Sources

Foreign Policy in Focus (FPIF): Progressive Response Index
http://fpif.org/progresp/index_body.html

This index is produced weekly by FPIF, a "think tank without walls," which is an international network of analysts and activists dedicated to "making the U.S. a more responsible global leader and partner by advancing citizen movements and agendas." This index lists volume and issue numbers, dates, and topics covered by the articles.

People & Planet
http://www.peopleandplanet.org

People & Planet is an organization of student groups at universities and colleges across the United Kingdom. Organized in 1969 by students at Oxford University, it is now an independent pressure group campaigning on world poverty, human rights, and the environment.

United Nations System Web Locator
http://www.unsystem.org

This is the Web site for all the organizations in the United Nations family. According to its brief overview, the United Nations, an organization of sovereign nations, provides the machinery to help find solutions to international problems or disputes and to deal with pressing concerns that face people everywhere, including the problems of the developing world, through the UN Development Program at *http://www.undp.org* and UNAIDS at *http://www.unaids.org*.

United States Census Bureau: International Summary Demographic Data
http://www.census.gov/ipc/www/idbsum.html

The International Data Base (IDB) is a computerized data bank containing statistical tables of demographic and socioeconomic data for all countries of the world.

World Health Organization (WHO)
http://www.who.ch

The WHO's objective, according to its Web site, is the attainment by all peoples of the highest possible level of health. Health, as defined in the WHO constitution, is a state of complete physical, mental, and social well-being and not merely the absence of disease or infirmity.

UNIT 1: Undertanding the Developing World

Africa Index on Africa
http://www.afrika.no/index/

A complete reference source on Africa is available on this Web site.

African Studies WWW (U. Penn)
http://www.sas.upenn.edu/African_Studies/AS.html

The African Studies Center at the University of Pennsylvania supports this ongoing project that lists online resources related to African Studies.

UNIT 2: Political Economy and the Developing World

Center for Third World Organizing
http://www.ctwo.org/

The Center for Third World Organizing (CTWO, pronounced "C-2") is a racial justice organization dedicated to building a social justice movement led by people of color. CTWO is a 20-year-old training and resource center that promotes and sustains direct action organizing in communities of color in the United States.

ENTERWeb
http://www.enterweb.org

ENTERWeb is an annotated meta-index and information clearinghouse on enterprise development, business, finance, international trade, and the economy in this age of cyberspace and globalization. The main focus is on micro-, small-, and medium-scale enterprises, cooperatives, and community economic development both in developed and developing countries.

International Monetary Fund (IMF)
http://www.imf.org

The IMF was created to promote international monetary cooperation, to facilitate the expansion and balanced growth of international trade, to promote exchange stability, to assist in the establishment of a multilateral system of payments, to make its general resources temporarily available under adequate safeguards to its members experiencing balance-of-payments difficulties, and to shorten the duration and lessen the degree of disequilibrium in the international balances of payments of members.

TWN (Third World Network)
http://www.twnside.org.sg/

The Third World Network is an independent, nonprofit international network of organizations and individuals involved in issues relating to development, the Third World, and North-South issues.

U.S. Agency for International Development (USAID)
http://www.info.usaid.gov

USAID is an independent government agency that provides economic development and humanitarian assistance to advance U.S. economic and political interests overseas.

The World Bank
http://www.worldbank.org

The International Bank for Reconstruction and Development, frequently called the World Bank, was established in July 1944 at the UN Monetary and Financial Conference in Bretton Woods, New Hampshire. The World Bank's goal is to reduce poverty and improve living standards by promoting sustainable growth and investment in people. The bank provides loans, technical

www.dushkin.com/online/

assistance, and policy guidance to developing country members to achieve this objective.

UNIT 3: Conflict and Instability

The Carter Center
http://www.cartercenter.org

The Carter Center is dedicated to fighting disease, hunger, poverty, conflict, and oppression through collaborative initiatives in the areas of democratization and development, global health, and urban revitalization.

Center for Strategic and International Studies (CSIS)
http://www.csis.org/

For four decades, the Center for Strategic and International Studies (CSIS) has been dedicated to providing world leaders with strategic insights on, and policy solutions to, current and emerging global issues.

Conflict Research Consortium
http://www.Colorado.EDU/conflict/

The site offers links to conflict- and peace-related Internet sites.

Institute for Security Studies
http://www.iss.co.za

This site is South Africa's premier source for information related to African security studies.

PeaceNet
http://www.igc.org/peacenet/

PeaceNet promotes dialogue and sharing of information to encourage appropriate dispute resolution, highlights the work of practitioners and organizations, and is a proving ground for ideas and proposals across the range of disciplines within the conflict-resolution field.

Refugees International
http://www.refintl.org

Refugees International provides early warning in crises of mass exodus. It seeks to serve as the advocate of the unrepresented—the refugee. In recent years, Refugees International has moved from its initial focus on Indochinese refugees to global coverage, conducting almost 30 emergency missions in the last 4 years.

UNIT 4: Political Change in the Developing World

Greater Horn Information Exchange (GHIE)
http://www.hydrosult.com/niledata/usaid/index.htm

The GHIE is a no-fee resource accessible via e-mail, Telnet, Gopher, and the World Wide Web. It expedites the sharing of site reports, fact sheets, activity summaries, data sets, scientific papers, and analyses on the countries of the Horn region of Africa.

Latin American Network Information Center—LANIC
http://www.lanic.utexas.edu

According to *Latin Trade*, LANIC is "a good clearing house for Internet-accessible information on Latin America."

ReliefWeb
http://www.reliefweb.int/w/rwb.nsf

ReliefWeb is the UN's Department of Humanitarian Affairs clearinghouse for international humanitarian emergencies.

World Trade Organization (WTO)
http://www.wto.org

The WTO is promoted as the only international body dealing with the rules of trade between nations. At its heart are the WTO agreements, the legal ground rules for international commerce and for trade policy.

UNIT 5: Population, Development, Environment, & Health

Earth Pledge Foundation
http://www.earthpledge.org

The Earth Pledge Foundation promotes the principles and practices of sustainable development—the need to balance the desire for economic growth with the necessity of environmental protection.

EnviroLink
http://envirolink.org

EnviroLink is committed to promoting a sustainable society by connecting individuals and organizations through the use of the World Wide Web.

Greenpeace
http://www.greenpeace.org

Greenpeace is an international NGO (nongovernmental organization) that is devoted to environmental protection.

Linkages on Environmental Issues and Development
http://www.iisd.ca/linkages/

Linkages is a site provided by the International Institute for Sustainable Development. It is designed to be an electronic clearinghouse for information on past and upcoming international meetings related to both environmental issues and economic development in the developing world.

Population Action International
http://www.populationaction.org

According to its mission statement, Population Action International is dedicated to advancing policies and programs that slow population growth in order to enhance the quality of life for all people.

The Worldwatch Institute
http://www.worldwatch.org

The Worldwatch Institute advocates environmental protection and sustainable development.

UNIT 6: Women and Development

African Women Global Network
http://www.ohio-state.edu/org/awognet/

African Women Global Network is a global organization that networks all men and women, organizations, institutions, and indigenous national organizations within Africa. The activities at this Web site target the improvement of the living conditions of women and children in Africa.

WIDNET: Women in Development NETwork
http://www.focusintl.com/widnet.htm

This site provides a wealth of information about women in development, including the Beijing '95 Conference, WIDNET statistics, and women's studies.

WomenWatch/Regional and Country Information
http://www.un.org/womenwatch/

The UN Internet Gateway on the Advancement and Empowerment of Women provides a rich mine of information.

We highly recommend that you review our Web site for expanded information and our other product lines. We are continually updating and adding links to our Web site in order to offer you the most usable and useful information that will support and expand the value of your Annual Editions. You can reach us at: *http://www.dushkin.com/annualeditions/*.

UNIT 1
Understanding the Developing World

Unit Selections

Key Points to Consider

- What factors account for the economic disparity both within developing countries and between them and the world's industrialized countries?

- How do immigration and trade policies in the industrialized countries affect the developing world?

- How can developing countries increase their influence at international conferences? Besides institutions, what other factors are crucial to development?

- What constitutes the Western model of development?

- Is the Western model of development transferable to the developing world?

- In what ways should the concept of development be broadened?

- Why does hunger persist in the developing world?

 **Links: www.dushkin.com/online/**
These sites are annotated in the World Wide Web pages.

Africa Index on Africa
http://www.afrika.no/index/

African Studies WWW (U. Penn)
http://www.sas.upenn.edu/African_Studies/AS.html

Understanding the diverse countries that make up the developing world has never been an easy task and it has become even more difficult as further differentiation among these countries has occurred. "Developing world" is a catch-all term that lacks precision and explanatory power. It encompasses a wide range of societies from traditional to modernizing. There is also controversy over what actually constitutes development. For some, it is economic growth or progress toward democracy; for others it involves greater empowerment and dignity. There are also differing views on why progress toward development has been uneven. The West tends to see the problem as stemming from institutional weakness and failure to embrace free-market principles. Critics from the developing world cite the legacy of colonialism and the international political and economic structures as the reasons for a lack of development. Lumping together the 100-plus nations that make up the developing world obscures the disparities in size, population, resources, forms of government, industrialization, distribution of wealth, ethnic diversity, and a host of other indicators that make it difficult to understand this large, diverse group of countries.

Despite their diversity, most nations of the developing world share some characteristics. Developing countries often have large populations, with annual growth rates between 2 and 4 percent. Poverty is widespread in both rural and urban areas, with rural areas often containing the poorest of the poor. While the majority of the developing world's inhabitants continue to live in the countryside, there is a massive rural-to-urban migration under way, and cities are growing rapidly. Wealth is unevenly distributed, making education, employment opportunities, and access to health care luxuries that few can enjoy. Corruption and mismanagement are widespread. With very few exceptions, these nations share a colonial past that has affected them both politically and economically. Moreover, critics charge that the neocolonial structure of the international economy and the West's political, military, and cultural links with the developing world amount to continued domination.

Developing countries continue to struggle to improve their citizens' living standards. Despite the economic success in some areas, poverty remains widespread, and more than a billion people live on less than a dollar a day. There is also growing economic inequality between the industrial countries and the developing world. This is especially true of the poorest countries, which have become further marginalized due to their fading strategic importance since the end of the cold war and their limited participation in the global economy. Inequality is also growing within developing countries where elite access to education, capital, and technology has significantly widened the gap between rich and poor.

Although the gap between rich and poor nations persists, some emerging markets saw significant growth during the 1990s. However, even these countries experienced the harsh realities of the global economy. The 1997 Asian financial crisis demonstrated the potential consequences of global finance and investment, and investors remain wary of investing in all but a few developing countries. Focus on emerging markets' problems further marginalizes the majority of developing countries

that have not experienced economic growth. The reasons for poor economic performance are complex. Both internal and external factors play a role in the inability of some developing countries to make economic progress. Among the factors are the colonial legacy, continued reliance on the export of primary products, stagnating or declining terms of trade for those primary products, protectionism in the industrialized countries, meager foreign aid contributions, debt, and the array of domestic social problems. Sluggish growth in the United States and lagging economies in the rest of the industrialized world have had a significant impact on the economic prospects of the developing world.

As colonies gained their independence in the post–World War II era, a combination of thought and action that constituted a loosely defined third world movement emerged. Besides shaping the domestic goals of many leaders, this perspective emphasized the revolutionary aspirations of developing world peoples and contributed to the view of a world divided between the industrialized North and the developing South. Based on this view, which portrayed the North as continuing to exploit the South, the developing countries challenged the North, calling for the establishment of a New International Economic Order and seeking to influence the agenda of various international organizations. Although this effort faded during the 1980s, elements of this perspective are still evident in a radical, non-Western worldview. Emphasizing the North's continuing domination of the developing world, its proponents view international relations as a struggle of the oppressed against their oppressors. Echoing the earlier efforts at solidarity, developing countries have coordinated efforts to extract concessions from the industrialized countries during the Doha round of trade talks. Such cooperative efforts promote solidarity and could enhance the developing world's ability to shape the international agenda.

In contrast to the developing world's criticism of the West, industrial countries continue to maintain the importance of institution-building and following the Western model that emphasizes a market-oriented approach to development. There is clearly a divergence of opinion between the industrialized countries and the developing world on issues ranging from human rights to governance and economic development.

The Great Divide
in the Global Village

Bruce R. Scott

INCOMES ARE DIVERGING

MAINSTREAM economic thought promises that globalization will lead to a widespread improvement in average incomes. Firms will reap increased economies of scale in a larger market, and incomes will converge as poor countries grow more rapidly than rich ones. In this "win-win" perspective, the importance of nation-states fades as the "global village" grows and market integration and prosperity take hold.

But the evidence paints a different picture. Average incomes have indeed been growing, but so has the income gap between rich and poor countries. Both trends have been evident for more than 200 years, but improved global communications have led to an increased awareness among the poor of income inequalities and heightened the pressure to emigrate to richer countries. In response, the industrialized nations have erected higher barriers against immigration, making the world economy seem more like a gated community than a global village. And although international markets for goods and capital have opened up since World War II and multilateral organizations now articulate rules and monitor the world economy, economic inequality among countries continues to increase. Some two billion people earn less than $2 per day.

At first glance, there are two causes of this divergence between economic theory and reality. First, the rich countries insist on barriers to immigration and agricultural imports. Second, most poor nations have been unable to attract much foreign capital due to their own government failings. These two issues are fundamentally linked: by forcing poor people to remain in badly governed states, immigration barriers deny those most in need the opportunity to "move up" by "moving out." In turn, that immobility eliminates a potential source of pressure on ineffective governments, thus facilitating their survival.

Since the rich countries are unlikely to lower their agricultural and immigration barriers significantly, they must recognize that politics is a key cause of economic inequality. And since most developing countries receive little foreign investment, the wealthy nations must also acknowledge that the "Washington consensus," which assumes that free markets will bring about economic convergence, is mistaken. If they at least admit these realities, they will abandon the notion that their own particular strategies are the best for all countries. In turn, they should allow poorer countries considerable freedom to tailor development strategies to their own circumstances. In this more pragmatic view, the role of the state becomes pivotal.

Why have economists and policymakers not come to these conclusions sooner?

Since the barriers erected by rich countries are seen as vital to political stability, leaders of those countries find it convenient to overlook them and focus instead on the part of the global economy that has been liberalized. The rich countries' political power in multilateral organizations makes it difficult for developing nations to challenge this self-serving world-view. And standard academic solutions may do as much harm as good, given their focus on economic stability and growth rather than on the institutions that underpin markets. Economic theory has ignored the political issues at stake in modernizing institutions, incorrectly assuming that market-based prices can allocate resources appropriately.

The fiasco of reform in Russia has forced a belated reappraisal of this blind trust in markets. Many observers now admit that the transition economies needed appropriate property rights and an effective state to enforce those rights as much as they needed the liberalization of prices. Indeed, liberalization without property rights turned out to be the path to gangsterism, not capitalism. China, with a more effective state, achieved much greater success in its transition than did Russia, even though Beijing proceeded much more slowly with liberalization and privatization.

Economic development requires the transformation of institutions as well as the freeing of prices, which in turn requires political and social modernization as well as economic reform. The state plays a key role in this process; without it, developmental strategies have little hope of succeeding. The creation of effective states in the developing world will not be driven by familiar market forces, even if pressures from capital markets can force fiscal and monetary discipline. And in a world still governed by "states rights," real progress in achieving accountable governments will require reforms beyond the mandates of multilateral institutions.

GO WITH THE FLOW

IN THEORY, globalization provides an opportunity to raise incomes through increased specialization and trade. This opportunity is conditioned by the size of the markets in question, which in turn depends on geography, transportation costs, communication networks, and the institutions that underpin markets. Free trade increases both the size of the market and the pressure to improve economic performance. Those who are most competitive take advantage of the enhanced market opportunities to survive and prosper.

Neoclassical economic theory predicts that poor countries should grow faster than rich ones in a free global market. Capital from rich nations in search of cheaper labor should flow to poorer economies, and labor should migrate from low-income areas toward those with higher wages. As a result, labor and capital costs—and eventually income—in rich and poor areas should eventually converge.

The U.S. economy demonstrates how this theory can work in a free market with the appropriate institutions. Since the 1880s, a remarkable convergence of incomes among the country's regions has occurred. The European Union has witnessed a similar phenomenon, with the exceptions of Greece and Italy's southern half, the *Mezzogiorno*. What is important, however, is that both America and the EU enjoy labor and capital mobility as well as free internal trade.

But the rest of the world does not fit this pattern. The most recent *World Development Report* shows that real per capita incomes for the richest one-third of countries rose by an annual 1.9 percent between 1970 and 1995, whereas the middle third went up by only 0.7 percent and the bottom third showed no increase at all. In the

Western industrial nations and Japan alone, average real incomes have been rising about 2.5 percent annually since 1950—a fact that further accentuates the divergence of global income. These rich countries account for about 60 percent of world GDP but only 15 percent of world population.

Why is it that the poor countries continue to fall further behind? One key reason is that most rich countries have largely excluded the international flow of labor into their markets since the interwar period. As a result, low-skilled labor is not free to flow across international boundaries in search of more lucrative jobs. From an American or European perspective, immigration appears to have risen in recent years, even approaching its previous peak of a century ago in the United States. Although true, this comparison misses the central point. Billions of poor people could improve their standard of living by migrating to rich countries. But in 1997, the United States allowed in only 737,000 immigrants from developing nations, while Europe admitted about 665,000. Taken together, these flows are only 0.04 percent of all potential immigrants.

Global markets offer opportunities for all, but opportunities do not guarantee results

---

The point is not that the rich countries should permit unfettered immigration. A huge influx of cheap labor would no doubt be politically explosive; many European countries have already curtailed immigration from poor countries for fear of a severe backlash. But the more salient issue is that rich nations who laud liberalism and free markets are rejecting those very principles when they restrict freedom of movement. The same goes for agricultural imports. Both Europe and Japan have high trade barriers in agriculture, while the United States remains modestly protectionist.

Mainstream economic theory does provide a partial rationalization for rich-country protectionism: Immigration barriers need not be a major handicap to poor nations because they can be offset by capital flows from industrialized economies to developing ones. In other words, poor people need not demand space in rich countries because the rich will send their capital to help develop the poor countries. This was

indeed the case before World War I, but it has not been so since World War II.

But the question of direct investment, which typically brings technologies and know-how as well as financial capital, is more complicated than theories would predict. The total stock of foreign direct investment did rise almost sevenfold from 1980 to 1997, increasing from 4 percent to 12 percent of world GDP during that period. But very little has gone to the poorest countries. In 1997, about 70 percent went from one rich country to another, 8 developing countries received about 20 percent, and the remainder was divided among more than 100 poor nations. According to the World Bank, the truly poor countries received less than 7 percent of the foreign direct investment to all developing countries in 1992–98. At the same time, the unrestricted opening of capital markets in developing countries gives larger firms from rich countries the opportunity for takeovers that are reminiscent of colonialism. It is not accidental that rich countries insist on open markets where they have an advantage and barriers in agriculture and immigration, where they would be at a disadvantage.

As for the Asian "tigers," their strong growth is due largely to their high savings rate, not foreign capital. Singapore stands out because it has enjoyed a great deal of foreign investment, but it has also achieved one of the highest domestic-savings rates in the world, and its government has been a leading influence on the use of these funds. China is now repeating this pattern, with a savings rate of almost 40 percent of GDP. This factor, along with domestic credit creation, has been its key motor of economic growth. China now holds more than $100 billion in low-yielding foreign-exchange reserves, the second largest reserves in the world.

In short, global markets offer opportunities for all, but opportunities do not guarantee results. Most poor countries have been unable to avail themselves of much foreign capital or to take advantage of increased market access. True, these countries have raised their trade ratios (exports plus imports) from about 35 percent of their GDP in 1981 to almost 50 percent in 1997. But without the Asian tigers, developing-country exports remain less than 25 percent of world exports.

Part of the problem is that the traditional advantages of poor countries have been in primary commodities (agriculture and minerals), and these categories have shrunk from about 70 percent of world trade in 1900 to about 20 percent at the end of the century. Opportunities for growth in the world market have shifted from raw or semiprocessed commodities toward manufactured goods and services—and, within these categories, toward more knowledge-intensive segments. This trend obviously favors rich countries over poor ones, since most of the latter are still peripheral players in the knowledge economy. (Again, the Asian tigers are the exception. In 1995, they exported as much in high-technology goods as did France, Germany, Italy, and Britain combined—which together have three times the population of the tigers.)

ONE COUNTRY, TWO SYSTEMS

W HY is the performance of poor countries so uneven and out of sync with theoretical forecasts? Systemic barriers at home and abroad inhibit the economic potential of poorer nations, the most formidable of these obstacles being their own domestic political and administrative problems. These factors, of course, lie outside the framework of mainstream economic analysis. A useful analogy is the antebellum economy of the United States, which experienced a similar set of impediments.

Like today's "global village," the U.S. economy before the Civil War saw incomes diverge as the South fell behind the North. One reason for the Confederacy's secession and the resulting civil war was Southern recognition that it was falling behind in both economic and political power, while the richer and more populous North was attracting more immigrants. Half of the U.S. population lived in the North in 1780; by 1860, this share had climbed to two-thirds. In 1775, incomes in the five original Southern states equaled those in New England, even though wealth (including slaves) was disproportionately concentrated in the South. By 1840, incomes in the northeast were about 50 percent higher than those in the original Southern states; the North's railroad mileage was about 40 percent greater (and manufacturing investment four times higher) than the South's. As the economist Robert Fogel has pointed out, the South was not poor—in 1860 it was richer than all European states except England—but Northern incomes were still much higher and increasing.

Why had Southern incomes diverged from those in the North under the same government, laws, and economy? Almost from their inception, the Southern colonies followed a different path from the North—specializing in plantation agriculture rather than small farms with diversified crops—due to geography and slavery. Thanks to slave labor, Southerners were gaining economies of scale and building comparative advantage in agriculture, exporting their goods to world markets and the North. Gang labor outproduced "free" (paid) labor. But the North was building even greater advantages by developing a middle class, a manufacturing sector, and a more modern social and political culture. With plans to complete transcontinental railroads pending, the North was on the verge of achieving economic and political dominance and the capacity to shut off further expansion of slavery in the West. The South chose war over Northern domination—and modernization.

Although the Constitution guaranteed free trade and free movement of capital and labor, the institution of slavery meant that the South had much less factor mobility than the North. It also ensured less development of its human resources, a less equal distribution of income, a smaller market for manufactures, and a less dynamic economy. It was less attractive to both European immigrants and external capital. With stagnant incomes in the older states, it was falling behind. In these respects, it was a forerunner of many of today's poor countries, especially those in Latin America.

What finally put the South on the path to economic convergence? Four years of civil war with a total of 600,000 deaths and vast destruction of property were only a start. Three constitutional amendments and twelve years of military "reconstruction" were designed to bring equal rights and due process to the South. But the reestablishment of racial segregation following Reconstruction led to sharecropping as former slaves refused to return to the work gangs. Labor productivity dropped so much that Southern incomes fell to about half of the North's in 1880. In fact, income convergence did not take off until the 1940s, when a wartime boom in the North's industrial cities attracted Southern migrants in search of better jobs. At the same time, the South began drawing capital as firms sought lower wages, an anti-union environment, and military contracts in important congressional districts. But this process did not fully succeed until the 1960s, as new federal laws and federal troops brought full civil rights to the South and ensured that the region could finally modernize.

THE GREAT DIVIDE

A LTHOUGH slavery is a rarity today, the traditional U.S. divide between North and South provides a good model for understanding contemporary circumstances in many developing countries. In the American South, voter intimidation, segregated housing, and very unequal schooling were the rule, not the exception—and such tactics are repeated today by the elites in today's poor countries. Brazil, Mexico, and Peru had abundant land relative to population when the Europeans arrived, and their incomes roughly approximated those in North America, at least until 1700. The economists Stanley Engerman and Kenneth Sokoloff have pointed out that these states, like the Confederacy, developed agricultural systems based on vast landholdings for the production of export crops such as sugar and coffee. Brazil and many Caribbean islands also adopted slavery, while Peru and Mexico relied on forced indigenous labor rather than African slaves.

History shows that the political development of North America and developing nations—most of which were colonized by Europeans at some point—was heavily influenced by mortality. In colonies with tolerable death rates (Australia, Canada, New Zealand, and the United States), the colonists soon exerted pressure for British-style protections of persons and property. But elsewhere (most of Africa, Latin America, Indonesia, and to a lesser degree, India), disease caused such high mortality rates that the few resident Europeans were permitted to exploit a disenfranchised laboring class, whether slave or free. When the colonial era ended in these regions, it was followed by "liberationist" regimes (often authoritarian and incompetent) that maintained the previous system of exploitation for the advantage of a small domestic elite. Existing inequalities within poor countries continued; policies and institutions rarely protected individual rights or private initiative for the bulk of the population and allowed elites to skim off rents from any sectors that could bear it. The economist Hernando de Soto has shown how governments in the developing world fail to recognize poor citizens' legal titles to their homes and businesses, thereby depriving them of the use of their assets for collateral. The losses in potential capital to these countries have dwarfed the cumulative capital inflows going to these economies in the last century.

The legacy of these colonial systems also tends to perpetuate the unequal distribution of income, wealth, and political power while limiting capital mobility. Thus major developing nations such as Brazil, China, India, Indonesia, and Mexico are experiencing a divergence of incomes by province within their economies, as labor and capital fail to find better opportunities. Even in recent times, local elites have fought to maintain oppressive conditions in Brazil, El Salvador, Guatemala, Mexico, Nicaragua, and Peru. Faced with violent intimidation, poor people in these countries have suffered from unjust law enforcement similar to what was once experienced by black sharecroppers in the American South.

Modernization and economic development inevitably threaten the existing distribution of power and income, and powerful elites continue to protect the status quo—even if it means that their society as a whole falls further behind. It takes more than a constitution, universal suffrage, and regular elections to achieve governmental accountability and the rule of law. It may well be that only the right of exit—emigration—can peacefully bring accountability to corrupt and repressive regimes. Unlike the U.S. federal government, multilateral institutions lack the legitimacy to intervene in the internal affairs of most countries. Europe's economic takeoff in the second half of the nineteenth century was aided by the emigration of 60 million people to North America, Argentina, Brazil, and Australia. This emigration—about 10 percent of the labor force—helped raise European wages while depressing inflated wages in labor-scarce areas such as Australia and the United States. A comparable out-migration of labor from today's poor countries would involve hundreds of millions of people.

Of course, Latin America has seen some success. Chile has received the most attention for its free market initiatives, but its reforms were implemented by a brutally repressive military regime—hardly a model for achieving economic reform through democratic processes. Costa Rica would seem to be a much better model for establishing accountability, but its economic performance has not been as striking as Chile's.

Italy, like the United States in an earlier era, is another good example of "one country, two systems." Italy's per capita income has largely caught up with that of its European neighbors over the past 20 years, even exceeding Britain's and equaling France's in 1990, but its *Mezzogiorno* has failed to keep up. Whereas overall Italian incomes have been converging toward those of the EU, *Mezzogiorno* incomes have been diverging from those in the north. Southern incomes fell from 65 percent of the northern average in 1975 to 56 percent 20 years later; in Calabria, they fell to 47 percent of the northern average. Southern unemployment rose from 8 percent in 1975 to 19 percent in 1995—almost three times the northern average. In short, 50 years of subsidies from Rome and the EU have failed to stop the *Mezzogiorno* from falling further behind. Instead, they have yielded local regimes characterized by greatly increased public-sector employment, patronage, dependency, and corruption—not unlike the results of foreign aid for developing countries. And the continuing existence of the Mafia further challenges modernization.

Democracy is not enough to ensure that the governed reap the gains of their own efforts.

Democracy, then, is not enough to ensure that the governed are allowed to reap the gains of their own efforts. An effective state requires good laws as well as law enforcement that is timely, evenhanded, and accessible to the poor. In many countries, achieving objective law enforcement means reducing the extralegal powers of vested interests. When this is not possible, the only recourse usually available is emigration. But if the educated elite manages to emigrate while the masses remain trapped in a society that is short of leaders, the latter will face even more formidable odds as they try to create effective institutions and policies. Although Italians still emigrate from south to north, the size of this flow is declining, thanks in part to generous transfer payments that allow them to consume almost as much as northerners. In addition, policymaking for the *Mezzogiorno* is still concentrated in Rome.

The immigration barriers in rich countries not only foreclose opportunities in the global village to billions of poor people, they help support repressive, pseudo-democratic governments by denying the citizens of these countries the right to vote against the regime with their feet. In effect, the strict dictates of sovereignty allow wealthy nations to continue to set the rules in their own favor while allowing badly governed poor nations to continue to abuse their own citizens and retard economic development. Hence the remedy for income divergence must be political as well as economic.

GETTING INSTITUTIONS RIGHT

ACCORDING TO ECONOMIC THEORY, developing nations will create and modernize the institutions needed to underpin their markets so that their markets and firms can gradually match the performance of rich countries. But reality is much more complex than theory. For example, de Soto's analysis makes clear that effectively mobilizing domestic resources offers a much more potent source of capital for most developing nations than foreign inflows do. Yet mainstream economists and their formal models largely ignore these resources. Western economic advisers in Russia were similarly blindsided by their reliance on an economic model that had no institutional context and no historical perspective. Economists have scrambled in recent years to correct some of these shortcomings, and the Washington consensus now requires the "right" institutions as well as the "right" prices. But little useful theory exists to guide policy when it comes to institutional analysis, and gaps in the institutional foundations in most developing countries leave economic models pursuing unrealistic solutions or worse.

The adjustment of institutions inevitably favors certain actors and disadvantages others. As a result, modernization causes conflict that must be resolved through politics as well as economics. At a minimum, successful development signifies that the forces for institutional change have won out over the status quo. Achieving a "level playing field" signifies that regulatory and political competition is well governed.

Economists who suggest that all countries must adopt Western institutions to achieve Western levels of income often fail to consider the changes and political risks involved. The experts who recommended that formerly communist countries apply "shock therapy" to markets and democracy disregarded the political and regulatory issues involved. Each change requires a victory in the "legislative market" and successful persuasion within the state bureaucracy for political approval. Countries with lower incomes and fewer educated people than Russia face even more significant developmental challenges just to achieve economic stability, let alone attract foreign investment or make effective use of it. Institutional deficiencies, not capital shortages, are the major impediment to development, and as such they must be addressed before foreign investors will be willing to send in capital.

Although price liberalization can be undertaken rapidly, no rapid process (aside from revolution) exists for an economy modernizing its institutions. Boris Yeltsin may be credited with a remarkable turnover, if not a coup d'état, but his erratic management style and the lack of parliamentary support ensured that his government would never be strong. In these circumstances, helping the new Russian regime improve law enforcement should have come ahead of mass privatization. Launching capitalism in a country where no one other than apparatchiks had access to significant amounts of capital was an open invitation to gangsterism and a discredited system. Naive economic models made for naive policy recommendations.

HOW THE WEST WON

THE STATE'S crucial role is evident in the West's economic development. European economic supremacy was forged not by actors who followed a "Washington consensus" model but by strong states. In the fifteenth century, European incomes were not much higher than those in China, India, or Japan. The nation-state was a European innovation that replaced feudalism and established the rule of law; in turn, a legal framework was formed for effective markets. Once these countries were in the lead, they were able to continuously increase their edge through technological advances. In addition, European settlers took their civilization with them to North America and the South Pacific, rapidly raising these areas to rich-country status as well. Thus Europe's early lead became the basis for accumulating further advantages with far-reaching implications.

Europe's rise to economic leadership was not rapid at first. According to the economist Angus Maddison, Europe's economy grew around 0.07 percent a year until 1700; only after 1820 did it reach one

percent. But the pace of technological and institutional innovation accelerated thereafter. Meanwhile, discovery of new markets in Africa, Asia, and the Americas created new economic opportunities. Secular political forces overthrew the hegemony of the Catholic Church. Feudalism was eroded by rising incomes and replaced by a system that financed government through taxes, freeing up land and labor to be traded in markets. Markets permitted a more efficient reallocation of land and labor, allowing further rises in incomes. Effective property rights allowed individuals to keep the fruits of their own labor, thereby encouraging additional work. And privatization of common land facilitated the clearing of additional acreage.

The nation-state helped forge all these improvements. It opened up markets by expanding territory; reduced transaction costs; standardized weights, measures, and monetary units; and cut transport costs by improving roads, harbors, and canals. In addition, it was the state that established effective property rights. The European state system thrived on flexible alliances, which constantly changed to maintain a balance of power. Military and economic rivalries prompted states to promote development in agriculture and commerce as well as technological innovation in areas such as shipping and weaponry. Absent the hegemony of a single church or state, technology was diffused and secularized. Clocks, for instance, transferred timekeeping from the monastery to the village clock tower; the printing press did much the same for the production and distribution of books.

Europe's development contrasts sharply with Asia's. In the early modern era, China saw itself as the center of the world, without real rivals. It had a much larger population than Europe and a far bigger market as well. But though the Chinese pioneered the development of clocks, the printing press, gunpowder, and iron, they did not have the external competitive stimulus to promote economic development. Meanwhile, Japan sealed itself off from external influences for more than 200 years, while India, which had continuous competition within the subcontinent, never developed an effective national state prior to the colonial era.

The Europeans also led in establishing accountable government, even though it was achieved neither easily nor peacefully. Most European states developed the notion that the sovereign (whether a monarch or a parliament) had a duty to protect subjects

and property in return for taxes and service in the army. Rulers in the Qing, Mughal, and Ottoman Empires, in contrast, never recognized a comparable responsibility to their subjects. During the Middle Ages, Italy produced a number of quasi-democratic city-states, and in the seventeenth century Holland created the first modern republic after a century of rebellion and warfare with Spain. Britain achieved constitutional monarchy in 1689, following two revolutions. After a bloody revolution and then dictatorship, France achieved accountable government in the nineteenth century.

Europe led in establishing accountable government, although it was not easy or peaceful.

Europe led the way in separating church and state—an essential precursor to free inquiry and adoption of the scientific method—after the Thirty Years' War. The secular state in turn paved the way for capitalism and its "creative destruction." Creative destruction could hardly become the norm until organized religion lost its power to execute as heretics those entrepreneurs who would upset the status quo. After the Reformation, Europeans soon recognized another fundamental tenet of capitalism: the role of interest as a return for the use of capital. Capitalism required that political leaders allow private hands to hold power as well as wealth; in turn, power flowed from the rural nobility to merchants in cities. European states also permitted banks, insurance firms, and stock markets to develop. The "yeast" in this recipe lay in the notion that private as well as state organizations could mobilize and reallocate society's resources—an idea with profound social, political, and economic implications today.

Most of Europe's leading powers did not rely on private initiative alone but adopted mercantilism to promote their development. This strategy used state power to create a trading system that would raise national income, permitting the government to enhance its own power through additional taxes. Even though corruption was sometimes a side effect, the system generally worked well. Venice was the early leader, from about 1000 to 1500; the Dutch followed in the sixteenth and seventeenth centuries; Britain became dominant in the eighteenth century. In Britain, as in the other cases, mercantilist export promotion was associated with a dramatic rise in state spending and employment (especially in the navy), as well as "crony capitalism." After World War II, export-promotion regimes were adopted by Japan, South Korea, Singapore, and Taiwan with similar success. Today, of course, such strategies are condemned as violations of global trade rules, even for poor countries.

Finally, geography played a pivotal role in Europe's rise, providing a temperate climate, navigable rivers, accessible coastline, and defensible boundaries for future states. In addition, Europe lacked the conditions for the production of labor-intensive commodities such as coffee, cotton, sugar, or tobacco—production that might have induced the establishment of slavery. Like in the American North, European agriculture was largely rain-fed, diversified, and small-scale.

Europe's rise, then, was partly due to the creation and diffusion of technological innovations and the gradual accumulation of capital. But the underlying causes were political and social. The creation of the nation-state and institutionalized state rivalry fostered government accountability. Scientific enlightenment and upward social mobility, spurred by healthy competition, also helped Europe achieve such transformations. But many of today's developing countries still lack these factors crucial for economic transformation.

PLAYING CATCH-UP

Globalization offers opportunities for all nations, but most developing countries are very poorly positioned to capitalize on them. Malarial climates, limited access to navigable water, long distances to major markets, and unchecked population growth are only part of the problem. Such countries also have very unequal income structures inherited from colonial regimes, and these patterns of income distribution are hard to change unless prompted by a major upheaval such as a war or a revolution. But as serious as these disadvantages are, the greatest disadvantage has been the poor quality of government.

If today's global opportunities are far greater and potentially more accessible than at any other time in world history, developing countries are also further behind than ever before. Realistic political logic suggests that weak governments need to show that they can manage their affairs much better before they pretend to have strategic ambitions. So what kind of catch-up models could they adopt?

Substituting domestic goods for imports was the most popular route to economic development prior to the 1980s. But its inward orientation made those who adopted it unable to take advantage of the new global opportunities and ultimately it led to a dead end. Although the United States enjoyed success with such a strategy from 1790 until 1940, no developing country has a home market large enough to support a modern economy today. The other successful early growth model was European mercantilism, namely export promotion, as pioneered by Venice, the Dutch republic, Britain, and Germany. Almost all of the East Asian success stories, China included, are modern versions of the export-oriented form of mercantilism.

For its part, free trade remains the right model for rich countries because it provides decentralized initiatives to search for tomorrow's market opportunities. But it does not necessarily promote development. Britain did not adopt free trade until the 1840s, long after it had become the world's leading industrial power. The prescription of lower trade barriers may help avoid even worse strategies at the hands of bad governments, but the Washington-consensus model remains best suited for those who are ahead rather than behind.

Today's shareholder capitalism brings additional threats to poor countries, first by elevating compensation for successful executives, and second by subordinating all activities to those that maximize shareholder value. Since 1970, the estimated earnings of an American chief executive have gone from 30 times to 450 times that of the average worker. In the leading developing countries, this ratio is still less than 50. Applying a similar "market-friendly" rise in executive compensation within the developing world would therefore only aggravate the income gap, providing new ammunition for populist politicians. In addition, shareholder capitalism calls for narrowing the managerial focus to the interests of shareholders, even if this means dropping activities that offset local market imperfections. A leading South African bank has shed almost a million small accounts—mostly held by blacks—to raise its earnings per share. Should this bank, like its American counterparts, have an obligation to serve its community, including its black members, in return for its banking license?

Poor nations must improve the effectiveness of their institutions and bureaucracies in spite of entrenched opposition and poorly paid civil servants. As the journalist Thomas Friedman has pointed out, it is true that foreign-exchange traders can dump the currencies of poorly managed countries, thereby helping discipline governments to restrain their fiscal deficits and lax monetary policies. But currency pressures will not influence the feudal systems in Pakistan and Saudi Arabia, the theocracies in Afghanistan and Iran, or the kleptocracies in Kenya or southern Mexico. The forces of capital markets will not restrain Brazilian squatters as they take possession of "public lands" or the slums of Rio de Janeiro or São Paulo, nor will they help discipline landlords and vigilantes in India's Bihar as they fight for control of their state. Only strong, accountable government can do that.

LOOKING AHEAD

Increased trade and investment have indeed brought great improvements in some countries, but the global economy is hardly a win-win situation. Roughly one billion people earn less than $1 per day, and their numbers are growing. Economic resources to ameliorate such problems exist, but the political and administrative will to realize the potential of these resources in poor areas is lacking. Developing-nation governments need both the pressure to reform their administrations and institutions, and the access to help in doing so. But sovereignty removes much of the external pressure, while immigration barriers reduce key internal motivation. And the Washington consensus on the universality of the rich-country model is both simplistic and self-serving.

The world needs a more pragmatic, country-by-country approach, with room for neomercantilist regimes until such countries are firmly on the convergence track. Poor nations should be allowed to do what today's rich countries did to get ahead, not be forced to adopt the laissez-faire approach. Insisting on the merits of comparative advantage in low-wage, low-growth industries is a sure way to stay poor. And continued poverty will lead to rising levels of illegal immigration and low-level violence, such as kidnappings and vigilante justice, as the poor take the only options that remain. Over time, the rich countries will be forced to pay more attention to the fortunes of the poor—if only to enjoy their own prosperity and safety.

Still, the key initiatives must come from the poor countries, not the rich. In the last 50 years, China, India, and Indonesia have led the world in reducing poverty. In China, it took civil war and revolution, with tens of millions of deaths, to create a strong state and economic stability; a de facto coup d'état in 1978 brought about a very fortunate change of management. The basic forces behind Chinese reform were political and domestic, and their success depended as much on better using resources as opening up markets. Meanwhile, the former Soviet Union and Africa lie at the other extreme. Their economic decline stems from their failure to maintain effective states and ensure the rule of law.

It will not be surprising if some of today's states experience failure and economic decline in the new century. Argentina, Colombia, Indonesia, and Pakistan will be obvious cases to watch, but other nations could also suffer from internal regional failures—for example, the Indian state of Bihar. Income growth depends heavily on the legal, administrative, and political capabilities of public actors in sovereign states. That is why, in the end, external economic advice and aid must go beyond formal models and conform to each country's unique political and social context.

BRUCE R. SCOTT is Paul W. Cherington Professor of Business Administration at Harvard Business School.

The Poor Speak Up

Leaders of the developing world are rising up with a strength not seen since Tito, Nasser and Nehru, challenging the rules of globalization as defined by both Western governments and Western activists

Rana Foroohar
With Ian MacKinnon in Delhi, Mac Margolis in Rio, Paul Mooney in Beijing and Mark Ashurst in London

They drew blanket press coverage and a watchful audience of New York police, but mainly on the strength of past protest performances. From Seattle to Davos, the riotous anti-globalization road show had pressed its case that rising trade and capital flows are bringing nothing but oppression and instability to the developing world. Their ranks boomed with the global economy, but faded with it as well. The thousands who descended on the World Economic Forum last year in Davos dwindled to a few hundred by the time the fete reconvened last week in New York.

Inside the party at the Waldorf-Astoria, the protesters still inspired a degree of soul-searching among the gathered stars of the global age—CEOs, prime ministers, princes and celebs—as if they had noticed that the era and the opposition have changed. But leadership of the effort to alter the course of globalization has shifted away from Western street rebels—and toward those they have long claimed to represent. The main actors now are leaders of the poor nations, who have united against the global trade elite and are demanding radical changes in the rules of international commerce.

This is now a far more powerful movement, particularly in the grim wake of an age of plenty. Unlike anarchists, greens and reds, the presidents and ministers of the developing world cannot be humored and dismissed as a bag of mixed nuts. They have a seat at the table, real influence. At the November trade summit in Doha, Qatar, they stepped forward in a show of unity and strength that, arguably, had not been seen from poor nations since the days of Tito, Nehru and Nasser. Led by India, Brazil and South Africa, this emerging front is opposed to globalization as defined by Western governments and Western activists. In short, they are for free trade (unlike the activists) but on their own terms (not those of the West). Born of long frustration with the postwar commercial order, the mystery is why this movement is erupting so publicly only now. "There was a clear improvement in the awareness of developing countries at Doha,

and in their capacity to deal," says EU Trade Commissioner Pascal Lamy, who now expects this front to "go on pushing us."

In Doha, a bloc of developing nations won unprecedented victories. They helped force Europe to consider phasing out subsidies for its farms, which elbow out produce from poor, farm-based economies. They compelled the United States to consider limits on anti-dumping laws, which are often used against cheap exports from emerging manufacturing powers. They won the right to ignore Western drug patents when necessary to fight developing world scourges like AIDS, and then kept right on fighting. They recently pushed for new rules that would give their own ambassadors the lead role in the new trade round launched at Doha, and lost. But they'll be back. Many want no less than major reform of global financial institutions that date to the Bretton Woods agreement of 1944, which they believe has failed crisis-torn emerging nations like Argentina. "Financial capital is globalized, but not policy decisions," Brazilian President Fernando Henrique Cardoso said recently. "The Bretton Woods system is obsolete."

The new alliance has distant roots in the Third World solidarity of the early postwar years. The Group of 77 was founded in 1955 as a united front against the injustices of the colonial era. Many of its members were still colonies, barred by their rulers from working as full partners in the emerging international trading regime. Once they won independence, the emerging nations set out to develop in solidarity with one another, as a union of the poor against the empires they had just thrown off. They shut out imports in the hope of developing local industry, sealing themselves off in a socialist Third World.

Many of the poor were still living in this state of partial isolation at the start of the 1980s, when Reaganism and Thatcherism set off a boom in trade and in the complexity of trade agreements. In 1986 the Uruguay Round expanded the talks beyond simple tariffs to subjects like financial services and intellectual property. By this time, many poor nations had begun to drop state capitalism and to prosper as export powers in the global markets—but few had the sophistication to analyze issues like the trade-related aspects of intellectual property rights, or TRIPS. "Many countries didn't understand what they were signing up for," says Leif Pagrotsky, Sweden's trade minister.

The poor would soon suffer the consequences of obliviousness. By 1995 the United States was demanding that poor nations live up to the detailed letter of Uruguay. Even those prospering as export powers couldn't always enforce new rules around issues like intellectual property. The United States wound up suing India and Pakistan for patent-law violations. "It became clear that no matter what we had thought, the other side would take all this quite seriously," says Rashid S. Kaukab, a former Pakistani diplomat now at the South Centre in Geneva. "We couldn't afford not to look at these issues very carefully."

The more they studied, the angrier they became. Even booming nations like India and Brazil came to believe that, while they benefit from global trade, the West was benefiting more. In fact, a growing body of research showed that the global tariff system favored the rich. According to the World Bank, the poor countries pay average Uruguay Round tariffs of more than 14 percent, a rate more than twice as high as everyone else. Worse, developing-country exports tend to be farm goods or simple manufactures, two of the markets most heavily protected in the West by nontariff barriers. For example, African sugar and Brazilian steel both face stiff quotas and heavy subsidies in the developed world. "Developing countries are encouraged to diversify their exports," says Brazilian Trade Minister Sergio Amaral. "But precisely when they do so, and become competitive, they hit barriers in the developed world."

By the mid- to late 1990s there was a growing perception among emerging countries that they had been bamboozled in Uruguay and that the West was still pressing its advantage in trade expertise and clout. "Throughout the 1980s and 1990s, it was easier for the U.S. and the EU to intimidate the developing countries," says Alan Winters, a fellow at the Centre for Economic Policy Research in London. "They'd pick a few at a time and work on them individually, getting them to agree to enforcement of trade issues, like patent laws." To a growing number of developing-world trade ministers, it looked like a Western strategy of divide and conquer.

The obvious response was a more united front, even if the interests of poor nations didn't always mesh. As early as a 1996 WTO meeting in Singapore, developing countries were beginning to unite behind a rough agenda: just say no to whatever the rich nations ask. By the time of the 1999 WTO summit in Seattle, trade alliances were emerging among the states of Africa, Asia and Latin America, with increasingly sophisticated demands for greater access to rich-country markets, and a greater say in trade talks. When several small nations were locked out of the "green room" negotiations of the big trade powers in Seattle, several coalitions of lesser powers threatened to walk out. Their backroom revolt had at least as much to do with the collapse of the Seattle talks as protests in the streets, and many were left more alienated than ever. After Seattle, says trade expert Kent Hughes, many developing nations were left wondering, "What were these meetings doing for them?"

In mounting frustration, various alliances of poor nations began plotting for the next big trade summit. Jolted by South African President Thabo Mbeki's critique of its economic failures, Africa began uniting. In July 2001 the ineffectual Organization for African Unity was replaced by the African Union, boasting a clear agenda for trade liberalization. Led by Tanzania, the least-developed countries met in Zanzibar to hash out a plan for Doha. Geneva ambassadors of the Like Minded Group, a coalition of 13 countries from Asia, Latin America and Africa, began dropping by each other's offices more frequently for tea, and targeted issues to press at the upcoming summit. China's Zhu Rongji led a pre-Doha relationship-building delegation to India. Leaders of five developing countries—India, Brazil, South Africa, Malaysia and Egypt—held informal talks to set common goals. And a bloc of 50 nations began lobbying on an old grievance: TRIPS. "There was constant, coordinated pressure by the various developing countries and missions in Geneva," says Egyptian trade negotiator Magdi Farahat. "Everyone kept everyone else in the loop."

The sweep of their victory changed the rules of the game. They pushed the envelope not only on intellectual property, farms and dumping, but also on their complaint that Uruguay left them with trade commitments they could not afford to enforce. The northern countries made concessions on 50 of these "implementation" issues and agreed to help poor nations build up the capacity to carry out their trade-related responsibilities. That could include everything from helping nations pay for technology to carry out stricter customs checks, to boosting their ability to carry out policy research and analysis. At the center of it was Murasoli Maran, who had worked as a screenwriter before becoming trade minister of India, and had scripted an eleventh-hour drama in Doha. Pressured by European and American negotiators to either sign a deal or bump the final agreement up to the presidential level for approval, Maran held out until he got his way. Asked whether he'd been intimidated by the West, Maran shot back, "No. I intimidated them."

It was a public declaration of victory for the new southern front, heralding a new era in the battle over globalization. The elite trade powers know it. Europe's lead negotiator, Lamy, says that trade has not only "risen on the agenda of these governments," but trade ministers are increasingly powerful figures in the developing world. Trade ministers like Maran and South Africa's Alec Erwin are "punching above their weight," says Lamy. This new strength will force Europe and America to spend more time courting opinion in the developing world. At the World Economic Forum in New York last week, U.S. Trade Representative Robert Zoellick stressed the importance of the new developing-nation alliances in the post-Doha world. "It is critical to look at the role that developing nations play and the networks they are creating," he said. In order to make the next round successful, Zoellick noted, "We need consensus-building."

But this is not the '50s, when one leader like Tito could claim to speak for the Third World. The movement was simpler then, united by animus against colonial rule, and asking only for a broad redistribution of wealth. It is no longer a unit bound by ideology, but a front of shifting alliances that change from issue to issue. "As long as you're opposed to something, you can hold together," says Columbia University economist Jagdish Bhagwati. "Once you move forward, interests are bound to diverge."

The splits are already appearing. While developing nations agree broadly on opening up richer markets, they differ on the details. India, Malaysia and Egypt want Europeans to cut agricultural subsidies while they keep on protecting their own farmers. Brazil and South Africa are opposed to protection for anyone. The recent European decision to grant preferred trade status to its former colonies pleased the former colonies, but no one else. "It's all a matter of what's convenient," says Tattamangalam Vishwanath, international trade adviser to the Confederation of Indian Industry. "In trade talks, there are no permanent friends and no permanent enemies."

The new battlelines don't necessarily pit rich versus poor anymore, either. In Doha, the United States backed the developing nations that protested against European farm subsidies. And the EU backed the stand against U.S. anti-dumping rules.

WTO director-general Michael Moore welcomes this new movement of pragmatic, nonideological developing nations as "healthy." It's no longer so much a revolt against colonial masters as a spat about stuff like steel quotas and banana tariffs. "There is no Third World anymore," says Brazil's chief trade negotiator, Sergio Amaral. "But there is a unanimity among the developing countries that protectionism is the common enemy."

This is a sharp departure from anti-globalization as championed by activists, whether from rich or poor nations. The activists aim to temper the destabilizing effects of rising trade and capital flows by protection, if necessary, and by requiring multinationals to raise wages and improve labor conditions in the developing world. This vision of managed trade would slow the pace of globalization, and was ridiculed at Davos 2000 by former Mexican president Ernesto Zedillo as a plan "to save the people of developing countries from developing." So the game is on. It pits the emerging front of poor nations against rich governments, who once made the rules, but also against Western activists, who once claimed to speak for them. The victors will shape the changing global order.

Institutions Matter, but Not for Everything

The role of geography and resource endowments in development shouldn't be underestimated.

Jeffrey D. Sachs

THE DEBATE over the role of institutions in economic development has become dangerously simplified. The vague concept of "institutions" has become, almost tautologically, the intermediate target for all efforts to improve an economy. If an economy is malfunctioning, the reasoning goes, something must be wrong with its institutions. In fact, recent papers have argued that institutions explain nearly everything about a country's level of economic development and that resource constraints, physical geography, economic policies, geopolitics, and other aspects of internal social structure, such as gender roles and inequalities between ethnic groups, have little or no effect. These papers have been written by such respected economists as Daron Acemoglu, Simon Johnson, and James Robinson; Dani Rodrik, Arvind Subramanian, and Francesco Trebbi; and William Easterly and Ross Levine.

Indeed, a single-factor explanation of something as important as economic development can be alluring, and the institutions-only argument has special allure for two additional reasons. First, it attributes high income levels in the United States, Europe, and Japan to allegedly superior social institutions; it even asserts that when incomes rise in other regions, they do so mainly because of the Western messages of freedom, property rights, and markets carried there by intrepid missionaries intent on economic development. Second, according to the argument, the rich world has little, if any, financial responsibility for the poor because development failures are the result of institutional failures and not of a lack of resources.

The problem is that the evidence simply does not support those conclusions. Institutions may matter, but they don't matter exclusively. The barriers to economic development in the poorest countries today are far more complex than institutional shortcomings. Rather than focus on improving institutions in sub-Saharan Africa, it would be wise to devote more effort to fighting AIDS, tuberculosis, and malaria; addressing the depletion of soil nutrients; and building more roads to connect remote populations to regional markets and coastal ports. In other words, sub-Saharan Africa and other regions struggling today for improved economic development require much more than lectures about good governance and institutions. They require direct interventions, backed by expanded donor assistance, to address disease, geographical isolation, low technological productivity, and resource limitations that trap them in poverty. Good governance and sound institutions would, no doubt, make such interventions more effective.

When economic growth fails

When Adam Smith, our profession's original and wisest champion of sound economic institutions, turned his eye to the poorest parts of the world in 1776, he did not so much as mention institutions in explaining their woes. It is worth quoting at length from Smith's *Wealth of Nations* on the plight of sub-Saharan Africa and central Asia, which remain the world's most troubled development hot spots:

> All the inland parts of Africa, and all that part of Asia which lies any considerable way north of the Euxine and Caspian seas, the ancient Scythia, the modern Tartary and Siberia, seem in all ages of the world to have been in the same barbarous and uncivilised state in which we find them at present. The Sea of Tartary is the frozen ocean which admits of no navigation, and though some of the greatest rivers in the world run through that country, they are at too great a distance from one another to carry commerce and communication through the greater part of it. There are in Africa none of those great inlets, such as the Baltic and Adriatic seas in Europe, the Mediterranean and Euxine seas in both Europe and Asia, and the gulfs of Arabia, Per-

sia, India, Bengal, and Siam, in Asia, to carry maritime commerce into the interior parts of that great continent: and the great rivers of Africa are at too great a distance from one another to give occasion to any considerable inland navigation. (Book I, Chapter III)

Smith's point is that Africa and central Asia could not effectively participate in international trade because transport costs were simply too high. And, without international trade, both regions were condemned to small internal markets, an inefficient division of labor, and continued poverty. These disadvantages of the hinterland exist to this day.

The ability of a disease to cut off economic development may seem surprising to some but reflects a lack of understanding of how disease can affect economic performance.

Smith couldn't know the half of it. The problems of African isolation went far beyond mere transport costs. Characterized by the most adverse malaria ecology in the world, Africa was as effectively cut off from global trade and investment by that killer disease. Although the disease ecology of malaria was not understood properly until two centuries after Adam Smith, what was known demonstrated that Africa's suffering was unique. It had a climate conducive to year-round transmission of malaria and was home to a species of mosquito ideally suited to transmitting malaria from person to person. When Acemoglu, Johnson, and Robinson find that the high mortality rates of British soldiers around 1820 in various parts of the world correlate well with the low levels of GNP per capita in the 1990s, they are discovering the pernicious effects of malaria in blocking long-term economic development.

The ability of a disease to cut off economic development may seem surprising to some but reflects a lack of understanding of how disease can affect economic performance. Thus, in writing that malaria has a limited impact in sub-Saharan Africa because most adults have some acquired immunity, Acemoglu, Johnson, and Robinson completely neglect the fact that the disease dramatically lowers the returns on foreign investments and raises the transaction costs of international trade, migration, and tourism in malarial regions. This is like claiming that the effects of the recent SARS (Severe Acute Respiratory Syndrome) outbreak in Hong Kong SAR can be measured by the number of deaths so far attributable to the disease rather than by the severe disruption in travel to and from Asia.

In an environment in which capital and people can move around with relative ease, the disadvantages of adverse geography—physical isolation, endemic disease, or other local problems (such as poor soil fertility)—are magnified. It is probably true that when human capital is high enough in any location,

physical capital will flow in as a complementary factor of production. Skilled workers can sell their outputs to world markets almost anywhere, over the Internet or by plane transport. Landlocked and at a high altitude, Denver can still serve as a high-tech hub of tourism, trade, and information technology. But when countries that are remote or have other problems related to their geography also have few skilled workers, these workers are much more likely to emigrate than to attract physical capital into the country. This is true even of geographically remote regions within countries. For example, China is having great difficulty attracting investments into its western provinces and is instead facing a massive shift of labor, including the west's few skilled workers, to the eastern and coastal provinces.

Recent history, then, confirms Smith's remarkable insights. Good institutions certainly matter, and bad institutions can sound the death knell of development even in favorable environments. But poor physical endowments may also hamper development. During the globalization of the past 20 years, economic performance has diverged markedly in the developing world, with countries falling into three broadly identifiable categories. First are the countries, and regions within countries, in which institutions, policies, and geography are all reasonably favorable. The coastal regions of east Asia (coastal China and essentially all of Korea, Taiwan Province of China, Hong Kong SAR, Singapore, Thailand, Malaysia, and Indonesia) have this beneficent combination and, as a result, have all become closely integrated with global production systems and benefited from large inflows of foreign capital.

Second are the regions that are relatively well endowed geographically but, for historical reasons, have had poor governance and institutions. These include the central European states, whose proximity to Western Europe brought them little benefit during the socialist regime. For such countries, institutional reforms are paramount. And, finally, there are impoverished regions with an unfavorable geography, such as most of sub-Saharan Africa, central Asia, large parts of the Andean region, and the highlands of Central America, where globalization has not succeeded in raising living standards and may, indeed, have accelerated the brain drain and capital outflows from the region. The countries that have experienced the severest economic failures in the recent past have all been characterized by initial low levels of income and small populations (and hence small internal markets) that live far from coasts and are burdened by disease, especially AIDS, tuberculosis, and malaria. These populations have essentially been trapped in poverty because of their inability to meet the market test for attracting private capital inflows.

When institutions *and* geography matter

It is a common mistake to believe—and a weak argument to make—that geography equals determinism. Even if good health is important to development, not all malarial regions are condemned to poverty. Rather, special investments are needed to fight malaria. Landlocked regions may be burdened by high transport costs but are not necessarily condemned to poverty.

Rather, special investments in roads, communications, rail, and other transport and communications facilities are even more important in those regions than elsewhere. Such regions may also require special help from the outside world to initiate self-sustaining growth.

A poor coastal region near a natural harbor may be able to initiate long-term growth precisely because few financial resources are needed to build roads and port facilities to get started. An equally poor landlocked region, however, may be stuck in poverty in the absence of outside help. A major project to construct roads and a port would most likely exceed local financing possibilities and may well have a rate of return far below the world market cost of capital. The market may be right: it is unlikely to pay a market return to develop the hinterland without some kind of subsidy from the rest of the world. Nor will institutional reforms alone get the goods to market.

In the short term, only three alternatives may exist for an isolated region: continued impoverishment of its population; migration of the population from the interior to the coast; or sufficient foreign assistance to build the infrastructure needed to link the region profitably with world markets. Migration would be the purest free market approach, yet the international system denies that option on a systematic basis; migration is systemically feasible only within countries. When populations do migrate from the hinterlands, the host country often experiences a political upheaval. The large migration from Burkina Faso to Côe d'Ivoire was one trigger of recent ethnic riots and civil violence.

A fourth and longer-term strategy that merits consideration is regional integration: a breaking down of artificial political barriers that limit the size of markets and condemn isolated countries to relative poverty. In this regard, the recent initiative to strengthen subregional and regional cooperation in Africa should certainly be supported. Yet, given political realities, this process will be too slow, by itself, to overcome the crisis of the poorest inland regions.

A good test of successful development strategy in these geographically disadvantaged regions is whether development efforts succeed in attracting new capital inflows. The structural adjustment era in sub-Saharan Africa, for example, was very disappointing in this dimension. Although the region focused on economic reforms for nearly two decades, it attracted very little foreign (or even domestic) investment, and what it did attract largely benefited the primary commodity sectors. Indeed, these economies remained almost completely dependent on a few primary commodity exports. The reform efforts did not solve the underlying fundamental problems of disease, geographical isolation, and poor infrastructure. The countries, unattractive to potential investors, could not break free from the poverty trap, and market-based infrastructure projects could not make up the difference.

Helping the poorest regions

Development thinking and policy must return to the basics: both institutions and resource endowments are critical, not just one or the other. That point was clear enough to Adam Smith but has been forgotten somewhere along the way. A crucial corollary is that poverty traps are real: countries can be too poor to find their own way out of poverty. That is, some locales are not favorable enough to attract investors under current technological conditions and need international help in even greater amounts than have been made available to them in recent decades.

An appropriate starting point for the international community would be to set actual developmental goals for such regions rather than "make do" with whatever economic results emerge. The best standards, by far, would be the Millennium Development Goals, derived from the international commitments to poverty alleviation adopted by all countries of the world at the UN Millennium Assembly of September 2000. The goals call for halving the 1990 rates of poverty and hunger by the year 2015 and reducing child mortality rates by two-thirds. Dozens of the poorest countries—those trapped in poverty—are too far off track to achieve these goals. Fortunately, at last year's UN Financing for Development Conference held in Monterrey, Mexico, and at the World Summit on Sustainable Development held in Johannesburg, South Africa, the industrial world reiterated its commitment to help those countries by increasing debt relief and official development assistance, including concrete steps toward the international target of 0.7 percent of donor GNP. The extra $125 billion a year that would become available if official development assistance were raised from the current 0.2 percent of GNP to 0.7 percent of GNP should easily be enough to enable all well-governed poor countries to achieve the Millennium Development Goals. Like official development assistance, debt-relief mechanisms have been wholly inadequate to date.

Armed with these goals and assurances of increased donor assistance, the international community, both donors and recipients, should be able to identify, for each country and in much greater detail than in the recent past, those obstacles—whether institutional, geographical, or other (including barriers to trade in the rich countries)—that are truly impeding economic development. For each of the Millennium Development Goals, detailed interventions—including their costs, organization, delivery mechanisms, and monitoring—can be assessed and agreed upon by stakeholders and donors. By freeing our thinking from one-factor explanations and understanding that poverty may have as much to do with malaria as with the exchange rate, we will become much more creative and expansive in our approach to the poorest countries. And, with this broader view, the international institutions can also be much more successful than past generations in helping to free these countries from their economic suffering.

References

Acemoglu, Daron, Simon Johnson, and James A. Robinson, 2001, "The Colonial Origins of Comparative Development: An Empirical Investigation," American Economic Review, Vol. 91 (December), pp. 1369–1401.

Bloom, David E., and Jeffrey D. Sachs, 1998, "Geography, Demography, and Economic Growth in Africa," *Brookings Papers on Economic Activity: 2,* Brookings Institution, pp.207–95.

Démurger, Sylvie, and others, 2002, "Geography, Economic Policy, and Regional Development in China," *Asian Economic Papers,* Vol. I (Winter), pp. 146–97.

Easterly, William, and Ross Levine, 2002, "Tropics, Germs and Crops: How Endowments Influence Economic Development," *NBER Working Paper 9106* (Cambridge, Massachusetts: National Bureau of Economic Research).

Gallup, John Luke, and Jeffrey D. Sachs with Andrew D. Mellinger, 1998, "Geography and Economic Development," paper presented at the Annual World Bank Conference on Development Economics, Washington, D.C., April.

Rodrik, Dani, Arvind Subramanian, and Francesco Trebbi, 2002, "Institutions Rule: The Primacy of Institutions over Geography and Integration in Economic Development," *NBER Working Paper 9305* (Cambridge, Massachusetts: National Bureau of Economic Research).

Sachs, Jeffrey D., 2002a , "A New Global Effort to Control Malaria," *Science,* Vol. 298 (October), pp. 122–24.

_____, 2002b, "Resolving the Debt Crisis of Low–Income Countries," *Brookings Papers on Economic Activity: 1,* Brookings Institution, pp. 257–86.

_____, 2003, "Institutions Don't Rule: Direct Effects of Geography on Per Capita Income," *NBER Working Paper 9490* (Cambridge, Massachusetts: National Bureau of Economic Research).

_____ and Pia Malaney, 2002, "The Economic and Social Burden of Malaria," *Nature Insight,* Vol. 415 (February), pp. 680–85.

United Nations Development Program, 2003, *Human Development Report* (New York), forthcoming.

Jeffrey D. Sachs is Director of the Earth Institute at Columbia University and is a special adviser to the UN secretary-general on the UN Millennium Development Goals.

From *Finance & Development,* June 2003, pp. 38-41. © 2003 by Finance & Development.

Development as Poison

Rethinking the Western Model of Modernity

STEPHEN A. MARGLIN

A*t the beginning of* Annie Hall, *Woody Allen tells a story about two women returning from a vacation in New York's Catskill Mountains. They meet a friend and immediately start complaining: "The food was terrible," the first woman says, "I think they were trying to poison us." The second adds, "Yes, and the portions were so small." That is my take on development: the portions are small, and they are poisonous. This is not to make light of the very* real gains that have come with development. In the past three decades, infant and child mortality have fallen by 66 percent in Indonesia and Peru, by 75 percent in Iran and Turkey, and by 80 percent in Arab oil-producing states. In most parts of the world, children not only have a greater probability of surviving into adulthood, they also have more to eat than their parents did—not to mention better access to schools and doctors and a prospect of work lives of considerably less drudgery.

Nonetheless, for those most in need, the portions are indeed small. Malnutrition and hunger persist alongside the tremendous riches that have come with development and globalization. In South Asia almost a quarter of the population is undernourished and in sub-Saharan Africa, more than a third. The outrage of anti-globalization protestors in Seattle, Genoa, Washington, and Prague was directed against the meagerness of the portions, and rightly so.

But more disturbing than the meagerness of development's portions is its deadliness. Whereas other critics highlight the distributional issues that compromise development, my emphasis is rather on the terms of the project itself, which involve the destruction of indigenous cultures and communities. This result is more than a side-effect of development; it is central to the underlying values and assumptions of the entire Western development enterprise.

The White Man's Burden

Along with the technologies of production, healthcare, and education, development has spread the culture of the modern West all over the world, and thereby undermined other ways of seeing, understanding, and being. By culture I mean something more than artistic sensibility or intellectual refinement. "Culture" is used here the way anthropologists understand the term, to mean the totality of patterns of behavior and belief that characterize a specific society. Outside the modern West, culture is sustained through community, the set of connections that bind people to one another economically, socially, politically, and spiritually. Traditional communities are not simply about shared spaces, but about shared participation and experience in producing and exchanging goods and services, in governing, entertaining and mourning, and in the physical, moral, and spiritual life of the community. The culture of the modern West, which values the market as the primary organizing principle of life, undermines these traditional communities just as it has undermined community in the West itself over the last 400 years.

The West thinks it does the world a favor by exporting its culture along with the technologies that the non-Western world wants and needs. This is not a recent idea. A century ago, Rudyard Kipling, the poet laureate of British imperialism, captured this sentiment in the phrase "White Man's burden," which portrayed imperialism as an altruistic effort to bring the benefits of Western rule to uncivilized peoples. Political imperialism died in the wake of World War II, but cultural imperialism is still alive and well. Neither practitioners nor theorists speak today of the white man's burden—no development expert of the 21st century hankers after clubs or golf courses that exclude local folk from membership. Expatriate development experts now work with local people, but their collaborators are themselves formed for the most part by Western culture and values and have more in common with the West than they do with their own people. Foreign advisers—along with their local collaborators—are still missionaries, missionaries for progress as the West defines the term. As our forbears saw imperialism, so we see development.

There are in fact two views of development and its relationship to culture, as seen from the vantage point of the modern West. In one, culture is only a thin veneer over a common, universal behavior based on rational calculation and maximization of individual self interest. On this view, which is probably the view of most economists, the Indian subsistence-oriented peasant is no less calculating, no less competitive, than the US commercial farmer.

Cultural imperialism is still alive and well. ... Foreign advisers ... are still missionaries, missionaries for progress as the West defines the term. As our forebears saw imperialism, so we see development.

There is a second approach which, far from minimizing cultural differences, emphasizes them. Cultures, implicitly or explicitly, are ranked along with income and wealth on a linear scale. As the West is richer, Western culture is more progressive, more developed. Indeed, the process of development is seen as the transformation of backward, traditional, cultural practices into modern practice, the practice of the West, the better to facilitate the growth of production and income.

What these two views share is confidence in the cultural superiority of the modern West. The first, in the guise of denying culture, attributes to other cultures Western values and practices. The second, in the guise of affirming culture, posits an inclined plane of history (to use a favorite phrase of the Indian political psychologist Ashis Nandy) along which the rest of the world is, and ought to be, struggling to catch up with us. Both agree on the need for "development." In the first view, the Other is a miniature adult, and development means the tender nurturing by the market to form the miniature Indian or African into a full-size Westerner. In the second, the Other is a child who needs structural transformation and cultural improvement to become an adult.

Both conceptions of development make sense in the context of individual people precisely because there is an agreed-upon standard of adult behavior against which progress can be measured. Or at least there was until two decades ago when the psychologist Carol Gilligan challenged the conventional wisdom of a single standard of individual development. Gilligan's book *In A Different Voice* argued that the prevailing standards of personal development were male standards. According to these standards, personal development was measured by progress

from intuitive, inarticulate, cooperative, contextual, and personal modes of behavior toward rational, principled, competitive, universal, and impersonal modes of behavior, that is, from "weak" modes generally regarded as feminine and based on experience to "strong" modes regarded as masculine and based on algorithm.

Drawing from Gilligan's study, it becomes clear that on an international level, the development of nation-states is seen the same way. What appear to be universally agreed upon guidelines to which developing societies must conform are actually impositions of Western standards through cultural imperialism. Gilligan did for the study of personal development what must be done for economic development: allowing for difference. Just as the development of individuals should be seen as the flowering of that which is special and unique within each of us—a process by which an acorn becomes an oak rather than being obliged to become a maple—so the development of peoples should be conceived as the flowering of what is special and unique within each culture. This is not to argue for a cultural relativism in which all beliefs and practices sanctioned by some culture are equally valid on a moral, aesthetic, or practical plane. But it is to reject the universality claimed by Western beliefs and practices.

Of course, some might ask what the loss of a culture here or there matters if it is the price of material progress, but there are two flaws to this argument. First, cultural destruction is not necessarily a corollary of the technologies that extend life and improve its quality. Western technology can be decoupled from the entailments of Western culture. Second, if I am wrong about this, I would ask, as Jesus does in the account of Saint Mark, "[W]hat shall it profit a man, if he shall gain the whole world, and lose his own soul?" For all the material progress that the West has achieved, it has paid a high price through the weakening to the breaking point of communal ties. We in the West have much to learn, and the cultures that are being destroyed in the name of progress are perhaps the best resource we have for restoring balance to our own lives. The advantage of taking a critical stance with respect to our own culture is that we become more ready to enter into a genuine dialogue with other ways of being and believing.

The Culture of the Modern West

Culture is in the last analysis a set of assumptions, often unconsciously held, about people and how they relate to one another. The assumptions of modern Western culture can be described under five headings: individualism, self interest, the privileging of "rationality," unlimited wants, and the rise of the moral and legal claims of the nation-state on the individual.

Individualism is the notion that society can and should be understood as a collection of autonomous individuals, that groups—with the exception of the nation-state—have no normative significance as groups; that all behavior, policy, and even ethical judgment should be reduced to their effects on individuals. All individuals play the game of life on equal terms, even if they start with different amounts of physical strength, in-

INSURANCE

Spending on Insurance Premiums

Region	Percent of Global Premium Market
NORTH AMERICA	**37.32**
Canada	1.91
United States	35.41
LATIN AMERICA	**1.67**
Brazil	0.51
Mexico	0.4
EUROPE	**31.93**
France	4.99
Germany	5.06
UK	9.7
ASIA	**26.46**
China	0.79
India	0.41
Japan	20.62
AFRICA	**1.03**
South Africa	0.87
OCEANIA	**1.59**
Australia	1.46

http://www.internationalinsurance.org

tellectual capacity, or capital assets. The playing field is level even if the players are not equal. These individuals are taken as given in many important ways rather than as works in progress. For example, preferences are accepted as given and cover everything from views about the relative merits of different flavors of ice cream to views about the relative merits of prostitution, casual sex, sex among friends, and sex within committed relationships. In an excess of democratic zeal, the children of the 20th century have extended the notion of radical subjectivism to the whole domain of preferences: one set of "preferences" is as good as another.

Self-interest is the idea that individuals make choices to further their own benefit. There is no room here for duty, right, or obligation, and that is a good thing, too. Adam Smith's best remembered contribution to economics, for better or worse, is the idea of a harmony that emerges from the pursuit of self-interest. It should be noted that while individualism is a prior condition for self-interest—there is no place for self-interest without the self—the converse does not hold. Individualism does not necessarily imply self-interest.

The third assumption is that one kind of knowledge is superior to others. The modern West privileges the algorithmic over the experiential, elevating knowledge that can be logically deduced from what are regarded as self-evident first principles over what is learned from intuition and authority, from touch and feel. In the stronger form of this ideology, the algorithmic is not only privileged but recognized as the sole legitimate form of knowledge. Other knowledge is mere belief, becoming legitimate only when verified by algorithmic methods.

Fourth is unlimited wants. It is human nature that we always want more than we have and that there is, consequently, never enough. The possibilities of abundance are always one step beyond our reach. Despite the enormous growth in production and consumption, we are as much in thrall to the economy as our parents, grandparents, and great-grandparents. Most US families find one income inadequate for their needs, not only at the bottom of the distribution—where falling real wages have eroded the standard of living over the past 25 years—but also in the middle and upper ranges of the distribution. Economics, which encapsulates in stark form the assumptions of the modern West, is frequently defined as the study of the allocation of limited resources among unlimited wants.

Finally, the assumption of modern Western culture is that the nation-state is the pre-eminent social grouping and moral authority. Worn out by fratricidal wars of religion, early modern Europe moved firmly in the direction of making one's relationship to God a private matter—a taste or preference among many. Language, shared commitments, and a defined territory would, it was hoped, be a less divisive basis for social identity than religion had proven to be.

An Economical Society

Each of these dimensions of modern Western culture is in tension with its opposite. Organic or holistic conceptions of society exist side by side with individualism. Altruism and fairness are opposed to self interest. Experiential knowledge exists, whether we recognize it or not, alongside algorithmic knowledge. Measuring who we are by what we have has been continually resisted by the small voice within that calls us to be our better selves. The modern nation-state claims, but does not receive, unconditional loyalty.

So the sway of modern Western culture is partial and incomplete even within the geographical boundaries of the West. And a good thing too, since no society organized on the principles outlined above could last five minutes, much less the 400 years that modernity has been in the ascendant. But make no mistake—modernity is the dominant culture in the West and increasingly so throughout the world. One has only to examine the assumptions that underlie contemporary economic thought—both stated and unstated—to confirm this assessment. Economics is simply the formalization of the assumptions of modern Western culture. That both teachers and students of economics accept these assumptions uncritically speaks volumes about the extent to which they hold sway.

It is not surprising then that a culture characterized in this way is a culture in which the market is the organizing principle of social life. Note my choice of words, "the market" and "social life," not markets and economic life. Markets have been with us since time out of mind, but the market, the idea of markets as a system for organizing production and exchange, is a distinctly modern invention, which grew in tandem with the cultural assumption of the self-interested, algorithmic individual who pursues wants without limit, an individual who owes allegiance only to the nation-state.

There is no sense in trying to resolve the chicken-egg problem of which came first. Suffice it to say that we can hardly have the market without the assumptions that justify a market system—and the market system can function acceptably only when the assumptions of the modern West are widely shared. Conversely, once these assumptions are prevalent, markets appear to be a "natural" way to organize life.

Markets and Communities

If people and society were as the culture of the modern West assumes, then market and community would occupy separate ideological spaces, and would co-exist or not as people chose. However, contrary to the assumptions of individualism, the individual does not encounter society as a fully formed human being. We are constantly being shaped by our experiences, and in a society organized in terms of markets, we are formed by our experiences in the market. Markets organize not only the production and distribution of things; they also organize the production of people.

The rise of the market system is thus bound up with the loss of community. Economists do not deny this, but rather put a market friendly spin on the destruction of community: impersonal markets accomplish more efficiently what the connections of social solidarity, reciprocity, and other redistributive institutions do in the absence of markets. Take fire insurance, for example. I pay a premium of, say, US$200 per year, and if my barn burns down, the insurance company pays me US$60,000 to rebuild it. A simple market transaction replaces the more cumbersome method of gathering my neighbors for a barn-raising, as rural US communities used to do. For the economist, it is a virtue that the more efficient institution drives out the less efficient. In terms of building barns with a minimal expenditure of resources, insurance may indeed be more efficient than gathering the community each time somebody's barn burns down. But in terms of maintaining the community, insurance is woefully lacking. Barn-raisings foster mutual interdependence: I rely on my neighbors economically—as well as in other ways—and they rely on me. Markets substitute impersonal relationships mediated by goods and services for the personal relationships of reciprocity and the like.

Why does community suffer if it is not reinforced by mutual economic dependence? Does not the relaxation of economic ties rather free up energy for other ways of connecting, as the English economist Dennis Robertson once suggested early in the 20th century? In a reflective mood toward the end of his life, Sir Dennis asked, "What does the economist economize?" His answer: "[T]hat scarce resource Love, which we know, just as well as anybody else, to be the most precious thing in the

world." By using the impersonal relationships of markets to do the work of fulfilling our material needs, we economize on our higher faculties of affection, our capacity for reciprocity and personal obligation—love, in Robertsonian shorthand—which can then be devoted to higher ends.

In the end, his protests to the contrary notwithstanding, Sir Dennis knew more about banking than about love. Robertson made the mistake of thinking that love, like a loaf of bread, gets used up as it is used. Not all goods are "private" goods like bread. There are also "public" or "collective" goods which are not consumed when used by one person. A lighthouse is the canonical example: my use of the light does not diminish its availability to you. Love is a *hyper* public good: it actually increases by being used and indeed may shrink to nothing if left unused for any length of time.

Economics is simply the formalization of the assumptions of modern Western culture. That both teachers and students of economics accept these assumptions uncritically speaks volumes about the extent to which they hold sway.

If love is not scarce in the way that bread is, it is not sensible to design social institutions to economize on it. On the contrary, it makes sense to design social institutions to draw out and develop the community's stock of love. It is only when we focus on barns rather than on the people raising barns that insurance appears to be a more effective way of coping with disaster than is a community-wide barn-raising. The Amish, who are descendants of 18th century immigrants to the United States, are perhaps unique in the United States for their attention to fostering community; they forbid insurance precisely because they understand that the market relationship between an individual and the insurance company undermines the mutual dependence of the individuals that forms the basis of the community. For the Amish, barn-raisings are not exercises in nostalgia, but the cement which holds the community together.

Indeed, community cannot be viewed as just another good subject to the dynamics of market supply and demand that people can choose or not as they please, according to the same market test that applies to brands of soda or flavors of ice cream. Rather, the maintenance of community must be a collective responsibility for two reasons. The first is the so-called "free rider" problem. To return to the insurance example, my decision to purchase fire insurance rather than participate in the give and take of barn raising with my neighbors has the side effect—the "externality" in economics jargon—of lessening my involvement with the community. If I am the only one to act this way, this effect may be small with no harm done. But when all of us opt for insurance and leave caring for the community to others, there will be no others to care, and the community will disintegrate. In the case of insurance, I buy insurance because it is more convenient, and—acting in isolation—I can reasonably say to myself that my action hardly undermines the community. But when we all do so, the cement of mutual obligation is weakened to the point that it no longer supports the community.

The free rider problem is well understood by economists, and the assumption that such problems are absent is part of the standard fine print in the warranty that economists provide for the market. A second, deeper, problem cannot so easily be translated into the language of economics. The market creates more subtle externalities that include effects on beliefs, values, and behaviors—a class of externalities which are ignored in the standard framework of economics in which individual "preferences" are assumed to be unchanging. An Amishman's decision to insure his barn undermines the mutual dependence of the Amish not only by making him less dependent on the community, but also by subverting the beliefs that sustain this dependence. For once interdependence is undermined, the community is no longer valued; the process of undermining interdependence is self-validating.

Thus, the existence of such externalities means that community survival cannot be left to the spontaneous initiatives of its members acting in accord with the individual maximizing model. Furthermore, this problem is magnified when the externalities involve feedback from actions to values, beliefs, and then to behavior. If a community is to survive, it must structure the interactions of its members to strengthen ways of being and knowing which support community. It will have to constrain the market when the market undermines community.

A Different Development

There are two lessons here. The first is that there should be mechanisms for local communities to decide, as the Amish routinely do, which innovations in organization and technology are compatible with the core values the community wishes to preserve. This does not mean the blind preservation of whatever has been sanctioned by time and the existing distribution of power. Nor does it mean an idyllic, conflict-free path to the future. But recognizing the value as well as the fragility of community would be a giant step forward in giving people a real opportunity to make their portions less meager and avoiding the poison.

The second lesson is for practitioners and theorists of development. What many Westerners see simply as liberating people from superstition, ignorance, and the oppression of tradition, is fostering values, behaviors, and beliefs that are highly problematic for our own culture. Only arrogance and a supreme failure of the imagination cause us to see them as universal rather than as the product of a particular history. Again, this is not to argue that "anything goes." It is instead a call for sensitivity, for entering into a dialogue that involves listening instead of dictating—not so that we can better implement our own agenda, but so that we can genuinely learn that which modernity has made us forget.

STEPHEN A. MARGLIN is Walter S. Barker Professor of Economics at Harvard University.

Why People Still Starve

The real crisis of hunger in Africa is that it is so widespread, chronic—and intractable.
From Malawi, a chronicle of starvation foretold.

By Barry Bearak

Late one afternoon, during the long melancholia of the hungry months, there was a burst of joyous delirium in Mkulumimba. Children began shouting the word *ngumbi*, announcing that winged termites were fluttering through the fields. These were not the bigger species of the insect, which can be fried in oil and sold as a delicacy for a good price. Instead, these were the smaller ones, far more wing than torso, which are eaten right away. Suddenly, most everyone was giddily chasing about; villagers were catching *ngumbi* with their fingers and tossing them onto their tongues, grateful for the unexpected gift of food afloat in the air.

Adilesi Faisoni was able to share in that happiness but not in the cavorting. For several years, old age had been catching up with her, until it had finally pulled even and then ahead. Her walk was unsteady now, her posture stooped, her eyesight dimmed. As the others ran about, she remained seated on the wet ground near the doorstep of her mud-brick hovel. It was the same place I always found her during my weeks in the villages of Malawi, weeks when I was examining the mechanisms of famine. "There is no way to get used to hunger," Adilesi told me once. "All the time something is moving in your stomach. You feel the emptiness. You feel your intestines moving. They are too empty, and they are searching for something to fill up on."

Hunger was the main topic of our talks. Most every year, Malawi suffers a food shortage during the so-called hungry months, December through March, the single growing season in a predominantly rural nation. Corn is this country's mainstay, what people mainly grow and what people mainly eat, usually as *nsima*, a thick porridge. Ideally, the yield from one harvest lasts until the next. But even in good times the food supply is nearing its end while the next crop is still rising from the ground. Families often endure this hungry period on a single meal a day, sometimes nothing more than a foraged handful of greens. Last year's food crisis was the worst in living memory. Hundreds, and probably thousands, of Malawians succumbed to the scythe of a hunger-related death.

Among those who perished were Adilesi's husband, Robert Mkulumimba, and their grown daughter Mdati Robert, herself the mother of four young sons. The two died within a month of each other, unable to subsist on the pumpkin leaves and wild vegetables that had become the family's only nourishment. "The first symptom was the swollen feet, and then the swelling started to move up his body," Adilesi said of her husband's illness.

It was strange the way Robert seemed to fade. Before the start of the hungry months, it had been he who had kept the family going, leaving before dawn each day to sell firewood or tend someone's fields. But then work became impossibly scarce, and Robert seemed to be using himself up in the search for it. "At the peak of the crisis, there was nothing to do but beg, and you were begging from others who needed to beg."

Even small jolts to the regular food supply can jar open the trapdoor between what is normal, which is chronic malnutrition, and what is exceptional, which is outright starvation.

As most people visualize it, famine is a doleful spectacle, the aftershock of some calamity that has left thousands of the starving flocked together, emergency food kept from their mouths by the perils of war or the callousness of despots or the impassibility of washed-away roads. But more often, in the nether regions of the developing world, famine is both less obvious and more complicated. Even small jolts to the regular food supply can jar open the trapdoor between what is normal, which is chronic malnutrition, and what is exceptional, which is outright starvation. Hunger and disease then malignly feed off each other, leaving the invisible poor to die in invisible numbers.

Nowhere is this truer than in sub-Saharan Africa, where President Bush was recently scheduled to travel. Each year, most nations in the region grow poorer, hungrier and sicker. Their share of global trade and investment has been collapsing. Average per capita income is lower now than in the 1960's, with half the population surviving on less than 65 cents a day. It is a situation seldom noticed, as wars on poverty are neglected for wars more animate. African countries now hold the 27 lowest places on the human-development index—a combined measure of health, literacy and income calculated by the United Nations. They occupy 38 spots in the bottom 50.

During the past decade or so, the poorest of Africa's poor have suffered as rarely before. Merely to survive, they have sold off their meager assets—household goods and farm animals and the tin roofs of their homes. Just now, the most urgent need is in Ethiopia, Eritrea and Zimbabwe. But hunger has become a chronic problem throughout the region, often occurring even under the best of weather conditions. The World Food Program warns that nearly 40 million Africans are struggling against starvation, a "scale of suffering" that is "unprecedented." Coincident with the hunger is H.I.V./AIDS, which has beset sub-Saharan Africa in a disproportionate way, cursing it with 29.4 million infections, nearly three-quarters of the world's caseload. Very few of the stricken can afford the drugs that forestall the virus's death work, and family after family is being purged of its breadwinning generations, leaving the very young and the very old to cope.

With survival so precarious, life is lived at the edge of nothingness, easily pushed over the side. Take Malawi, I was told again and again—for this land-locked, overpopulated nation in southeastern Africa seems to be a favored specimen of researchers. There is a relative innocence to Malawi's impoverishment: no tyrannical dictator currently in power, no army of goons marauding in civil war, no disastrous weather wiping out the harvest. And yet last year, the nation was nudged into starvation. It happened while there was grain in the stores, if only the poor had the money to buy it. It happened while well-meaning people were arguing about whether it was happening at all.

Three years ago, Adilesi's daughter, Lufinenti, separated from her husband, who later died after a long illness. Adilesi lost her husband, Robert, to starvation last year.

To track the origins of the crisis, my plan was this: to find a family that had lost someone to last year's hunger and then work my way back through the hows and whys. Though I mostly shuttled over the narrow and soggy mud roads between Mkulumimba, Adilesi's village, and Lilongwe, the capital, the actual route of causality reaches beyond Malawi's borders. It extends toward wealthier nations and their shared institutions—the World Bank and the International Monetary Fund. It travels the uncertain ups and downs of global commodity prices and currency valuations—and of course passes into the limited access roads of humanity's conscience.

"THE NEW GENERATION are the unfortunates, because now there is a food shortage every year," Adilesi said. "Things began getting bad when I was done with my childbearing years. If they had been this bad before, all my children would have died."

In a Malawian village, guests are customarily greeted outside and offered a mat as a seat. Adilesi's front door faced a broad clearing that slanted down, allowing a splendid view of cornfields and acacias and the Dzalanyama Mountains in the distance. About 20 yards to the left was a banana tree and the ruins of an outhouse, its mud walls half-collapsed by a recent storm. The main house itself was a single room, about 9 feet by 12 feet, with a roof of bundled grass. In one corner were the ashy remnants of a small fire. Otherwise, the room was empty but for a tin pail, two pots, a few baskets and plastic bowls and some empty grain sacks that could be used as blankets. Adilesi lived in this hut with her daughter Lufinenti and 10 grandchildren. It was hard to imagine the geometry that would allow them all to sleep in so spare a space.

"We squeeze like worms," Adilesi said, explaining rather than complaining.

Thin-faced and withered, the old woman owned only one set of clothes, a colorful wrap that went around her waist and a faded T-shirt that showed a San Francisco street scene and advertised Levi's 501 jeans. The lettering proclaimed, "Quality Never Goes Out of Style." She had no idea what the words meant. "Is this something offensive?" she asked in Chichewa, Malawi's main language, bending her head down so she could examine the thinned cloth. The villagers all bought such used clothing, the discards of people from richer nations. Children who had never seen television unwittingly sported apparel that allied them with Ninja Turtles and Power Rangers. Often, holes in these shirts rivaled the size of the remaining garment. This shamed the children, and some refused to go to school. "I have no idea what San Francisco is," Adilesi said with a smile, repeating the words she had just been told. "I couldn't tell you whether it's an animal or a man."

She did not know her age, either, but she could remember the historic famine of 1949. She was a youngster then, that year when the skies cruelly withheld the rains. The undisturbed sun not only parched the cornstalks; it seemed to melt the glue that held the village together. Neighbors, once generous, hid away what food they had, afraid of theft. Women sang prayers of apology to their ancestors for any conceivable wrongdoing and begged them to reopen the clouds. Men wandered far from their homes, disappearing for weeks in a desperate search for work. "We refer to it simply as '49," Adilesi said.

She married Robert soon after. She can't recall exactly when. Robert was a nephew of the village chief, and their wedding was preceded by an evening of dancing, with the entire village sharing in a feast of two goats, several chickens and homemade beer. Vows were recited at the African Abraham Church nearby. Adilesi would later bear 12 children, including 8 who lived to be adults—an average rate of survival in a country where half the children suffer stunted growth and one in four die before age 5.

Robert, tall and stout, was a good provider. As a young man, he went to South Africa and toiled in the mines. Then, back in Malawi, he worked for the forestry department, slashing away underbrush with his panga knife. In a village of farmers, he was one of the few men who carried home monthly paychecks. But that job ended four years ago, when the government, under pressure from foreign lenders, drastically reduced its payroll. Robert then spent more time farming and doing *ganyu*, day labor.

Toward the end of 2001, after an overabundance of rain and a disappointing harvest, corn prices leapt as high as 40 *kwacha* per kilo, about 50 cents, a forbidding sum for people used to paying a tenth as much. Foraging became necessary, as it had been in '49, as it was last year, as it is even now. The toil was not unproductive. In the openness of the plain, with the daily rain slapping hard at the mud, edible leaves reached out for the taking from small stems. They held vitamin C, some iron, some beta carotene. Occasionally there were tubers. People could eat, just not a balanced diet, just not enough.

Hunger, like many diseases, is often an abettor of death rather than an absolute cause. Who really knew: was it the tuberculosis or the malnutrition that came first, and which of them delivered the fatal blow? But symptomatically, starvation usually arrives with anemia and extreme wasting and swelling from fluid in the tissue. There can be loss of appetite, and there can be diarrhea. Robert suffered all these symptoms. That a grown man would be among those to succumb to the hunger was not so uncommon. Men, it was explained to me, used up more of themselves in the unceasing search for *ganyu*.

Whatever the undertow, Robert grew too weak to work. He and Adilesi went to the government hospital, where he was treated for malnutrition, then later treated for malaria, then sent home. When they released him, the doctors said he needed to eat better or he would die. Inevitably, there was little food, so he began his capitulation, imparting final goodbyes. "He told me we needed to remain united as a family," said Kiniel, 16, the youngest of his children.

ROBERT'S DAUGHTER MDATI fell ill soon after he went into his decline. She was about 30. Her husband had been a philanderer to whom she had said good riddance, and now she was suddenly incapable of caring for their four sons herself. The entire family had always depended on her. Mdati was the only one who could read and write. "She went to school up to the second grade," said her sister Lufinenti. "She was very smart."

Unlike her father, Mdati couldn't keep food down when she found something to eat. This raised a suspicion that she had somehow been bewitched.

The family regularly attended the African Abraham Church, a tiny red-brick building with pews and an altar molded out of mud. As with most Christians in the area, they found ways to blend witchcraft into their beliefs. "Some people protect their fields with charms, but we can't afford such things," Adilesi told me. This safeguard against thievery required the interces-

sion of someone with magical powers, a *sing'anga*. (My interpreter—the daily intermediary between my English and the villagers' Chichewa—used the word "witch doctor," though a more respectful term would be "traditional healer.") The family had great hopes that a *sing'anga* could break the spell that gripped Mdati. They took her to two of them.

The first, Bomba Kamchewere, is a tall, bony man with a missing front tooth. When I visited him, he spread out a mat of tightly stitched reeds so we could sit together beneath his favorite tree. He had been tutored, in dreams, by Jesus himself, he said. But even with divine insight into the curative uses of roots and herbs, his powers had limits. While he claimed to cure stomachaches, venereal disease and tuberculosis, he confessed that other sicknesses baffled him. AIDS was particularly confounding, as was *njala*, or hunger, which had been Mdati's problem. "With her case, the spirits told me I could not do anything," he said. Then, somewhat shamefully, he confessed that around that time he himself had endured *njala*, quite frighteningly so. "I went three weeks without any solid food, and I developed some strange swelling." At a hospital, the doctors recommended that he eat more, which was advice that struck the *sing'anga* as less than a revelation.

Mose Chinkhombe, a young, self-confident man with a spacious smile, was the second healer. His home was hours away in the village of Chiseka. To get there, the starving Mdati, limp as a rag doll, had to be placed on a borrowed bicycle and guided over the roads by five companions, who took turns keeping both her and the wheels steady.

A year later, the healer still remembered her. "My diagnosis was anemia," he said as he sat in a dark room on a half-sack of dried lime, all the while shooing flies with an oxtail. He was in his vocational attire, a spotless white frock and floppy hat. "She was so weak from lack of food," the healer said. "I could treat her for this anemia. But I told her she needed to eat enough food to recharge her body. When she left, she had improved slightly. But then I heard she died." He nodded rather forcefully as he said this. Then, perhaps in defense of his medical craft, he apparently felt he needed to tell me the obvious, that the "big hunger" had taken a great many lives during those dismal months.

There is a belief that when a stray black dog crosses your path, terrible times will come, he said. "Last year," he explained, "a black dog walked across the entire country."

SOME 11 MILLION people live in Malawi, though far too few live especially long. Average life expectancy from birth has fallen to 36, one of the lowest figures anywhere, according to the World Health Organization. Tuberculosis cases have doubled in the past 10 years, and an estimated 16 percent of people ages 15 to 49 have H.I.V./AIDS, though with little testing going on, few of them know it. Nearly 500,000 children have lost one or both parents to the virus, according to the United Nations. In the villages, where AIDS is seldom discussed, people call it "government disease," because it seems to spring from the city.

For the poor, conditions began rapidly deteriorating in the early 90's, during the last days of the "Lion of Malawi," Dr.

Hastings Kamuzu Banda, the Western-trained physician who was the nation's dictator for 40 years. Tobacco is Malawi's only major cash crop, and the doctor amassed a fortune by granting himself valuable licenses to grow it. At the same time, his government benefited from the foreign aid of prosperous friends. The West applauded his steadfast anti-Communism; South Africa admired his tolerance of apartheid. Banda ruled—ruthlessly and myopically—until 1994, long enough to take himself well into his 90's and senility.

By then, geopolitical necessities had changed, as had theories on how to develop the third world. Benefactors began attaching tighter strings to their money, first during the final decade of the Banda regime, then with the subsequent elected government. The World Bank and the International Monetary Fund had entered their "structural adjustment" period. Austerity in government spending was preached, the overriding principle being that the poor were best served through the efficiency of free markets. The fine print in most loan agreements committed governments to reduce subsidies, curtail spending and sell off monopolies.

Whatever eventual benefit there might be in such reforms, the immediate impact on Malawian farmers—paupers during the best of times—was distress. Corn prices, no longer set by the government, became unpredictable. Given the risk caused by instability, the private sector did not mature as expected. Worse yet, the *kwacha* was repeatedly devalued. Falling prices in the world tobacco market had strained already thin foreign-currency reserves. At the urging of the I.M.F., the government instituted small devaluations in 1990 and 1991 and two larger ones in 1992. Finally, in 1994, Malawi moved from a fixed exchange rate for the *kwacha* to one that floated. For farmers, that meant the cost of fertilizer, an imported good, ballooned as the *kwacha* shriveled.

Before, fertilizer had been subsidized. Loans had been, too. Farmers now found themselves adrift "in the worst of both worlds, a Bermuda Triangle," deprived of the benefits of a regulated economy while yet to gain the benefits of a free market, said Lawrence Rubey, the United States Agency for International Development's chief of agriculture in Malawi. He gave an example with some dismal arithmetic: in dollar terms, the price of a bag of fertilizer had actually gone down. But in devalued *kwachas*, the cost had risen fivefold. This was devastating to farmers with badly leached soil. "The past arrangement of high state control of the economy was inefficient, but at least it was stable," Rubey said.

A YEARNING FOR the stability of the Banda days—though rarely for the doctor himself—is a commonly expressed sentiment throughout the nation. Malawi's youthful democracy has lacked equilibrium, even though Bakili Muluzi, a onetime Banda protégé who fell out of favor, has been its only president. The unsteadiness results in large part from the pull-and-tug of two parallel sources of power, the elected government and the international patrons who finance it. Like many poor, heavily indebted countries, Malawi operates something like a business in receivership. Lenders and donors—among them the World

Bank, the I.M.F., the British, the Americans and the European Union—carefully monitor fiscal policy and budget expenditures. Their approvals are necessary, or their generosity is withdrawn. The spigot of aid goes on, off, on, off.

The World Bank's chief economist points out that each day, the average European cow receives $2.50 in subsidies while 75 percent of the people in Africa are scrimping by on less than $2.

Understandably, this has made for a peevish relationship. The Malawians quite correctly contend that the donors are hypocrites: while opposing state subsidies elsewhere, wealthy nations hand out $1 billion a day to their own farmers, about six times what they give in development aid to the globe's poor. (Nicholas Stern, the World Bank's chief economist, once pointed out that each day, the average European cow receives $2.50 in subsidies while 75 percent of the people in Africa are scrimping by on less than $2.) These subsidies also depress commodity prices, undercutting the ability of developing nations to compete in world markets and get their nations off the dole.

The benefactors, on the other hand, quite correctly contend that the government is persistently wasteful and inconsistently honest, prone to overspending on frivolous travel and lax in underwriting programs for the poor. So they give their aid with a chiding finger and chastening attitude: Malawi needs better government!

And yet good government, like good deeds, is most often a complicated matter.

With fertilizer so unaffordable, the government in 1998 cooperated with the British, the World Bank and others to furnish beleaguered farmers with a "starter pack," enough free seed and fertilizer to grow a healthy quarter-acre, about 15 percent of a typical family plot. Soon, the cornfields of Malawi took on the look of a bad haircut, with one cluster of stalks as high as an elephant's eye and the rest barely above the knee.

The starter-pack program combined with favorable weather to produce a bumper crop in the 1998–99 season. In fact, the surplus was so great that a newly established government safeguard, the National Food Reserve Agency, was able to purchase huge amounts, fattening its grain stockpile to about 190,000 tons. For Malawi, so often hungry, this cache was a comforting buffer against famine. But it was also costly. The reserve agency, without capital of its own, bought the corn with loans at a staggering domestic interest rate, above 50 percent. Storage itself was expensive, and after a second bumper crop in 1999–2000, there were concerns that the untapped grain reserves would rot.

The donors felt adjustments were necessary. The starter-pack program now seemed too benevolent. Were farmers to be given fertilizer forever? they asked. If so, that would create a dependency that in the jargon of development economics was "unsustainable." So the program was reconstructed as "targeted inputs," cut in half to reach only the "neediest" of the destitute. As for the grain reserve, the donors, like the government, fretted over those gargantuan interest payments. To satisfy the debt, they suggested that the corn be sold, with future stocks kept to a modest 35,000 to 65,000 tons, considered enough to meet most emergencies until more grain could be imported.

MALAWI HAS STUNNING skies, with a blue so bright and clouds so shapely that they seem to be the work of a cartoonist. Those skies are also fickle, suddenly exchanging a sunny disposition for an angry pout and unleashing thunderstorms that seem to hurl water rather than drop it: In February and March of 2001, as the harvest approached, the skies were angrier than usual, causing regional flooding. That brought the first clue of the trouble to come. This accumulating water kept the fields a hearty green, but the cornstalks, standing uncomfortably in shallow pools, failed to mature fully. Many a farmer was fooled by the deepness of color. They had thought they were going to have a good harvest and did not face the disappointing truth until it came time to pick.

Farmers were not the only ones fooled. Crop forecasters from the Agriculture Ministry underestimated the impact of the flooding on the corn harvest. These errors were compounded by a grossly wayward overestimation of the cassava and sweet potato yield, a false optimism perhaps bred from a false pride. The Agriculture Ministry was involved in a multimillion-dollar United States aid program for crop diversification. Field agents who reported high expectations for these roots and tubers were the same people whose job it was to encourage their planting. The net effect was to delay concern about the total food supply. Indeed, a surplus was predicted: if people ran out of corn, let them eat cassava.

By late summer, when the final estimates came in, the corn harvest was put at 1.9 million tons, about a third lower than the excellent crop of the year before. This was considered a bad though not a terrible yield. And yet to those attentive enough, something ominous was happening. In some areas, corn prices had increased more than fivefold, a sign that stocks were perilously low. Though neither the nation's preoccupied president, Bakili Muluzi, nor the donors were yet alarmed, the Agriculture Ministry was fretful enough to advocate a grain purchase. In September, a decision was announced to import corn from South Africa, Tanzania and Uganda.

As usual, the government asked the donors to foot the bill. The donors, in reply, inquired about the national grain reserves—and were shocked to learn that they were entirely gone. Yes, the donors acknowledged, they had recommended the selling of most of it. But if the entire storehouse was empty, who bought it? And where was the money?

Answers were not forthcoming. That infuriated the always suspicious donors, so much so that the mystery of the missing grain overshadowed the unfolding fate of the fragile poor for several crucial months. This delay was tragic. By year's end, people were lining up at clinics to plead for food. Jos Kuppens, an activist Dutch priest, said he saw starving people resorting to meals of fodder, with one woman even thickening her gruel with sawdust. He recalled a hopelessly thin little girl he had seen at a Catholic hospital. He asked her where she was from. "I had to bend close to hear her," he said. "She could barely speak, and she said, 'Njala,' which is the word for hunger. The sister told me: 'There's nothing we can do for a child like this. It's too late.'"

Despite the donors' unwillingness to pay the bill, the government tried to proceed with its plans to import 220,000 tons of corn. But a freak coincidence of disasters stalled deliveries: washed-out bridges, a derailed train, a port on fire. It also became clear that Malawi's neighbors faced a similar shortage. They began to vie to buy the same food.

A British agronomist named Harry Potter, dean of the foreign agricultural experts among the donors in Lilongwe, is a man far less inclined toward wizardry than his fictional namesake. Nevertheless, he saw a supernatural hand dispensing punishment.

"It was almost as if somebody up there had decided, Malawi, you have to be taught a lesson," he said. But cosmic reprimand or not, the consequences were of course grave only to the very poor and not to their nefarious officials.

Little grain would arrive in time for the start of the hungry months in December.

FAMINE IS VARIOUSLY defined by scholars, though its common usage most often implies a large surge in hunger-related mortality. Some would even designate a minimum body count. Malawians, at least linguistically, make little distinction between "hunger" and "famine" except to say that famine is "the hunger that kills." Perspective, of course, changes depending on whether you are studying the situation or starving within it. Stephen Devereux of the University of Sussex—an expert on African food security—talks about "outsider" and "insider" views of such crises. Outsiders presume famine to be an extraordinary event, while insiders see it as an intensification of what they already suffer.

The Nobel laureate Amartya Sen is the world's best-known authority on famine. He argues that such catastrophes do not occur in a "functioning democracy," where a free press is a roving sentinel and elected governments have strong incentives to take preventive measures. If so, it is hard to say whether Malawi would be an exception. First, there is the matter of how well the nation's nine-year-old democracy qualifies as functioning. Then there is the question of whether the hunger that killed Robert Mkulumimba and Mdati Robert qualifies as famine.

By the beginning of the hungry months, starvation deaths were already being recorded within miles of the capital. Lilongwe seems a very un-Malawian city, with a scattering of big office buildings and fancy shopping malls. Notable on some thoroughfares are dozens of roadside "coffin workshops," their carpenters steadily at work in the open air, taking morbid advantage of a rare growth industry. Capital Hill—home to the government ministries—is an isolated slope of manicured land, best reachable by car, far away from the masses.

For weeks, I tried to interview President Muluzi. Finally, I was informed that he saw no reason to see me since I had already spoken with his friend Friday Jumbe, the finance minister.

That conversation had seemed cordial enough, though on reflection, perhaps I was considered impertinent for asking about the missing grain reserves. At the time of the sale, Jumbe headed a quasi-governmental corporation that had physical control of the stored corn. A parliamentary committee had investigated the deals, and while the probe turned up little solid evidence, its report concluded that the reserves were "mainly sold to politicians and individuals politically connected." Jumbe was faulted for failing to explain his recent financing of a luxury hotel in the city of Blantyre.

"It is true that some people bought the maize at a price of 4 or 5 *kwacha*, and then one day maize became scarce," the finance minister told me one morning in his office. "That's business!" Elegantly dressed in a three-piece suit, he was dismissive, and occasionally sarcastic, about the allegations. He said it was "foolish" to assume that he was a thief just because he owned a hotel and pointed out that the author of the damning report had been fired and that a revised investigation of the facts was in the works. "If there was any theft of money, maybe, I don't know what, 100 or 200 or 300 *kwacha*—but not big money. But the donors, that's where the bone of contention is. They are assuming that certain politicians bought this grain at a small price and made a killing."

Indeed, that is—and was—the assumption. To this day—after two years and five investigations—the donors know little more than that some traders bought corn for very little and sold it for quite a lot, yet another scandalous story about people in the right place at the right time with the right friends, a Malawian Enron. Unfortunately, the donors' righteous pique played a part in the tardy recognition of the deadly hunger, as did Muluzi's willful ignorance. In February of last year, the president was still refuting any reports of a food crisis. "Nobody has died of hunger," he insisted.

But by then, firm statistical proof had begun to appear in repeated health surveys. Also, hospital admissions for severe malnutrition were swiftly escalating far beyond the normal. At month's end, even Muluzi did a turnabout. On Feb. 27, he declared a state of emergency in a national broadcast—11 days after the death of Robert Mkulumimba.

Mkulumimba is one in a cluster of five villages, each with 35 to 60 households. Their boundaries reach deeply into one another, and it was sometimes possible to move between them simply by walking from hut to hut. Any visit required the permission of the village headman—the chief—who customarily then assumed the role of official escort. I was fortunate to be befriended by Elias Mitengo, a 50-year-old headman widely held in high regard. A poor farmer, he combined the gravitas of a diplomat with the easy humor of a kibitzer. He was chief of the village of Mdauma. And he was also Adilesi's son-in-law.

Elias said he had always tried to help his in-laws whenever he could. But last year, he often did not have enough food for even his own children. "It was an impossible time," he said.

Being headman, an inherited job, included sensitive duties. The chief allocated land and mediated disputes and sometimes judged criminals. He was also the village marriage counselor and arbiter of property settlements in the event of divorce. Elias prided himself on absolute fairness. When I wanted to observe church on Sunday mornings, he went along and even addressed the congregations. But generally, he said, he no longer attended services. With five churches in the area, he worried about favoritism.

One morning, Elias invited me to a meeting under a stand of trees. Several chiefs were discussing a proper punishment for stealing food. Instant justice was being applied in some places, with the hungry crook forfeiting a hand. This, the headmen agreed, was too harsh. Yet when thieves were simply turned over to the police, they were released after a few days, and that seemed too lenient. A better penalty, the chiefs agreed, would be stiff fines—perhaps three goats or chickens. But this was impractical. Thieves usually belonged to the poorest of families. The discussion went on and on without resolution.

For Elias, hungry thieves seemed less a problem than organized crime. In recent years, it was not uncommon for a truckload of men to descend during the night. Usually, they rustled livestock. But sometimes they emptied entire cornfields of ripe cobs. Elias decided to levy a 5 *kwacha* tax on each household, enough to pay vigilantes to guard the roads, waiting with axes and pangas and bows and arrows.

"These are not the best of days," he told me, laughing at his own understatement.

Actually, Elias had plunged into deep pessimism, something he was more likely to express after a few liters of *chibuku*, a cheap beer made from corn and millet. He said the *njala*, the hunger, was prying apart families, turning husbands and wives against each other. "These women carry their vegetables to the trading center, and if they can't sell them, they sell themselves," he said. "It's the poverty." Sometimes, he told me, he would call village meetings to speak out against prostitution, particularly addressing the young women. "I tell them: 'You see your elders. They got to live this long because they kept themselves clean of promiscuity.'" He would invoke the threat of "government disease."

Men were hardly blameless, Elias said. Some were lazy. And of those who were not, some occasionally found work in other districts and then returned home triumphantly with a second wife. Fear of AIDS was also making men do crazy things. Some thought sex with a virgin cured the virus. In parts of Malawi, this superstition had led to rapes.

"Things were never like this before," Elias said.

While I had done no scientific sampling, it seemed that more than half the adults I met were not living with spouses. Women usually said they had thrown their husbands out. Adilesi's daughter Lufinenti was typical. Her husband, the father of her three children, had been gone from the village for three years. "He used to be a hard worker, but when he made a lot of money, he wanted to marry more women, and I didn't want to be in a two-wife marriage," she said. At any rate, there was now no chance of his return. He had moved to the south and then died after a long illness, one of his friends had told her. Maybe it was

AIDS, maybe tuberculosis, maybe both. "They said he was always coughing," she said.

In the case of Mdati, Adilesi's deceased daughter, her husband was also described as a good worker, a man who could get construction jobs. But his "sleeping around" had become intolerable. "He'd even take other women into their house," Lufinenti said. While she described these dalliances, at one point she mentioned parenthetically that the "last" of Mdati's sons was now living with his father in a different village. This unexpected detail confused me. For weeks, I had assumed that all four of Mdati's sons were among the many small boys I always saw nearby, playing with tops and homemade pull toys.

It was only then that I learned that the three youngest had died during these past months, a tragedy so within the parameters of the commonplace that it had not merited any special mention. The first boy died of malaria, the second of a rash and high fever.

The third, Legina Robert, perished only in December. He was 3 years old and very tiny, just another Malawian youngster with growth so stunted that he had yet to walk. Children warming themselves by a fire dropped him as they carelessly passed his body across the flames. The adults were in the fields when they heard frenzied calls for help. The little boy needed to be rushed to the hospital—a three-hour journey by foot but only half that if they had been able to borrow one of the village's few bicycles. "We looked, but we didn't find one," Lufinenti recalled. "He died while I was carrying him on my back."

PARADOXICALLY, food was growing everywhere, the slender cornstalks nudging up against the roads, leaning along the hillsides, creeping down to the river's edge. Crops were wedged into the spare emptiness of the cities, spreading in half-ovals around office buildings, stores and saloons, reaching even the back entrances of the government ministries. To be in Malawi during the hot and wet growing season was to be embraced by a landscape of tantalizing abundance—a perplexing sight in view of all the hunger.

But in Malawi, a visitor soon realizes that the lush color is a cruel tease. All those planted fields, all that greenness, are merely symbols of desperation. Crops appear in unlikely places because farmers feel a need to stretch their holdings. Those inescapable cornstalks are telltale of a nation growing food to fill its belly rather than to compete in world markets. Little economic development is taking place.

Always on arrival at Adilesi's house, my first question was "What are you going to eat today?" And each time I listened to the family's contingencies, that day's single meal dependent on some relative being lucky enough to get *ganyu* or Lufinenti being able to sell her foraged greens in the city. When a little money was earned, the family spent it frugally, never splurging on cornmeal but instead cooking *gaga*, a dish made from only the outer husks of the corn kernels, a sifted residue more commonly employed as fodder.

Adilesi's field is only a short walk from the house. The family's corn plants were puny for that late in the growing season, and many of the leaves were yellowish and droopy.

Children broke off the least promising stalks and chewed them like sugar cane.

On one visit, I was accompanied by Ellard Malindi, the country's chief technical adviser for agriculture. He was a big-hearted man who talked easily with the farmers. A few months later, tragically, he would come down with cerebral malaria and die. But on that sunny day, he was at his vigorous best, and during a stroll through Adilesi's field, he moved from row to row, examining the sorry plants and shaking his head in frustration. "We're standing on the richest soils in the entire country, maybe in southern Africa," he said as his arm cut an arc through the air. "We're in the medium-altitude plateau. But the soil has become depleted by continuous planting, and it has lost its organic nutrients."

Goliati Faisoni, Adilesi's oldest son, was with us. A skinny, sickly man with bloodshot eyes, he agreed with Malindi. The family would not be reaping much this year. And however much there was, they intended to eat it early, when the cobs were still new and green. The family was often too hungry to allow their crops to ripen fully. Last year, in fact, they used their green corn to feed the mourners at Mdati's funeral.

"Didn't you get a starter pack?" Malindi asked Goliati.

The family had. But in their desperation they used the fertilizer incorrectly. They were supposed to apply the bag of mixed nutrients—nitrogen, phosphate, potash and sulfur—when they planted, then a second bag with urea when the stalks were knee high. Instead, the family combined the bags. Then, rather than concentrating on one quarter-acre, they spread it thinly over their entire field, hoping to outsmart science.

"There were also beans in the pack," Malindi said. "Did you plant a bean crop?"

No, he confessed. The family had been unable to resist. They had eaten the beans.

FAMILIES STARVE because families lack money. In most cases, it is that simple.

But last year, while most of Mkulumimba went hungry, its chief, Daniel Mkulumimba, was getting by all right. I had often wondered about him. He was the one man in all five villages whom others considered wealthy. His tobacco field was only a 10-minute walk from Adilesi's door. "He could easily help people, and sometimes he does," Elias, the headman, told me. "But it's not his nature. He mostly takes care of his own."

By American standards, Daniel was hardly rich. He, too, lived in a one-room house, though it was a bigger room, and it was covered with a roof that did not leak. He owned an ox cart, a bicycle, six donkeys and four goats. Besides corn and tobacco, he grew cabbage, lettuce, turnips and sugar cane on his 15 acres of land. He fertilized his fields, and the healthy cornstalks towered above him. He had two wives and a pot belly.

I wanted to understand why Daniel had not done more to help. When I posed the question, he considered it for a few seconds before saying, "When the food situation is very serious, the rich and the poor are the same, and it's everyone for himself." I reminded him that his cousin Robert had actually died of starvation. "As the chief, I'm not supposed to help one particular family," he replied. "I'm chief of the whole village."

For me, Daniel came to represent the "haves" of the world. They do assist Africa, though not with a strenuous effort, certainly not in proportion to the hunger and the disease and the benumbing poverty. Indeed, in relation to their gross domestic product, donor nations are now spending considerably less on foreign assistance than they were a decade ago. Among wealthy countries, the United States spends the lowest percentage of all, something President Bush is understandably reluctant to mention when he talks about Africa.

Many experts debate whether aid does more harm than good. Certainly Africa's problems are immense and confounding: paralyzing debt, sorry infrastructure, depleted soil, meager exports, bad government and ethnic and tribal warfare. The majority of Africa's poorest countries have average incomes below the level of Western Europe at the start of the 17th century, according to the distinguished economic historian Angus Maddison.

Unlike the days when structural adjustments were seen as direct routes to poverty reduction, now there seems to be little consensus on what to try next. Proposals tend to be modest. In Lilongwe, I heard one idea after another: soil renourishment, manufacturing schemes, public-service jobs, small-scale irrigation. Lawrence Rubey, the American booster of free enterprise, showed me a bag of chilies grown for export to Germany. "Niche marketing," he said, in much the same way "plastics" was advised in "The Graduate."

Maybe chili peppers can be one of the answers, maybe not. But in the meantime, even if poverty and hunger seem unconquerable, famine surely can be overcome. Only our indifference—only our neglect—allows it to persevere. In Malawi, the timely distribution of fertilizer ought to be preferable to the inevitability of emergency food. That is what every farmer in the villages asked for: if you give us fertilizer, or a reasonable way to buy it, we'll manage for ourselves from one hungry season to the next.

As I heard these sentiments, I would always nod sympathetically, writing the anguished words in my notebook. The people I met were invariably gracious, even though I knew an unspoken tension existed between us. After all, they were hungry, and I had the means to change that in my wallet, as easily as I handed my kids lunch money back in America. But I wanted to understand how people coped with hunger, and handouts would have made that impossible. So I had made a decision in advance: if I met people who seemed gravely ill, I would take them to the hospital and pay their expenses. (That happened twice.) Otherwise, I'd give no one money until the reporting was done.

And that day gave me as much relief as it gave them—perhaps more. It is awful to be with the hungry, to watch them ebb and falter and scrounge.

"You ask so many questions about death!" Adilesi said to me during one of my final visits. "It is hard on us. We believe that when you talk about the dead, you get visits by their spirits at night. When are the questions about the dead going to stop?"

I apologized but said that I had come to learn about hunger and that I had learned a lot.

I thanked her for that. And that is when she and her daughter thought to share with me their "ingenious" trick, something she thought every human being ought to know.

If I were ever so hungry I could no longer work, they advised me, there was a way for a determined mind to outfox a hollow stomach. "Tie some cloth tightly around your waist right at the navel," Lufinenti said. "Make it as tight as you can."

For a few hours, you can fool your belly into thinking that it's full.

Barry Bearak is a staff writer for the magazine. His last article was about the reconstruction of Afghanistan.

From the *New York Times* Magazine, July 13, 2003, pp. 32-37, 52, 60-61. © 2003 by Barry Bearak. Distributed by The New York Times Special Features. Reprinted by permission.

Putting a human face on development

Rubens Ricupero

There are better ways of ending a century (indeed, a millennium) than with a war superimposed on a major economic crisis amidst recurring bouts of food panic.

Wars, crises, food panics, share a common effect: they produce fear, anxiety, insecurity. They do it not only by inflicting pain but also by threatening to take away the possibility of having a future. As we live almost as much for the future as the present, without that perspective it is hard to contemplate life itself.

If we want to avoid Eliot's terrible conclusion; that "The cycles of Heaven in twenty centuries // Bring us farther from God and nearer to the Dust",[1] we have to go back to where we started. At a minimum, the security of human beings has to be assured in all of life's dimensions.

But security is not enough: no one can live exclusively from security or stability. These only provide human beings with the possibility of having a future. We need the dream that tomorrow will be better than today and yesterday; that our children, and their children, shall be free from fear and want, that they will not only be secure but that they will be able to fulfil their lives through productive, creative work and through love, affection, solidarity, and cooperation.

However, the century is ending with failure to solve two major threats to that future: mass unemployment and growing inequality. No system of organising production has ever been able to provide a productive job for every man and woman who wanted to work. Moreover, disparities in the distribution of wealth and income are on the rise, both within and between nations.

In the poor parts of the world, covering much of the planet, the very possibility of sustainable development has been called into question by the economic crisis that began in Asia in 1997. The fifth serious monetary and financial crisis of the last 20 years, it truly deserves to be called a "crisis of development", for three main reasons. First, it hit almost exclusively developing countries, at the same time sparing and even benefiting the industrial economies through falling prices for commodities, capital flight, and cheap manufactured imports, because of currency devaluations. Second, and paradoxically, it was much more destructive of the most advanced of the developing nations, raising serious doubts about whether, as had long been assumed, development is a process that reduces the vulnerability of economies to external shocks. Third, it created uncertainty whether it would be possible, once the crisis was over, to regain the levels of economic performance that constituted the only convincing demonstration so far of the possibility of development over several decades; that is, the experience of that group of countries once dubbed "the Asian tigers".

In the course of the crisis, millions of people lost their jobs in the affected countries; 30 years of progress in fighting poverty wiped out in a matter of weeks; and anguish, desperation, insecurity and, in some cases, political disintegration and violence returned in force. For the first time in many years, in 1998 and 1999 economic growth in the rich countries has been significantly higher than in poor nations, widening rather than narrowing the gap between the two.

Contrasts in the impact of crises resulting from differences in power and knowledge have reappeared in the economy, as they have in the areas of political or environmental security. Currency devaluations in the UK or Italy in the early 1990s did not st off a financial meltdown or an investors' stampede, as they did in Thailand or the Republic of Korea in 1997. Was it because the two European industrial economies had more economic power and better "fundamentals"? Or was it because they had more knowledge and skills about how to regulate and supervise financial markets?

This situation forces us to reflect on the whole concept and experience of development in the last few years, and to examine closely the recipes or formulas that have been advanced for economic development.

Of course, it is not only a problem of economic development or of poverty; the impact of this crisis has been much broader. In reality, it affected the world economy in general. It also raised questions regarding financial liberalisation even in industrialised countries. However, this article will concentrate mostly on questions related to economic development and its social implications, while keeping in mind that there is an underlying link connecting economic development, poverty, and globalisation.

We are witnessing a search for an alternative to the paradigm of development that has been hegemonic in the last 12 years, the so-called "Washington Consensus". This took its name from a famous article by John Williamson, an economist, who tried to codify the paradigm in a series of 10 principles. His approach to development was based on three major areas. First, sound macroeconomic policies, that is, low inflation, minimal budget deficits, and balanced external accounts. Second, the advice that countries should open up and follow the path of trade and financial liberalisation. (Although the Washington Consensus did not make a major distinction between the concepts of trade and financial liberalisation, financial liberalisation is much more difficult to deal with than trade liberalisation, as the Asian experience has shown. The Asian countries were very successful in trade liberalisation, but not in financial liberalisation.) Finally, the third element was to promote the role of the market much more than that of the state, through privatisation and reducing the action of the state to essential tasks, deregulation, and related matters.

These three categories of measure have been enforced by the IMF and the World Bank over the last 12 to 13 years in a top-down approach imposed through the conditionalities of the loans of the two institutions; and the underlying principles were the inspiration for the so-called structural adjustment programmers, applied over many years in many different countries. Those assumptions are now coming under increasing scrutiny.

We at UNCTAD are trying to look beyond the Washington Consensus in order to reach one that can be widely shared, a consensus built on the need for balance, equilibrium, and a sense of proportion, that will not reopen the old ideological battles of the 1970s or the 1980s, but will integrate more fully the complexity and diversity of conditions influencing development. There is still a great deal to be done in trying to reconcile apparently contradictory approaches, such as the role of the state and the market, price stability and economic growth, flexibility of the labour market and job security, or integration into the world economy and the building up of a national industrial base. These have often been presented as antagonistic, mutually exclusive positions, but the search now under way for alternatives is inspired precisely by the need to take a more multidisciplinary approach, to see to what extent we can make these goals mutually reinforcing and complementary.

We need, then, a thorough and comprehensive study of the experience of development over the last few decades, with three basic objectives. The first is to take stock of what went right and what went wrong in development. The second is to identify what was missing in the original approaches and concepts. In the 1950s and 1960s, the approach was much too macroeconomic; it emphasised aspects such as economic growth, capital accumulation, and productivity increases, but it did not give sufficient attention to the quality of development, the quality of life and social aspects such as the distribution of income or wealth. Other aspects that were totally ignored in the 1960s were the environmental dimension, the so-called sustainable quality of development, the role of women in the economy and the role of minorities and indigenous communities. The third objective of the study should be to identify the challenges ahead: what is the challenge facing development in the next century, in the next millennium? To realise these objectives is of course a tall order, and I would refrain from giving the false impression that anybody in the world has a ready recipe.

In a world of globalised finance, one of the biggest challenges is to reach a harmonious complementarity between social and economic development. One should beware of thinking that as soon as the developing Asian economies recover, their societies will immediately and automatically reach the social level they enjoyed in pre-crisis times. That has not been the experience in Latin America, where, even today, 17 years after the beginning of the Mexican foreign debt crisis, followed by crises in Argentina, Brazil, Peru and others, the continent has yet to recapture the pre-crisis level in social indicators.

An excellent recent report by the Economic Commission for Latin America and the Caribbean (ECLAC) entitled "The social panorama of Latin America" shows that even now the level of poverty in Latin America stands at 39% of the population—in other words, 209 million people in Latin America live below the poverty line—4 percentage points above the 1982 precrisis level of 35%. For some countries, such as Chile, the results are better, but ECLAC refers to the average of the continent as a whole.

One of the aspects of the emerging paradigm for development, which will only materialise if we work together to produce it, is the central importance of knowledge and information in the economy of the future. Professor Stiglitz, Chief Economist of the World Bank, and a strong dissenter from some of the basic premises of the Washington Consensus, has made an important contribution to a new branch of economics called "information economics". This does not refer, as some people think, to electronics or the way we transmit information through telecommunications. It refers to information in economic terms. The classical economists tended to consider that information had a zero or negligible cost; that is, every actor in the market had equal access to information about the market, and so the cost of acquiring it could be considered to be zero or negligible. The main contribution of the information economists has been to show that this is not true; information does have a cost, the so-called "transaction cost", which sometimes makes the difference between success and failure. Those who have had a good education or training tend to be better at dealing with information and to succeed where others fail. The problem is, what to do with the failures—the legion of unskilled workers in industrial countries, or the poor countries that are not able to compete in the market place because they do not have appropriate access to information.

This is a particularly acute and important problem today, for the simple reason that we are moving towards a new kind of economy and a new form of development, where the decisive factors are no longer capital, cheap labour, or an abundance of natural resources. More and more, the central, crucial factor is knowledge, information, patents; how to deal with the knowledge that is being constantly generated. As we move towards a knowledge-intensive economy, access to information becomes the difference between prosperity and poverty, and between domination and liberation. This is why information and knowledge will have to be increasingly considered in the rule-making negotiations on trade and on investment and in relation to economic life in general.

Competition of the intensified kind we are seeing as a result of globalisation has many analogies to a game; and it is not by accident that game theory, with its mathematical formulation, is nowadays frequently applied to competition. As in every game, competition certainly needs fair rules, such as the norms of the World Trade Organization, and it also needs an impartial arbiter, as in the WTO's dispute-settlement mechanism. Governments and trade negotiators tend to think that once we have fair rules and an impartial arbiter, the ideal conditions for competition will be in place. They forget a third and fundamental element of competition. In order to play a game, it is not enough to know the rules and to obey the arbiter, you have to learn how to play it; you need to be educated and trained. Nobody can run the 100 metres in the Olympic Games just because there are rules and umpires! So how can we include this element of learning and training as an integral part of competition in order to have a truly level playing field?

At the very least, it is necessary to provided each and every beginner with an equal opportunity to learn to play the game, with training time, during which the newcomer would not be crushed by veterans. Even a relatively level playing field may not be enough when inequality and poverty are such that countries and individuals are starting from such widely disparate levels.

As R. H. Tawney, the British historian, wrote: "… opportunities to rise are not a substitute for a large measure of practical equality of income and social condition. The existence of such opportunities… depends not only upon an open road but upon an equal start". In the *New Statesman* essay by the former Labour Minister Roy Hattersley,[2] in which I found the above quotation, the author comments that, 130 years after William Gladstone removed the institutional barriers to civil service appointments and military commissions in Great Britain, the same people are still getting most of the jobs!

Recognising that in an unequal society, families below the poverty line are doomed to remain poor in absolute as well as relative terms, some countries have resorted to US-style affirmative action, equal opportunity laws and other pro-active measures to correct gross disparities at the start. The same rationale, I would suggest, holds true in terms of the continuous need for special and differential treatment for developing countries, redefined from its previous form in a more concrete and up-to-date way.

Federico Mayor, former Director-General of UNESCO, has said that "trade not aid" should be the instrument of development. Everybody agrees with this. Thus, you would think that trade-related technical cooperation would be a very significant part of what happens in the field of technical cooperation. But that is not the case. OECD figures show that in fact only 2% of technical cooperation is trade-related. Nobody is really trying

hard to teach countries how to produce, how to trade, or how to compete. This is why information economics should be an important element in the revision of the rules concerning development, conceived and recast as a continuous learning process.

At the root of all movement towards globalisation, there has always been a revolution in ideas, science, and technology. It was so at the beginning of the expansion of the West with the Galilean revolution in the sixteenth century; again with the Newtonian revolution in the eighteenth century leading up to the Industrial Revolution; and once more today. The difference this time is that the revolution is about time and space: the previous revolutions were about matter and energy. This time it is the very concept of distance and time that is being changed by telecommunications and by informatics, and this is why the problem of access to information becomes central.

The fact that information, technology, and science are fundamental components of human development does not in itself, however, guarantee that these elements will not again be used for oppression or domination. This we must guard against. In the past, scientific knowledge has too often been used for oppression and domination. We should not be so naive as to think that access to information is just a matter of pedagogy, of learning, of education. There is an element of power—both market and political power—in controlling information. But access to information will remain the crucial condition for development.

Norbert Wiener, the founder of cybernetics, used to say that to be informed is to be free. He meant freedom to make choices, to choose among options. But in order to be able to choose, one needs to have knowledge about what the options are, if indeed there are options, and about the relative costs and benefits of each of them, because in political life, as in culture and in the economy, you always have a trade-off—you win some and lose some. In order to choose an option, you need information. Let us hope that, this time, information will serve not to oppress and to exploit humankind, but to liberate it and promote true human development.

Notes

1. T. S. Eliot, "The Rock", in *The Complete Poems and Plays 1909–1950*. Harcourt Brace World Inc, New York, 1971, p. 96.
2. R. Hattersley, "Up and down the social ladder". *New Statesman* (London), 22 January 1999.

Rubens Ricupero is Secretary-General of the United Nations Conference on Trade and Development (UNCTAD). Previously, he was Brazil's Minister of Environment and Amazonian Affairs, and Minister of Finance. He has also represented his country at the UN, World Bank, IMF, and GATT. He has been Professor of International Relations at the University of Brasilia, and Professor at the Rio Branco Institute. He is author of several books and essays on international relations.

From *International Social Science Journal*, UNESCO 2000, pp. 441–446. © 2000 by Blackwell Publishers Ltd.

UNIT 2

Political Economy and the Developing World

Unit Selections

Key Points to Consider

- In what ways can globalization be regulated to help the poor?

- What steps can developing countries themselves take to improve their benefits from trade? How can changes in the trade policy of the industrial countries benefit the developing world?

- Are the IMF and World Bank unaccountable? How does the IMF answer its critics?

- What industrialized countries are the most committed to development in poor countries?

- How can foreign aid help the developing world? In what ways can foreign aid be made more effective?

- What caused the breakdown in the 2003 WTO ministerial talks in Cancun? How are the developing countries excluded from the WTO?

 **Links: www.dushkin.com/online/**
These sites are annotated in the World Wide Web pages.

Center for Third World Organizing
  http://www.ctwo.org/
ENTERWeb
  http://www.enterweb.org
International Monetary Fund (IMF)
  http://www.imf.org
TWN (Third World Network)
  http://www.twnside.org.sg/
U.S. Agency for International Development (USAID)
  http://www.info.usaid.gov
The World Bank
  http://www.worldbank.org

Economic issues are among the developing world's most pressing concerns. Economic growth and stability are essential to progress on the variety of problems confronting developing countries. The developing world's position in the international economic system contributes to the difficulty of achieving consistent economic growth. From their incorporation into the international economic system during colonialism to the present, the majority of developing countries have been primarily suppliers of raw materials, agricultural products, and inexpensive labor. Dependence on commodity exports has meant that developing countries have had to deal with fluctuating, and frequently declining, prices for their exports. At the same time, prices for imports have remained constant or have increased. At best, this decline in the terms of trade has made development planning difficult; at worst, it has led to economic stagnation and decline.

With some exceptions, developing nations have had limited success in breaking out of this dilemma by diversifying their economies. Efforts at industrialization and export of light manufactures have led to competition with less efficient industries in the industrialized world. The response of industrialized countries has often been protectionism and demands for trade reciprocity, which can overwhelm markets in developing countries. Although the World Trade Organization (WTO) was established to standardize trade regulations and increase international trade, critics charge that the WTO continues to disadvantage developing countries and remains dominated by the wealthy industrial countries. The developing world also asserts that they are often shut out of trade negotiations, that they must accept deals dictated by the wealthy countries, and that they lack sufficient resources to effectively participate in the wide range of forums and negotiations that take place around the world. Moreover, developing countries see a selective dismantling of trade barriers; delegates from poor countries recently walked out of the WTO ministerial meeting in Cancun, Mexico, to protest rich countries' reluctance to eliminate agricultural subsidies.

The economic situation in the developing world, however, is not entirely attributable to colonial legacy and protectionism on the part of industrialized countries. Developing countries have sometimes constructed their own trade barriers and have relied on preferential trade relationships. Industrialization schemes involving heavy government direction were often ill-conceived or resulted in corruption and mismanagement. Industrialized countries frequently point to these inefficiencies in calling for market-oriented reforms, but the emphasis on privatization does not adequately recognize the role of the state in developing countries and may result in foreign control of important sectors of the economy as well as a loss of jobs.

Debt has further compounded economic problems for many developing countries. During the 1970s, developing countries' prior economic performance and the availability of petrodollars encouraged extensive commercial lending. Developing countries sought these loans to fill the gap between revenues from exports and foreign aid and development expenditures. The second oil price hike in the late 1970s, declining export earnings, and worldwide recession in the early 1980s left many developing countries unable to meet their debt obligations. The commercial banks weathered the crisis, and some actually showed a profit. Commercial lending declined and international financial institutions became the lenders of last resort for many developing countries. Access to World Bank and International Monetary Fund financing became conditional on the adoption of structural adjustment programs that involved steps such as reduced public expenditures, devaluation of currencies, and export promotion, all geared to debt reduction. The consequences of these programs have been painful for developing countries. Fewer public services, higher prices, and greater exploitation of resources have resulted.

The poorest countries in particular continue to struggle with heavy debt burdens, and the IMF and World Bank have come under increasing criticism for their programs in the developing world. Though they have made some efforts to shift the emphasis to poverty reduction, some critics charge that the reforms are superficial and that the international financial institutions lack accountability. Furthermore, advocates for developing countries condemn debt repayments that result in more capital flowing from poor countries to rich countries than in the opposite direction.

Globalization has emerged as a controversial issue with differing views regarding benefits and costs of this trend for the developing world. Advocates claim that closer economic integration, especially trade and financial liberalization, increases economic prosperity in developing countries and encourages good governance, transparency, and accountability. Critics respond that globalization's requirements, as determined by the powerful nations and reinforced through the international financial institutions, impose difficult and perhaps counterproductive policies on struggling economies. They also charge that globalization undermines workers' rights and encourages environmental degradation. Moreover, most of the benefits of globalization have gone to those countries that were already growing.

In part due to the realization that poverty in the developing world contributes to the despair and resentment that lead some to terrorism, there has been increased attention focused on foreign aid. Although the United States has recently committed to increasing foreign aid, a recently developed Commitment to Development Index shows that the smaller industrialized countries provide more aid to the developing world, suggesting that the G-7 countries could do more to help the poor. While aid has often been criticized, it does produce benefits. Those benefits could be enhanced by a more effective aid bureaucracy and efforts to enhance local capacity-building.

The Free-Trade Fix

So far, **globalization** has failed the world's poor.
But it's not trade that has hurt them. It's a rigged system.

By Tina Rosenberg

Globalization is a phenomenon that has remade the economy of virtually every nation, reshaped almost every industry and touched billions of lives, often in surprising and ambiguous ways. The stories filling the front pages in recent weeks—about economic crisis and contagion in Argentina, Uruguay and Brazil, about President Bush getting the trade bill he wanted—are all part of the same story, the largest story of our times: what globalization has done, or has failed to do.

Globalization is meant to signify integration and unity—yet it has proved, in its way, to be no less polarizing than the cold-war divisions it has supplanted. The lines between globalization's supporters and its critics run not only between countries but also through them, as people struggle to come to terms with the defining economic force shaping the planet today. The two sides in the discussion—a shouting match, really—describe what seem to be two completely different forces. Is the globe being knit together by the Nikes and Microsofts and Citigroups in a dynamic new system that will eventually lift the have-nots of the world up from medieval misery? Or are ordinary people now victims of ruthless corporate domination, as the Nikes and Microsofts and Citigroups roll over the poor in nation after nation in search of new profits?

The debate over globalization's true nature has divided people in third-world countries since the phenomenon arose. It is now an issue in the United States as well, and many Americans—those who neither make the deals inside World Trade Organization meetings nor man the barricades outside—are perplexed.

When I first set out to see for myself whether globalization has been for better or for worse, I was perplexed, too. I had sympathy for some of the issues raised by the protesters, especially their outrage over sweatshops. But I have also spent many years in Latin America, and I have seen firsthand how protected economies became corrupt systems that helped only those with clout. In general, I thought the protesters were simply being sentimental; after all, the masters of the universe must know what they are doing. But that was before I studied the agreements that regulate global trade—including this month's new law granting President Bush a free hand to negotiate trade agreements, a document redolent of corporate lobbying. And it was before looking at globalization up close in Chile and Mexico, two nations that have embraced globalization especially ardently in the region of the third world that has done the most to follow the accepted rules. I no longer think the masters of the universe know what they are doing.

The architects of globalization are right that international economic integration is not only good for the poor; it is essential. To embrace self-sufficiency or to deride growth, as some protesters do, is to glamorize poverty. No nation has ever developed over the long term without trade. East Asia is the most recent exam-ple. Since the mid-1970's, Japan, Korea, Taiwan, China and their neighbors have lifted 300 million people out of poverty, chiefly through trade.

But the protesters are also right—no nation has ever developed over the long term under the rules being imposed today on third-world countries by the institutions controlling globalization. The United States, Germany, France and Japan all became wealthy and powerful nations behind the barriers of protectionism. East Asia built its export industry by protecting its markets and banks from foreign competition and requiring investors to buy local products and build local know-how. These are all practices discouraged or made illegal by the rules of trade today.

The World Trade Organization was designed as a meeting place where willing nations could sit in equality and negotiate rules of trade for their mutual advantage, in the service of sustainable international development. Instead, it has become an unbalanced institution largely controlled by the United States and the nations of Europe, and especially the agribusiness, pharmaceutical and financial-services industries in these countries. At W.T.O. meetings, important deals are hammered out in negotiations attended by the trade ministers of a couple dozen powerful nations, while those of poor countries wait in the bar outside for news.

The International Monetary Fund was created to prevent future Great Depressions in part by lending countries in recession money and pressing them to adopt expansionary policies, like deficit spending and low interest rates, so they would continue to buy their neighbors' products. Over time, its mission has evolved into the reverse: it has become a long-term manager of the economies of developing countries, blindly committed to the bitter medicine of contraction no matter what the illness. Its formation was an acknowledgment that markets sometimes work imperfectly, but it has become a champion of market supremacy in all situations, echoing the voice of Wall Street and the United States Treasury Department, more interested in getting wealthy creditors repaid than in serving the poor.

It is often said that globalization is a force of nature, as unstoppable and difficult to contain as a storm. This is untrue and misleading. Globalization is a powerful phenomenon—but it is not irreversible, and indeed the previous wave of globalization, at the turn of the last century, was stopped dead by World War I. Today it would be more likely for globalization to be sabotaged by its own inequities, as disillusioned nations withdraw from a system they see as indifferent or harmful to the poor.

Globalization's supporters portray it as the peeling away of distortions to reveal a clean and elegant system of international commerce, the one nature intended. It is anything but. The accord creating the W.T.O. is 22,500 pages long—not exactly a *free* trade agreement. All globalization, it seems, is local, the rules drawn up by, and written to benefit, powerful nations and powerful interests within those nations. Globalization has been good for the United States, but even in this country, the gains go disproportionately to the wealthy and to big business.

It's not too late for globalization to work. But the system is in need of serious reform. More equitable rules would spread its benefits to the ordinary citizens of wealthy countries. They would also help to preserve globalization by giving the poor of the world a stake in the system—and, not incidentally, improve the lives of hundreds of millions of people. Here, then, are nine new rules for the

global economy—a prescription to save globalization from itself.

1. Make the State a Partner

If there is any place in Latin America where the poor have thrived because of globalization, it is Chile. Between 1987 and 1998, Chile cut poverty by more than half. Its success shows that poor nations can take advantage of globalization—if they have governments that actively make it happen.

> The previous wave of globalization was stopped dead by World War I. Today it would be more likely for globalization to be sabotaged by its own inequities, as disillusioned nations withdraw from a system they see as indifferent or harmful to the poor.

Chile reduced poverty by growing its economy—6.6 percent a year from 1985 to 2000. One of the few points economists can agree on is that growth is the most important thing a nation can do for its poor. They can't agree on basics like whether poverty in the world is up or down in the last 15 years—the number of people who live on less than $1 a day is slightly down, but the number who live on less than $2 is slightly up. Inequality has soared during the last 15 years, but economists cannot agree on whether globalization is mainly at fault or whether other forces, like the uneven spread of technology, are responsible. They can't agree on how to reduce inequality—growth tends not to change it. They can't agree on whether the poor who have not been helped are victims of globalization or have simply not yet enjoyed access to its benefits—in other words, whether the solution is more globalization or less. But economists agree on one thing: to help the poor, you'd better grow.

For the rest of Latin America, and most of the developing world except China (and to a lesser extent India), globalization as practiced today is failing,

and it is failing because it has not produced growth. Excluding China, the growth rate of poor countries was 2 percent a year lower in the 1990's than in the 1970's, when closed economies were the norm and the world was in a recession brought on in part by oil-price shocks. Latin American economies in the 1990's grew at an average annual rate of 2.9 percent—about half the rate of the 1960's. By the end of the 1990's, 11 million more Latin Americans lived in poverty than at the beginning of the decade. And in country after country, Latin America's poor are suffering—either from economic crises and market panics or from the day-to-day deprivations that globalization was supposed to relieve. The surprise is not that Latin Americans are once again voting for populist candidates but that the revolt against globalization took so long.

When I visited Eastern Europe after the end of Communism, a time when democracy was mainly bringing poverty, I heard over and over again that the reason for Chile's success was Augusto Pinochet. Only a dictator with a strong hand can put his country through the pain of economic reform, went the popular wisdom. In truth, we now know that inflicting pain is the easy part; governments democratic and dictatorial are all instituting free-market austerity. The point is not to inflict pain but to lessen it. In this Pinochet failed, and the democratic governments that followed him beginning in 1990 have succeeded.

What Pinochet did was to shut down sectors of Chile's economy that produced goods for the domestic market, like subsistence farming and appliance manufacturing, and point the economy toward exports. Here he was following the standard advice that economists give developing countries—but there are different ways to do it, and Pinochet's were disastrous. Instead of helping the losers, he dismantled the social safety net and much of the regulatory apparatus that might have kept privatization honest. When the world economy went into recession in 1982, Chile's integration into the global marketplace and its dependence on foreign capital magnified the crash. Poverty soared, and unemployment reached 20 percent.

Pinochet's second wave of globalization, in the late 1980's, worked better, because the state did not stand on the side. It regulated the changes effectively and aggressively promoted exports. But Pinochet created a time bomb in Chile: the country's exports were, and still are, nonrenewable natural resources. Chile began subsidizing companies that cut down native forests for wood chips, for example, and the industry is rapidly deforesting the nation.

Chile began to grow, but inequality soared—the other problem with Pinochet's globalization was that it left out the poor. While the democratic governments that succeeded Pinochet have not yet been able to reduce inequality, at least it is no longer increasing, and they have been able to use the fruits of Chile's growth to help the poor.

Chile's democratic governments have spread the benefits of economic integration by designing effective social programs and aiming them at the poor. Chile has sunk money into revitalizing the 900 worst primary schools. It now leads Latin America in computers in schools, along with Costa Rica. It provides the very low-income with housing subsidies, child care and income support. Open economy or closed, these are good things. But Chile's government is also taking action to mitigate one of the most dangerous aspects of global integration: the violent ups and downs that come from linking your economy to the rest of the world. This year it created unemployment insurance. And it was the first nation to institute what is essentially a tax on short-term capital, to discourage the kind of investment that can flood out during a market panic.

The conventional wisdom among economists today is that successful globalizers must be like Chile. This was not always the thinking. In the 1980's, the Washington Consensus—the master-of-the-universe ideology at the time, highly influenced by the Reagan and Thatcher administrations—held that government was in the way. Globalizers' tasks included privatization, deregulation, fiscal austerity and financial liberalization. "In the 1980's and up to 1996 or 1997, the state was considered the devil," says Juan Martin, an Argentine economist at the

United Nations' Economic Commission for Latin America and the Caribbean. "Now we know you need infrastructure, institutions, education. In fact, when the economy opens, you need *more* control mechanisms from the state, not fewer."

And what if you don't have these things? Bolivia carried out extensive reforms beginning in 1985—a year in which it had inflation of 23,000 percent—to make the economy more stable and efficient. But in the words of the World Bank, "It is a good example of a country that has achieved successful stabilization and implemented innovative market reforms, yet made only limited progress in the fight against poverty." Latin America is full of nations that cannot make globalization work. The saddest example is Haiti, an excellent student of the rules of globalization, ranked at the top of the I.M.F.'s index of trade openness. Yet over the 1990's, Haiti's economy *contracted*; annual per capita income is now $250. No surprise—if you are a corrupt and misgoverned nation with a closed economy, becoming a corrupt and misgoverned nation with an open economy is not going to solve your problems.

Import Know-How Along With the Assembly Line

2. If there is a showcase for globalization in Latin America, it lies on the outskirts of Puebla, Mexico, at Volkswagen Mexico. Every New Beetle in the world is made here, 440 a day, in a factory so sparkling and clean that you could have a baby on the floor, so high-tech that in some halls it is not evident that human beings work here. Volkswagen Mexico also makes Jettas and, in a special hall, 80 classic Beetles a day to sell in Mexico, one of the last places in the world where the old Bug still chugs.

The Volkswagen factory is the biggest single industrial plant in Mexico. Humans do work here—11,000 people in assembly-line jobs, 4,000 more in the rest of the factory—with 11,000 more jobs in the industrial park of VW suppliers across the street making parts, seats, dashboards and other components. Perhaps 50,000 more people work in other companies around Mexico that supply VW. The average monthly wage in the

plant is $760, among the highest in the country's industrial sector. The factory is the equal of any in Germany, the product of a billion-dollar investment in 1995, when VW chose Puebla as the exclusive site for the New Beetle.

Ahhh, globalization.

Except… this plant is not here because Mexico has an open economy, but because it had a closed one. In 1962, Mexico decreed that any automaker that wanted to sell cars here had to produce them here. Five years later, VW opened the factory. Mexico's local content requirement is now illegal, except for very limited exceptions, under W.T.O. rules; in Mexico the local content requirement for automobiles is being phased out and will disappear entirely in January 2004.

The Puebla factory, for all the jobs and foreign exchange it brings Mexico, also refutes the argument that foreign technology automatically rubs off on the local host. Despite 40 years here, the auto industry has not created much local business or know-how. VW makes the point that it buys 60 percent of its parts in Mexico, but the "local" suppliers are virtually all foreign-owned and import most of the materials they use. The value Mexico adds to the Beetles it exports is mainly labor. Technology transfer—the transmission of know-how from foreign companies to local ones—is limited in part because most foreign trade today is intracompany; Ford Hermosillo, for example, is a stamping and assembly plant shipping exclusively to Ford plants in the United States. Trade like this is particularly impenetrable to outsiders. "In spite of the fact that Mexico has been host to many car plants, we don't know how to build a car," says Huberto Juárez, an economist at the Autonomous University of Puebla.

Volkswagen Mexico is the epitome of the strategy Mexico has chosen for globalization—assembly of imported parts. It is a strategy that makes perfect sense given Mexico's proximity to the world's largest market, and it has given rise to the maquila industry, which uses Mexican labor to assemble foreign parts and then re-export the finished products. Although the economic slowdown in the United States is hurting the *maquila* industry, it still employs a million people and brings the country $10 billion a year in foreign exchange. The factories have turned Mexico into one of the developing world's biggest exporters of medium- and high-technology products. But the maquila sector remains an island and has failed to stimulate Mexican industries—one reason Mexico's globalization has brought disappointing growth, averaging only 3 percent a year during the 1990's.

In countries as varied as South Korea, China and Mauritius, however, assembly work has been the crucible of wider development. Jeffrey Sachs, the development economist who now directs Columbia University's Earth Institute, says that the maquila industry is "magnificent." "I could cite 10 success stories," he says, "and every one started with a maquila sector." When Korea opened its export-processing zone in Masan in the early 1970's, local inputs were 3 percent of the export value, according to the British development group Oxfam. Ten years later they were almost 50 percent. General Motors took a Korean textile company called Daewoo and helped shape it into a conglomerate making cars, electronic goods, ships and dozens of other products. Daewoo calls itself "a locomotive for national economic development since its founding in 1967." And despite the company's recent troubles, it's true—because Korea made it true. G.M. did not tutor Daewoo because it welcomed competition but because Korea demanded it. Korea wanted to build high-tech industry, and it did so by requiring technology transfer and by closing markets to imports.

Maquilas first appeared in Mexico in 1966. Although the country has gone from assembling clothing to assembling high-tech goods, nearly 40 years later 97 percent of the components used in Mexican maquilas are still imported, and the value that Mexico adds to its exports has actually declined sharply since the mid-1970's.

Mexico has never required companies to transfer technology to locals, and indeed, under the rules of the North American Free Trade Agreement, it cannot. "We should have included a technical component in Nafta," says Luis de la Calle, one of the treaty's negotiators and later Mexico's under secretary of economy for foreign trade. "We should be getting a significant transfer of technology from the United States, and we didn't really try."

Without technology transfer, maquila work is marked for extinction. As transport costs become less important, Mexico is increasingly competing with China and Bangladesh—where labor goes for as little as 9 cents an hour. This is one reason that real wages for the lowest-paid workers in Mexico dropped by 50 percent from 1985 to 2000. Businesses, in fact, are already leaving to go to China.

Sweat the Sweatshops—But Sweat Other Problems More

3. When Americans think about globalization, they often think about sweatshops—one aspect of globalization that ordinary people believe they can influence through their buying choices. In many of the factories in Mexico, Central America and Asia producing American-brand toys, clothes, sneakers and other goods, exploitation is the norm. The young women who work in them—almost all sweatshop workers are young women—endure starvation wages, forced overtime and dangerous working conditions.

In Chile, I met a man who works at a chicken-processing plant in a small town. The plant is owned by Chileans and processes chicken for the domestic market and for export to Europe, Asia and other countries in Latin America. His job is to stand in a freezing room and crack open chickens as they come down an assembly line at the rate of 41 per minute. When visitors arrive at the factory (the owners did not return my phone calls requesting a visit or an interview), the workers get a respite, as the line slows down to half-speed for show. His work uniform does not protect him from the cold, the man said, and after a few minutes of work he loses feeling in his hands. Some of his colleagues, he said, are no longer able to raise their arms. If he misses a day he is docked $30. He earns less than $200 a month.

Is this man a victim of globalization? The protesters say that he is, and at one point I would have said so, too. He—and all workers—should have dignified conditions and the right to organize. All companies should follow local labor laws, and activists should pressure companies to pay their workers decent wages.

But today if I were to picket globalization, I would protest other inequities. In a way, the chicken worker, who came to the factory when driving a taxi ceased to be profitable, is a beneficiary of globalization. So are the millions of young women who have left rural villages to be exploited gluing tennis shoes or assembling computer keyboards. The losers are those who get laid off when companies move to low-wage countries, or those forced off their land when imports undercut their crop prices, or those who can no longer afford life-saving medicine—people whose choices in life *diminish* because of global trade. Globalization has offered this man a hellish job, but it is a choice he did not have before, and he took it; I don't name him because he is afraid of being fired. When this chicken company is hiring, the lines go around the block.

Get Rid of the Lobbyists

4. The argument that open economies help the poor rests to a large extent on the evidence that closed economies do not. While South Korea and other East Asian countries successfully used trade barriers to create export industries, this is rare; most protected economies are disasters. "The main tendency in a sheltered market is to goof off," says Jagdish Bhagwati, a prominent free-trader who is the Arthur Lehman professor of economics at Columbia University. "A crutch becomes a permanent crutch. Infant-industry protection should be for infant industries."

Anyone who has lived or traveled in the third world can attest that while controlled economies theoretically allow governments to help the poor, in practice it's usually a different story. In Latin America, spending on social programs largely goes to the urban middle class. Attention goes to people who can organize, strike, lobby and contribute money. And in a closed economy, the "state" car factory is often owned by the dictator's son and the country's forests can be chopped down by his golf partner.

Free trade, its proponents argue, takes these decisions away from the government and leaves them to the market, which punishes corruption. And it's true that a system that took corruption and undue political influence out of economic decision-making could indeed benefit the poor. But humans have not yet invented such a system—and if they did, it would certainly not be the current system of globalization, which is soiled with the footprints of special interests. In every country that negotiates at the W.T.O. or cuts a free-trade deal, trade ministers fall under heavy pressure from powerful business groups. Lobbyists have learned that they can often quietly slip provisions that pay big dividends into complex trade deals. None have been more successful at getting what they want than those from America.

The most egregious example of a special-interest provision is the W.T.O.'s rules on intellectual property. The ability of poor nations to make or import cheap copies of drugs still under patent in rich countries has been a boon to world public health. But the W.T.O. will require most of its poor members to accept patents on medicine by 2005, with the very poorest nations following in 2016. This regime does nothing for the poor. Medicine prices will probably double, but poor countries will never offer enough of a market to persuade the pharmaceutical industry to invent cures for their diseases.

The intellectual-property rules have won worldwide notoriety for the obstacles they pose to cheap AIDS medicine. They are also the provision of the W.T.O. that economists respect the least. They were rammed into the W.T.O. by Washington in response to the industry groups who control United States trade policy on the subject. "This is not a trade issue," Bhagwati says. "It's a royalty-collection issue. It's pharmaceuticals and software throwing their weight around." The World Bank calculated that the intellectual-property rules will result in a transfer of $40 billion a year from poor countries to corporations in the developed world.

No Dumping

5. Manuel de Jesús Gómez is a corn farmer in the hills of Puebla State, 72 years old and less than five feet tall. I met him in his field of six acres, where he was trudging behind a plow pulled by a burro. He farms the same way *campesinos* in these hills have been farming for thousands of years. In Puebla, and in the poverty belt of Mexico's southern states—Chiapas, Oaxaca, Guerrero—corn growers plow with animals and irrigate by praying for rain.

Before Nafta, corn covered 60 percent of Mexico's cultivated land. This is where corn was born, and it remains a symbol of the nation and daily bread for most Mexicans. But in the Nafta negotiations, Mexico agreed to open itself to subsidized American corn, a policy that has crushed small corn farmers. "Before, we could make a living, but now sometimes what we sell our corn for doesn't even cover our costs," Gómez says. With Nafta, he suddenly had to compete with American corn—raised with the most modern methods, but more important, subsidized to sell overseas at 20 percent less than the cost of production. Subsidized American corn now makes up almost half of the world's stock, effectively setting the world price so low that local small farmers can no longer survive. This competition helped cut the price paid to Gómez for his corn by half.

Because of corn's importance to Mexico, when it negotiated Nafta it was promised 15 years to gradually raise the amount of corn that could enter the country without tariffs. But Mexico voluntarily lifted the quotas in less than three years—to help the chicken and pork industry, Mexican negotiators told me unabashedly. (Eduardo Bours, a member of the family that owns Mexico's largest chicken processor, was one of Mexico's Nafta negotiators.) The state lost some $2 billion in tariffs it could have charged, and farmers were instantly exposed to competition from the north. According to ANEC, a national association of campesino cooperatives, half a million corn farmers have left their land and moved to Mexican cities or to America. If it were not for a weak peso, which keeps the price of imports relatively

high, far more farmers would be forced off their land.

The toll on small farmers is particularly bitter because cheaper corn has not translated into cheaper food for Mexicans. As part of its economic reforms, Mexico has gradually removed price controls on tortillas and tortilla flour. Tortilla prices have nearly tripled in real terms even as the price of corn has dropped.

Is this how it was supposed to be? I asked Andrés Rosenzweig, a longtime Mexican agriculture official who helped negotiate the agricultural sections of Nafta. He was silent for a minute. "The problems of rural poverty in Mexico did not start with Nafta," he said. "The size of our farms is not viable, and they get smaller each generation because farmers have many children, who divide the land. A family in Puebla with five hectares could raise 10, maybe 15, tons of corn each year. That was an annual income of 16,000 pesos," the equivalent of $1,600 today. "Double it and you still die of hunger. This has nothing to do with Nafta.

"The solution for small corn farmers," he went on, "is to educate their children and find them jobs outside agriculture. But Mexico was not growing, not generating jobs. Who's going to employ them? Nafta."

One prominent antiglobalization report keeps referring to farms like Gómez's as "small-scale, diversified, self-reliant, community-based agriculture systems." You could call them that, I guess; you could also use words like "malnourished," "undereducated" and "miserable" to describe their inhabitants. Rosenzweig is right—this is not a life to be romanticized.

But to turn the farm families' malnutrition into starvation makes no sense. Mexico spends foreign exchange to buy corn. Instead, it could be spending money to bring farmers irrigation, technical help and credit. A system in which the government purchased farmers' corn at a guaranteed price—done away with in states like Puebla during the free-market reforms of the mid-1990's—has now been replaced by direct payments to farmers. The program is focused on the poor, but the payments are symbolic—

$36 an acre. In addition, rural credit has disappeared, as the government has effectively shut down the rural bank, which was badly run, and other banks won't lend to small farmers. There is a program—understaffed and poorly publicized—to help small producers, but the farmers I met didn't know about it.

FREE TRADE IS a religion, and with religion comes hypocrisy. Rich nations press other countries to open their agricultural markets. At the urging of the I.M.F. and Washington, Haiti slashed its tariffs on rice in 1995. Prices paid to rice farmers fell by 25 percent, which has devastated Haiti's rural poor. In China, the tariff demands of W.T.O. membership will cost tens of millions of peasants their livelihoods. But European farmers get 35 percent of their income from government subsidies, and American farmers get 20 percent. Farm subsidies in the United States, moreover, are a huge corporate-welfare program, with nearly 70 percent of payments going to the largest 10 percent of producers. Subsidies also depress crop prices abroad by encouraging overproduction. The farm bill President Bush signed in May—with substantial Democratic support—provides about $57 billion in subsidies for American corn and other commodities over the next 10 years.

Wealthy nations justify pressure on small countries to open markets by arguing that these countries cannot grow rice and corn efficiently—that American crops are cheap food for the world's hungry. But with subsidies this large, it takes chutzpah to question other nations' efficiency. And in fact, the poor suffer when America is the supermarket to the world, even at bargain prices. There is plenty of food in the world, and even many countries with severe malnutrition are food exporters. The problem is that poor people can't afford it. The poor *are* the small farmers. Three-quarters of the world's poor are rural. If they are forced off their land by subsidized grain imports, they starve.

Help Countries Break the Coffee Habit

6. Back in the 1950's, Latin American economists made a simple calculation. The products their nations exported—copper, tin, coffee, rice and other commodities—were buying less and less of the high-value-added goods they wanted to import. In effect, they were getting poorer each day. Their solution was to close their markets and develop domestic industries to produce their own appliances and other goods for their citizens.

The strategy, which became known as import substitution, produced high growth—for a while. But these closed economies ultimately proved unsustainable. Latin American governments made their consumers buy inferior and expensive products—remember the Brazilian computer of the 1970's? Growth depended on heavy borrowing and high deficits. When they could no longer roll over their debts, Latin American economies crashed, and a decade of stagnation resulted.

At the time, the architects of import substitution could not imagine that it was possible to export anything but commodities. But East Asia—as poor or poorer than Latin America in the 1960's—showed in the 1980's and 1990's that it can be done. Unfortunately, the rules of global trade now prohibit countries from using the strategies successfully employed to develop export industries in East Asia.

American trade officials argue that they are not using tariffs to block poor countries from exporting, and they are right—the average tariff charged by the United States is a negligible 1.7 percent, much lower than other nations. But the rules rich nations have set—on technology transfer, local content and government aid to their infant industries, among other things—are destroying poor nations' abilities to move beyond commodities. "We are pulling up the ladder on policies the developed countries used to become rich," says Lori Wallach, the director of Public Citizen's Global Trade Watch.

The commodities that poor countries are left to export are even more of a dead end today than in the 1950's. Because of oversupply, prices for coffee, cocoa, rice, sugar and tin dropped by more than 60 percent between 1980 and 2000. Because of the price collapse of commodi-

ties and sub-Saharan Africa's failure to move beyond them, the region's share of world trade dropped by two-thirds during that time. If it had the same share of exports today that it had at the start of the 1980's, per capita income in sub-Saharan Africa would be almost twice as high.

Let the People Go

7. Probably the single most important change for the developing world would be to legalize the export of the one thing they have in abundance—people. Earlier waves of globalization were kinder to the poor because not only capital, but also labor, was free to move. Dani Rodrik, an economist at Harvard's Kennedy School of Government and a leading academic critic of the rules of globalization, argues for a scheme of legal short-term migration. If rich nations opened 3 percent of their work forces to temporary migrants, who then had to return home, Rodrik says, it would generate $200 billion annually in wages, and a lot of technology transfer for poor countries.

Free the I.M.F.

8. Globalization means risk. By opening its economy, a nation makes itself vulnerable to contagion from abroad. Countries that have liberalized their capital markets are especially susceptible, as short-term capital that has whooshed into a country on investor whim whooshes out just as fast when investors panic. This is how a real-estate crisis in Thailand in 1997 touched off one of the biggest global conflagrations since the Depression.

The desire to keep money from rushing out inspired Chile to install speed bumps discouraging short-term capital inflows. But Chile's policy runs counter to the standard advice of the I.M.F., which has required many countries to open their capital markets. "There were so many obstacles to capital-market integration that it was hard to err on the side of pushing countries to liberalize too much," says Ken Rogoff, the I.M.F.'s director of research.

Prudent nations are wary of capital liberalization, and rightly so. Joseph

Stiglitz, the Nobel Prize-winning economist who has become the most influential critic of globalization's rules, writes that in December 1997, when he was chief economist at the World Bank, he met with South Korean officials who were balking at the I.M.F.'s advice to open their capital markets. They were scared of the hot money, but they could not disagree with the I.M.F., lest they be seen as irresponsible. If the I.M.F. expressed disapproval, it would drive away other donors and private investors as well.

In the wake of the Asian collapse, Prime Minister Mahathir Mohamad imposed capital controls in Malaysia—to worldwide condemnation. But his policy is now widely considered to be the reason that Malaysia stayed stable while its neighbors did not. "It turned out to be a brilliant decision," Bhagwati says.

Post-crash, the I.M.F. prescribed its standard advice for nations—making loan arrangements contingent on spending cuts, interest-rate hikes and other contractionary measures. But balancing a budget in recession is, as Stiglitz puts it in his new book, "Globalization and Its Discontents," a recommendation last taken seriously in the days of Herbert Hoover. The I.M.F.'s recommendations deepened the crisis and forced governments to reduce much of the cushion that was left for the poor. Indonesia had to cut subsidies on food. "While the I.M.F. had provided some $23 billion to be used to support the exchange rate and bail out creditors," Stiglitz writes, "the far, far, smaller sums required to help the poor were not forthcoming."

IS YOUR INTERNATIONAL financial infrastructure breeding Bolsheviks? If it does create a backlash, one reason is the standard Bolshevik explanation—the I.M.F. really is controlled by the epicenter of international capital. Formal influence in the I.M.F. depends on a nation's financial contribution, and America is the only country with enough shares to have a veto. It is striking how many economists think the I.M.F. is part of the "Wall Street-Treasury complex," in the words of Bhagwati. The fund serves "the interests of global finance," Stiglitz says. It listens to the "voice of the markets," says

Nancy Birdsall, president of the Center for Global Development in Washington and a former executive vice president of the Inter-American Development Bank. "The I.M.F. is a front for the U.S. government—keep the masses away from our taxpayers," Sachs says.

I.M.F. officials argue that their advice is completely equitable—they tell even wealthy countries to open their markets and contract their economies. In fact, Stiglitz writes, the I.M.F. told the Clinton administration to hike interest rates to lower the danger of inflation—at a time when inflation was the lowest it had been in decades. But the White House fortunately had the luxury of ignoring the I.M.F.: Washington will only have to take the organization's advice the next time it turns to the I.M.F. for a loan. And that will be never.

Let the Poor Get Rich
the Way the Rich Have

9. The idea that free trade maximizes benefits for all is one of the few tenets economists agree on. But the power of the idea has led to the overly credulous acceptance of much of what is put forward in its name. Stiglitz writes that there is simply no support for many I.M.F. policies, and in some cases the I.M.F. has ignored clear evidence that what it advocated was harmful. You can always argue—and American and I.M.F. officials do—that countries that follow the I.M.F.'s line but still fail to grow either didn't follow the openness recipe precisely enough or didn't check off other items on the to-do list, like expanding education.

Policy makers also seem to be skipping the fine print on supposedly congenial studies. An influential recent paper by the World Bank economists David Dollar and Aart Kraay is a case in point. It finds a strong correlation between globalization and growth and is widely cited to support the standard rules of openness. But in fact, on close reading, it does not support them. Among successful "globalizers," Dollar and Kraay count countries like China, India and Malaysia, all of whom are trading and growing but still have protected economies and could not be doing more to misbehave by the received wisdom of globalization.

Dani Rodrik of Harvard used Dollar and Kraay's data to look at whether the single-best measure of openness—a country's tariff levels—correlates with growth. They do, he found—but not the way they are supposed to. High-tariff countries grew *faster*. Rodrik argues that the countries in the study may have begun to trade more because they had grown and gotten richer, not the other way around. China and India, he points out, began trade reforms about 10 years after they began high growth.

When economists talk about many of the policies associated with free trade today, they are talking about national averages and ignoring questions of distribution and inequality. They are talking about equations, not what works in messy third-world economies. What economic model taught in school takes into account a government ministry that stops work because it has run out of pens? The I.M.F. and the World Bank—which recommends many of the same austerity measures as the I.M.F. and frequently conditions its loans on I.M.F.-advocated reforms—often tell countries to cut subsidies, including many that do help the poor, and impose user fees on services like water. The argument is that subsidies are an inefficient way to help poor people—because they help rich people too—and instead, countries should aid the poor directly with vouchers or social programs. As an equation, it adds up. But in the real world, the subsidies disappear, and the vouchers never materialize.

The I.M.F. argues that it often saves countries from even more budget cuts. "Countries come to us when they are in severe distress and no one will lend to them," Rogoff says. "They may even have to run surpluses because their loans are being called in. Being in an I.M.F. program means less austerity." But a third of the developing world is under I.M.F. tutelage, some countries for decades, during which they must remodel their economies according to the standard I.M.F. blueprint. In March 2000, a panel appointed to advise Congress on international financial institutions, named for its head, Allan Meltzer of Carnegie Mellon University, recommended unanimously that the I.M.F. should undertake only short-term crisis assistance and get out of the business of long-term economic micromanagement altogether.

The standard reforms deprive countries of flexibility, the power to get rich the way we know can work. "Most Latin American countries have had deep reforms, have gone much further than India or China and haven't gotten much return for their effort," Birdsall says. "Many of the reforms were about creating an efficient economy, but the economic technicalities are not addressing the fundamental question of why countries are not growing, or the constraint that all these people are being left out. Economists are way too allergic to the wishy-washy concept of fairness."

THE PROTESTERS IN the street, the Asian financial crisis, criticism from respected economists like Stiglitz and Rodrik and those on the Meltzer Commission and particularly the growing realization in the circles of power that globalization is sustainable for wealthy nations only if it is acceptable to the poor ones are all combining to change the rules—slightly. The debt-forgiveness initiative for the poorest nations, for all its limitations, is one example. The Asian crisis has modified the I.M.F.'s view on capital markets, and it is beginning to apply less pressure on countries in crisis to cut government spending. It is also debating whether it should be encouraging countries to adopt Chile's speed bumps. The incoming director of the W.T.O. is from Thailand, and third-world countries are beginning to assert themselves more and more.

But the changes do not alter the underlying idea of globalization, that openness is the universal prescription for all ills. "Belt-tightening is not a development strategy," Sachs says. "The I.M.F. has no sense that its job is to help countries climb a ladder."

Sachs says that for many developing nations, even climbing the ladder is unrealistic. "It can't work in an AIDS pandemic or an endemic malaria zone. I don't have a strategy for a significant number of countries, other than we ought to help them stay alive and control dis-ease and have clean water. You can't do this purely on market forces. The prospects for the Central African Republic are not the same as for Shanghai, and it doesn't do any good to give pep talks."

China, Chile and other nations show that under the right conditions, globalization can lift the poor out of misery. Hundreds of millions of poor people will never be helped by globalization, but hundreds of millions more could be benefiting now, if the rules had not been rigged to help the rich and follow abstract orthodoxies. Globalization can begin to work for the vast majority of the world's population only if it ceases to be viewed as an end in itself, and instead is treated as a tool in service of development: a way to provide food, health, housing and education to the wretched of the earth.

Tina Rosenberg writes editorials for The Times. Her last article for the magazine was about human rights in China.

Trading for development:
The poor's best hope

Removing trade barriers is not just a job for the rich. The poor must do the same in order to prosper, says Jagdish Bhagwati

WASHINGTON, DC

THE launch of a new round of multilateral trade negotiations (MTN) at Doha dealt a needed blow to the anti-globalisers who triumphed at Seattle just two years ago. But it was also important for a different reason. The word "development" now graces the name of the new round. This is unconventional, but it underlines the fact that development of the poor countries will be the round's central objective.

Pleasing rhetoric aside, however, we must ask: What does this mean? The question is not idle. For if the current thinking among policymakers and NGOs is any guide, the answer they would give is not the right one. And that is cause for alarm.

Of course, proponents of trade have always considered that trade is the policy and development the objective. The experience of the post-war years only proves them right. The objections advanced by a handful of dissenting economists, claiming that free-traders exaggerate the gains from trade or forget that good trade policy is best embedded within a package of reforms, are mostly setting up and knocking down straw men.

But if trade is indeed good for the poor countries, what can be done to enhance its value for them? A great deal. But not until we confront and discard several misconceptions. Among them:

- the world trading system is "unfair": the poor countries face protectionism that is more acute than their own;
- the rich countries have wickedly held on to their trade barriers against poor countries, while using the Bretton Woods institutions to force down the poor countries' own trade barriers; and

- it is hypocritical to ask poor countries to reduce their trade barriers when the rich countries have their own.

In fact, asymmetry of trade barriers goes the other way. Take industrial tariffs. As of today, rich-country tariffs average 3%; poor countries' tariffs average 13%. Nor do peaks in tariffs—concentrated in textiles and clothing, fisheries and footwear, and clearly directed at the poor countries—change the picture much: the United Nations Council for Trade, Aid and Development (UNCTAD) has estimated that they apply to only a third of poor-country exports. Moreover, the trade barriers of the poor countries against one another are more significant restraints on their own development than those imposed by the rich countries.

The situation is little different when it comes to the use of anti-dumping actions, the classic "fair trade" instrument that has ironically been used "unfairly" to undermine free trade. The "new" users, among them Argentina, Brazil, India, South Korea, South Africa and Mexico, are now filing more anti-dumping complaints than the rich countries (see chart 1). Between July and December 2001 alone, India carried out more anti-dumping investigations than anywhere else.

The wicked rich?

These facts fly in the face of the populist myth that the rich countries, often acting through the conditionality imposed by the World Bank and the International Monetary Fund (IMF), have demolished the trade barriers of the poor countries while holding on to their own. Indeed, both the omnipotence of the Bretton Woods institutions, and the wickedness of the rich countries, have been grossly exaggerated.

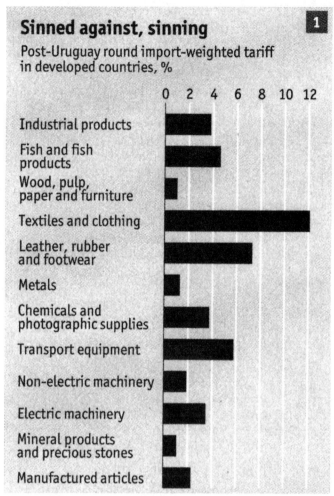

Sinned against, sinning

Post-Uruguay round import-weighted tariff
in developed countries, %

Source: UNCTAD

was reversed in 1979, the year in which another arrangement was negotiated with the IMF.

Moreover, the comparatively higher trade barriers against labour-intensive products are not usually the result of wickedness, but of simple political economy. Unilateral reductions of trade barriers are in fact not uncommon, and I document them for many countries and several sectors in the post-war period in my new book, "Going Alone: The Case for Relaxed Reciprocity in Freeing Trade" (MIT Press, July). But the fact remains that the developing countries were exempted by the economic ideology of the time, which embraced "Special and Differential" treatment for them, from having to make trade concessions of their own at the successive multilateral trade negotiations that reduced trade barriers after the war. The rich countries, denied reciprocal concessions from the poor countries, wound up concentrating on liberalising trade in products of interest largely to themselves, such as machinery, chemicals and manufactures, rather than textiles and clothing.

The situation changed when the poor countries became full participants. In 1995 in Marrakesh, where the Uruguay round was concluded, action was taken at last to dismantle the infamous Multi-fibre Arrangement (MFA), which—from its birth in 1961 as the Short-term Cotton Textile Arrangement—had grown by 1974 into a Frankenstein monster incorporating several separate agreements restricting world trade in all textiles. At Marrakesh the MFA was put on the block, and was scheduled to end in ten years.

But even if rich-country protectionism were asymmetrically higher, it would be dangerous to argue that it is therefore hypocritical to suggest that poor countries should reduce their own trade barriers. Except in the few cases of oligopolistic competition, such as that between Fuji and Eastman Kodak (hardly applicable to poor countries) where strategic tit-for-tat is credible, the net effect of matching other people's protection with one's own is to hurt oneself twice over. But there is ample evidence that many leaders of the poor countries have predictably made the wrong inference: that rich-country protectionism excuses, and justifies, going easy on relaxing their own barriers to trade.

In fact, the protectionism of the poor and the rich countries must be viewed together symbiotically to ensure effective exports by the poor countries. Thus, even if the doors to the markets of the rich countries were fully open to imports, exports from the poor countries would have to get past their own doors.

We know from numerous case studies dating back to the 1970s (which only corroborated elementary economic logic) that protection is often the cause of dismal export, and hence economic, performance. It creates a "bias against exports" by sheltering domestic markets that then become more lucrative. Just ask yourself why, though India and the far-eastern countries faced virtually the same external trade barriers in the quarter-century after the

The World Bank's conditionality is so extensive and diffused, and its need to lend so compelling, that it can in fact be bypassed. Many client states typically satisfy some conditions while ignoring others. Besides, countries go to the IMF when there is a stabilisation crisis. Since stabilisation requires that the excess of expenditures over income be brought into line, the IMF has often been reluctant to suggest tariff reductions. These could reduce revenues, exacerbating the crisis.

Then again, since countries are free to return to their bad ways once the crisis is past and the loans repaid, tariff reforms can be reversed. Countries do not "bind" their tariff reductions under the IMF programmes, as they do at the World Trade Organisation (WTO). Equally, tariff reductions may be reversed when a stabilisation crisis recurs and the tariffs are reimposed to increase revenues. My student Ravi Yatawara, who has studied what he calls "commercial policy switches", documents several instances of such tariff-reduction reversals by countries borrowing from the IMF. For instance, Uruguay in 1971 increased trade protection during an IMF programme that began the year before, and even managed to get another credit tranche the year after. Kenya's 1977 liberalisation

1960s, inward-looking India registered a miserable export performance while outward-looking South Korea, Taiwan, Singapore and Hong Kong chalked up spectacular exports. Just as charity begins at home, so exports begin with a good domestic policy. In the near-exclusive focus on rich-country protectionism, this dramatic lesson has been lost from view.

A strategy for change

Rich-country protectionism matters too, of course. And it must be assaulted effectively. But here, too, we witness folly. The current fashion is to shame the rich countries by arguing that their protection hurts the poor countries, whose poverty is the focus of renewed international efforts. And where action is actually undertaken, the preference is for granting preferences to the poorer countries, with yet deeper preferences for the poorest among them (the least-developed countries, or LDCs, as they are now called). But the former solution is woefully inadequate, and the latter is downright wrong.

If shame were sufficient, there would be no rich-country protectionism left. Trade economists and international institutions such as UNCTAD and the General Agreement on Tariffs and Trade (the GATT) have denounced the rich countries on this count over three decades. Added support, from charities such as Oxfam, could help in principle. But these charities need both expertise and a talent for strategy, not simply a conscience and a voice. They fall short. By subscribing to the counterproductive language of "hypocrisy" and the rhetoric of "unfair trade" to attack protection by the rich, a charity such as Oxfam, splendid at fighting plagues and famines, does more harm than good.

The argument to rich countries should be made in quite a different way: If you hold on to your own protection, no matter how much smaller, and in fact even raise it as the United States did recently with steel tariffs and the farm bill, you are going to undermine seriously the efforts of those poor-country leaders who have turned to freer trade in recent decades. It is difficult for such countries to reduce protection if others, more prosperous and fiercer supporters of free trade, are breaking ranks.

Beyond this, an effective tariff-reduction strategy requires that we handle labour-intensive goods such as textiles separately from agriculture. The differences between them dwarf the commonalities. Labour-intensive manufactures in the rich countries typically employ their own poor, the unskilled. To argue that we should eliminate protection, harming them simply because it helps yet poorer folk abroad, runs into evident ethical (and hence political) difficulties. The answer must be a gradual, but certain, phase-out of protection coupled with a simultaneous and substantial adjustment and retraining programme. That way, we address the problems of the poor both at home and abroad.

Once this is done, church groups and charities can be asked to endorse a programme that is balanced and just. Such a strategy is morally more compelling than either marching against free trade to protect workers in the labour-intensive industries of the rich nations—while forgetting the needs of poor workers in poor countries—or asking for trade restrictions to be abolished without providing for workers in such industries in the rich countries.

The removal of agricultural protection does not raise the same ethical problems; production and export subsidies in the United States and the European Union go mainly to large farmers. That should make it easier to dismantle farm protection on the grounds of helping the poor. At the same time, however, agricultural protectionism is energetically defended as necessary for preserving greenery and the environment. With the greens in play, protectionism becomes more difficult to remove. But, just as income support can be de-linked from increasing production and exports, so measures to support greenery can be de-linked too. Such new measures, and other environmental protections added as sweeteners, must be part of the strategic assault on agricultural protection.

The target date of Jubilee 2000 helped greatly to focus efforts on the objective of debt relief. Following that example, I and Arvind Panagariya of the University of Maryland suggested well over a year ago—with a nod from Kofi Annan, the UN's secretary general—a Jubilee 2010 movement to eliminate protection on labour-intensive products by 2010. Since agricultural protection is politically a harder nut to crack, 2020, rather than 2010, is probably a more realistic date for its demise. Leaders of rich and poor countries could endorse both targets at the mammoth UN Conference on Sustainable Development in Johannesburg in August.

The perils of preferences

A final word is necessary on the efforts to open rich-country doors. This is often done not by dismantling barriers on a most-favoured-nation (MFN) basis, which reduces them in a non-discriminatory manner, but through grants of preferences to the poor countries. This approach goes back to the Generalised System of Preferences (GSP), introduced in 1971 through a waiver and then granted legal status in 1979 with an enabling clause at the GATT. Under this, the eligible poor countries were granted entry at preferentially lower tariff rates.

GSP did little for the poor countries. The eligible products often excluded those on which poor countries had pinned their hopes of increasing exports. Thus the United States' GSP scheme excluded textiles, clothing and footwear. Upper caps were also introduced. The United States imposed a $100m limit on exports per tariff line, per year, per country; beyond this limit, the preferential rate vanished. Even the benefits granted were not "bound", and could be varied at a rich country's displea-

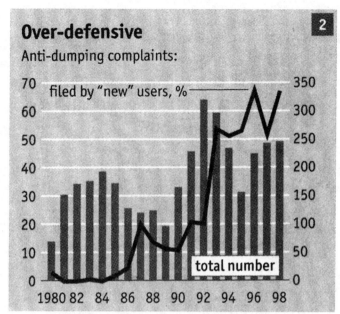

Over-defensive

Anti-dumping complaints:

Source: "Worldwide Use of Anti-Dumping, 1980–1998", by Thomas Prusa and Susan Skeath. NBER Working Paper no. W8424, Aug. 2001

local-content specifications (for example, shoes had to have uppers, soles and laces made locally) to qualify for GSP benefits.

The rich countries are still going down this preferential route today. The United States has introduced the Africa Growth and Opportunity Act (AGOA), while the EU has brought in the "everything but arms" initiative, properly known as the EBE, to eliminate trade barriers for the 49 LDCs. Yet virtually every drawback of GSP applies to these schemes as well. If anything, they are worse. Under the AGOA, for example, preferences for African garments are tightly linked to reverse preferences for American fabrics.

Since preferences typically divert trade away from non-preferred countries, they tend to pitch poor nations against each other. They are also a wasting asset, since they are relative to an MFN tariff that will probably decline with further multilateral liberalisation. And since they are non-binding and can be readily withdrawn for political reasons, investors are not likely to be impressed by them.

Preferences sound attractive and generous, and the poor countries have accepted them as such. But this has been a mistake. There is no good substitute for the MFN reduction of trade barriers in the rich countries. It should go hand in hand with enhanced technical and financial assistance. By focusing this help preferentially on the poor nations, the poor should be able to exploit the trade opportunities that are opened up for them by non-preferential treatment. This is the only way ahead.

sure. Thus, when India was put on the Special 301 list in 1991 and the United States trade representative determined unilaterally that India's intellectual-property protection was "unreasonable", President George Bush senior suspended duty-free privileges under GSP for $60m in trade from India in April 1992.

Preferences were also often dropped for commodities when they began to be successfully exported, a fact documented in a forthcoming study by Caglar Ozden and Eric Reinhardt of Emory University. Rules of origin served to curb exports, too. Exported items had to satisfy stringent

Jagdish Bhagwati is University Professor at Columbia University and Andre Meyer Senior Fellow in International Economics at the Council on Foreign Relations. His most recently published book is "Free Trade Today" (Princeton).

Rich Nations' Tariffs and Poor Nations' Growth

by Shweta Bagai and Richard Newfarmer

In an attempt to curb terrorism, international policymakers are now focusing on helping poor countries raise their standard of living, thus hopefully diminishing the appeal of radical ideologies.

Beginning with the Conference on Financing for International Development, held in Monterrey, Mexico, last year, the United States and other industrialized nations pledged to increase aid to help developing countries reach goals for reducing poverty, raising literacy, and lengthening life expectancy by 2015. Unfortunately, however, one potent method for helping poor nations grow has been underutilized, namely, tearing down trade barriers in rich countries that keep out exports from poor states.

On average, trade barriers are higher for developing countries than for rich ones. Exporters in impoverished countries generally face tariffs that are greater than duties in industrial countries, a difference that widens if other trade restrictions are taken into account. Several reasons explain this. Until the 1990s, successive rounds of world trade liberalization focused solely on manufactured products, as these were of greatest interest to the relatively few, mainly industrial countries that historically dominated the negotiations. Today, although poor countries tend to export a much higher percentage of agricultural products than wealthy nations do, tariffs on agriculture are five or more times higher than those on manufactures.

Countries such as Brazil, which are big producers of fruits and sugar, are hindered in their development by wealthy nation's tariffs.

Moreover, industrial states have sought to protect employment in certain domestic sectors and so have agreed to limit competition in labor-intensive products, particularly textiles and clothing. For these reasons, products of poor Asian nations such as Sri Lanka and Nepal entering the United States are taxed at an average rate of 11 percent, while Japanese and the European products pay tariffs that average below 2 percent. This means that poor countries pay more in tariffs for their total exports than rich countries do. For instance, according to Edward Gresser of the Progressive Policy Institute, even though France exports 12 times more to the United States than Bangladesh does, American tariff collections on imports from Bangladesh were roughly the same as those from France.

POOR NATIONS' TARIFFS AGAINST EACH OTHER

Since the world's poor make their living in agriculture or industries that are labor-intensive, this system of protection tends to work against them. In fact, the average poor person—someone living on less than $1 or $2 per day—selling into the global marketplace confronts an average tariff that is more than twice as much as for those who are not poor.

Though here we are concerned with ways for the United States and other rich countries to use their levers of policy to promote development, it should be noted that barriers to trade erected in middle- and low-income developing countries are also part of the problem. Most countries in South Asia and Africa, wrote Jagdish Bhagwati in the *Economist* last year, have tariff levels higher than the average for both developing and developed countries. According to recent World Bank figures, developing country tariffs in manufacturing average four times higher for imports from fellow developing countries than the tariffs imposed by industrial countries on imports from developing countries (12.8 as opposed to 3.4 percent). Agricultural subsidies in the countries that form the industrialized-nation club of the Organization for Economic Cooperation and Devel-

Free Traders' Trade Barriers	
	The world's industrialized nations--those that push most vocally for free trade--still impose high tariffs on a broad range of goods, hitting developing nations especially hard.
	Poor countries' export strength lies particularly in agricultural commodities, clothing, and textiles—the very items on which rich states pile their heaviest duties.
	The reason is that world free-trade talks have been led by industrialized countries, most of whose exports are manufactured goods and specialized services.
	Part of the trade problem, though, is that the developing nations themselves set high tariffs on imports from fellow poor states.
$	If world trade were fully liberalized, poor countries could have their annual income boosted by over $500 billion, dwarfing the $50 billion a year they currently get in foreign aid.

opment (OECD) are of course much greater than those in non-OECD countries. Even so, developing countries have as much—or more—to gain from reducing protection on trade among themselves as they do from rich countries' liberalization.

The round of trade negotiations launched at the World Trade Organization (WTO) ministerial meeting in Doha, Qatar, in November 2001 represents an opportunity to provide poor people with access to markets around the world. If both rich and poor countries were to phase out their merchandise trade barriers over the five years beginning in 2005, developing countries would experience a faster growth rate that would accelerate poverty reduction throughout the developing world—and at the same time benefit rich countries.

The World Bank simulated the effects of these policies in its global economic model and found that some 300 million poor people would be lifted above the $2 per day international poverty line—over and above the number that normal assumptions of growth would produce. Said differently, reducing protectionism would mean some 13 percent *fewer* people living in poverty in 2015. Developing countries could receive income gains of over $500 billion from full trade liberalization. This implies up to a 5 percent boost in their incomes. These numbers tend to dwarf the approximately $50 billion a year that developing countries currently receive in foreign aid.

MANY FACES OF PROTECTIONISM

U.S. protectionism is biased against imports from least-developed countries, while the European Union (EU) is unfair to imports from middle-income countries. World Bank research has found that developing-country exports generally confront trade barriers that are higher than on industrial-country exports. Barriers can assume different forms: tariffs, tariff peaks, and subsidies.

Average tariff levels mask selectively high tariffs—called *tariff peaks*—on particular products. These high tariffs (that is, ad valorem tariffs of 15 percent and greater) are more common in rich than in developing countries, and more so in the areas of textiles and clothing. Japan and the EU commonly have very high tariffs in agriculture, footwear, and food products. Tariff peaks in rich countries are 40 times higher than average tariffs.

Another distortion in rich countries that slows development is *tariff escalation*, in which tariffs increase in proportion to the value added to the product. Chilean firms, for example, can export fresh tomatoes to the United States, paying a tariff of 2.2 percent; however, if they dry the tomatoes and put them in a package, the U.S. tariff is 8.7 percent; and if they make salsa out of the tomatoes and export it to America, the duty is 11.6 percent. As in this example, protection is low for primary products and increases as the goods become more processed. Developing countries thus end up producing goods that require minimal technological sophistication.

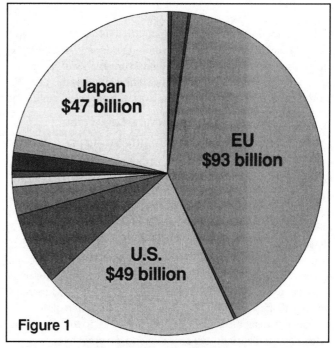

Figure 1

Crop props: **OECD countries spent $230 billion in 2001 on direct subsidies for their agricultural producers.**

Subsidies are even more pernicious in agriculture. Lacking the capital and skill base common in rich countries, developing countries rely on low-cost labor to compete in global markets. Yet it is these labor-intensive products that commonly encounter the greatest obstacles in world markets. The highest protection by rich countries covers products in which developing countries have comparative advantage: sugar, rice, fruits,

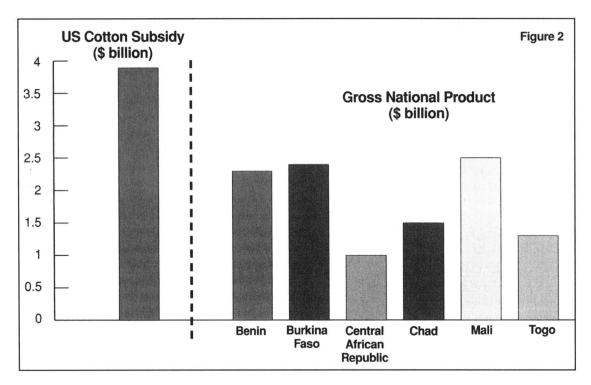

Cotton "candy": U.S. cotton subsidies dwarfed the gross national incomes of many African countries in 2000.

and vegetables. The American peanut market is restrictive and virtually shuts out the big developing-country peanut producers, such as China and India and even smaller African nations such as Senegal. High tariffs in the United States and EU on imported orange juice prevents Brazil from fully exploiting its comparative advantage in citrus.

In high-income countries, according to a 2001 study by the OECD and another report last year by the World Bank and International Monetary Fund, hefty subsidies to agriculture can aggravate the effects of tariffs. Support to agricultural producers has remained high—$311 billion in 2001—many times the level of international development assistance. One measure of this support is the extra income farmers get through tariff-assisted prices and subsidies, the producer support estimate. EU farmers get the most protection by this measure in absolute terms (see figure 1). Prices received by OECD farmers for their produce were much higher—31 percent above world prices (measured at the border). Beyond the $230 billion in direct subsidies, additional billions of *indirect* subsidies go to agricultural extensions and agricultural research and development.

EUROPE, U.S., JAPAN COULD DO MORE

Most of the subsidies are directed to large corporate and individual farms. Two-thirds of the subsidies in America go to farms with sales of more than $100,000. The same is true in Europe. The largest 17 percent of European farms, with earnings two or three times the average domestic wage, receive half of the public support, whereas the bottom 40 percent receive 8 percent, according to Patrick Messerlin, writing in *Foreign Policy* magazine last year. If rich countries were to cut back these subsidies, target them on small farmers, and decouple them from production, the benefits to U.S. and other rich-country consumers would be great—and at the same time would help raise standards of living the world over.

The European Union provides substantial budgetary support to farmers under its Common Agricultural Policy (CAP). Reforms to the CAP have been supported by various members but face opposition from those with strong farm lobbies. In particular, President Jacques Chirac of France and Chancellor Gerhard Schröder of Germany have drawn the EU toward a deal that keeps the present system largely intact until 2013. The CAP already costs $40 billion a year and the proposed admittance into the EU of central and eastern European countries with large agricultural sectors could make the costs soar.

In Japan, the new Basic Agricultural Law, passed in 1999, was intended to make Japanese agriculture more market oriented. In fact, the law is primarily directed at bolstering self-sufficiency through import restrictions.

Similarly, the 2002 U.S. farm bill increases support spending by a projected $45 billion, or 21 percent during fiscal years 2002–07, which is likely to cause a further decline in world prices, lowering the price developing countries can get for their exports. A large fraction of the money will be allocated to market assistance loans and special programs for dairy, sugar, and peanuts. The new peanut program alone will cost American taxpayers $4.9 billion.

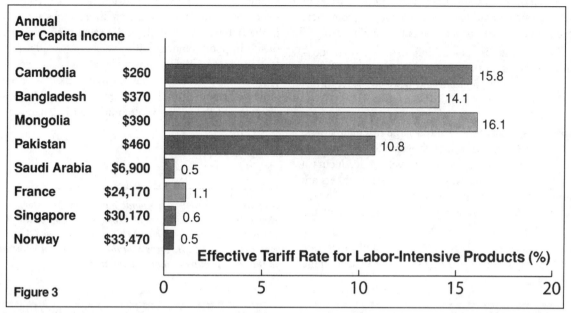

Keeping them down on the farm: Countries with a lower per capita income face higher tariffs in labor-intensive products like shoes and clothing.

U.S. sugar policy is one of the more egregious examples of antidevelopment policies. The program of U.S price supports and quotas only benefits a small number of domestic beet sugar and corn syrup producers at the expense of American consumers. It is estimated that about 9 percent of the 1999–2000 domestic crop (800,000 tons) is sitting in warehouses, paid for by U.S. taxpayers. Sugar protection causes a loss to the U.S. economy of about $1 billion annually.

High protection has also prompted the relocation of U.S. plants outside American borders to take advantage of lower trade barriers. Kraft has moved production of Life Savers candy to Canada, where imported sugar is cheaper. Liberalizing the sugar industry could therefore benefit rich countries; at the same time, says a World Bank study, it would help developing nations by increasing their annual export earnings by $1.5 billion.

CLOTHING AND TEXTILES

The same is true for U.S. cotton subsidies. The payments to 25,000 domestic cotton producers cost U.S. taxpayers nearly $4 billion a year (figure 2). This is three times the entire U.S. aid budget for Africa. To put this figure in perspective, it exceeds the entire gross domestic product of Burkina Faso, where 2 million people depend on cotton as a principal source of revenue. Even though the United States is an inefficient, high-cost producer by global standards, U.S. cotton production has doubled over the last 20 years.

The result is a glut that costs African economies an estimated $250 million a year in lost sales. The economic losses to Africa inflicted by the subsidy program exceed the aid that Americans provide. For example, the aid organization Oxfam estimates

that Mali received $37 million in assistance in 2001 but lost $43 million due to lower export earnings.

Support to Japanese rice growers equaled more than seven times the world price for their output. Huge government transfer payments ensure that the domestic market is shielded from foreign competitors. This also applies to support for wheat farmers in rich countries. Removal of wheat protection is estimated to lead to a rise in the price of wheat by 10 percent. In fact, many developing countries could become potentially large exporters of food grains if it were not for trade policy distortions, according to a study last year in the *Journal of African Economies*.

Agriculture is not the only major area in which trade is biased against the poor. Clothing and textiles also offer major potential sources of job creation and affect the low-skilled workers in developing countries. The World Bank and IMF say that about 50 percent of world textile exports and 70 percent of world clothing exports are accounted for by developing countries. Under the WTO's Uruguay Round Agreement on Textiles and Clothing, quota restrictions are to be abolished gradually over the period 1995–2005. However, the elimination of restrictions has been modest so far, with the exception of Norway. The United States has eliminated only 1 percent of quota arrangements, the EU 7 percent, and Canada 14 percent.

The tariffs on labor-intensive consumer goods such as shoes and clothing confronted by less-developed countries are four to five times higher than those facing the developed countries, even though developing countries' share in U.S. imports is only 6.7 percent. Cambodia, where the average income is only $260 a year, faces one of the highest effective tariff rates in the world, at 15.8 percent, whereas its industrialized neighbor, Singapore, which has a per capita income more than 100 times higher (see figure 3), faces a 0.6 percent overall tariff rate on its exports. In fact, textiles and clothing represent 87 percent of Cambodia's merchandise exports. Industrial-country restrictions in trade in textiles have cost some 20 million jobs in developing countries, many of which would constitute a move out of rural poverty.

Poor Americans generally pay a higher share of their consumer dollar for trade taxes than do the rich. The reason is that the poor spend more on agriculture and labor-intensive products, and cheap imports often are taxed at a higher rate. Consider, for example, the case of water glasses. A 30-cent imported water glass, commonly sold in large discount stores, faces a tariff of 38 percent, while a glass worth $5 or more, undoubtedly destined for the tables of the well-to-do, incurs a tariff of only 5 percent. Similarly, tariffs on cheap sneakers are proportionally higher than tariffs on leather golf shoes or ski boots—discriminating against low-income consumers in the United States.

All in all, then, policymakers in America and other rich countries who desire to help raise living standards in developing countries have a powerful instrument at their command. Moreover, liberalizing trade would create new jobs and raise incomes in the rich countries while at the same time raising the quality of life in poor countries. Still, vested interests supporting protection are politically powerful all around the world, and progress to date in multilateral trade talks has been glacially slow. Continuing—and indeed intensifying—U.S. leadership is necessary if the trade negotiations are to realize their development promise.

Richard Newfarmer is economic adviser in the International Trade Department and Development Prospects Group of the World Bank. He is lead author of the bank's annual report, **Global Economic Prospects**. *Shweta Bagai is a junior professional associate working in the same group on international trade.*

UNELECTED
Government

MAKING THE IMF AND THE WORLD BANK MORE ACCOUNTABLE

Accused of being secretive, unaccountable, and ineffective, both the International Monetary Fund and the World Bank are seeking to become more transparent, more participatory, and more accountable. Yet few attempts have been made to dissect the existing structure of accountability within the organizations. Applying the concept of accountability as understood and refined in political science can illuminate both the existing accountability of the institutions and some of the ways in which they could better be held to account.

By Ngaire Woods

Accountability to Governments

Like many international organizations, the IMF and the World Bank face complex problems of accountability that originate in a simple question: to whom should they be accountable and how? Democratic political systems ensure accountability not just by elections but by mechanisms such as transparent decisionmaking, judicial review, and the use of ombudsmen. The aim is to ensure that political actions are predictable, nonarbitrary, and procedurally fair, that decisionmakers are answerable, and that rules and parameters on the exercise of power are enforced.

Unlike a democratically elected government, international institutions cannot claim that voters elect them and can vote them out of office. Nor have they been subjected to the normal restraints politicians face from the checks and balances of government, including the role played by judges, ombudsmen, and the like.

Rather, international organizations grapple with an unwieldy structure of government representation that makes ensuring their own accountability extremely difficult. Until roughly the 1980s, the mission of the IMF and the World Bank was narrowly technical, and accountability, accordingly, less of an issue. Today, however, both the Fund and the Bank are being pushed by their most powerful members to perform a much wider range of tasks directly affecting a broad range of people. The need for accountability has thus become more critical.

The IMF and the World Bank were established after World War II as mutual assistance organizations through which all member countries could help each other with postwar reconstruction and development as well as with balance-of-payments problems. The voting and governance structures of both organizations reflect that early vision. Today, however, both lend only to developing

and transition countries, and both condition their lending heavily on broad changes in borrowers' economic policies. Indeed, "conditionality" has widened dramatically over the past couple of decades, increasing the intrusiveness of the institutions' work and changing the nature of their relationship with stakeholders. Yet their structure of accountability has remained in many ways unchanged

In both the Fund and the Bank, the basic structure of accountability works through representatives of governments. At the top of the system, Boards of Governors meet just once a year and are supposed to maintain overall oversight and control of the institutions. The day-to-day work is overseen by executive-directors, representatives of member states who sit on the Executive Boards and perform a dual role. Individually each executive-director represents a country or a group of countries; collectively they manage the organization. They appoint and can dismiss the head of each organization, who in turn controls the management and staff.

The links between most member governments and the IMF and the World Bank are extremely weak.

The chain of representative accountability is long and imperfect. The links between most member governments and the IMF and the World Bank are extremely weak. Most member governments (with the obvious exception of the United States) are too far removed from the workings of the Executive Board, which in turn exercises too little control over the staff and management. Four elements of this are worth elaborating.

Unequal Representation of Member States

The Executive Boards are the vital link from countries to the IMF and the World Bank. Yet only the largest member countries—the United States, Germany, France, Japan, the United Kingdom, Saudi Arabia, Russia, and China—are directly represented by their own executive-director. All other economies are grouped within constituencies represented collectively by just one executive-director. Most national governments thus have only the weakest link to formal deliberations and decisionmaking. In the IMF, for example, 21 Anglophone African countries, at least 11 of which have an intensive-care relationship with the institution and all of which are deeply affected by its work, are represented by a single executive-director and have a voting share of 3.26 percent. In the World Bank, the same group of countries plus the Seychelles is represented by one executive-director and has a voting share of 4.07 percent. This small formal share of votes reflects a rationale that was appropriate to the organizations' original mandate and membership. Each, at its inception, had far fewer member states; membership has since more than tripled, with decolonization creating a host of

new, independent states. Originally, members were mutually assisting, and each had a vote proportional to its stake. Now that both the Fund and the Bank make highly conditional development loans only to developing and transition economies, the traditional voting structure looks anachronistic and inadequate as a formal mechanism of accountability.

Inadequate Oversight of Staff and Management

Although the Executive Boards appoint and oversee the senior management and work of both institutions, several practical impediments complicate this role. Often both Boards are perceived as simply nodding through decisions already made by the staff and management in consultation with the most powerful member countries. Several organizational features play into this. The Board is not independent of the staff and management. Executive-directors are paid by and housed within each institution and have a dual role as officials of the organization as well as representatives of countries; many flit from Board to staff and back again. Many executive-directors are ill placed to follow closely and to prepare positions on all issues in front of them. Unlike their counterparts in the G-7 (the finance ministers and central bankers of the seven major industrialized countries), few have the advantage of resources and staff working on Fund and Bank issues in their constituency countries. Furthermore, when proposals come before the Board, executive-directors are not privy to the disagreements and alternatives that have been debated among staff and senior management. The latter present just one proposal to the Board, leaving executive-directors either to accept or to reject. Fine-tuning is seldom entertained.

Flawed Appointment Process

Neither the Bank nor the Fund has an open and transparent process by which to appoint its head—to whom all staff are accountable. Rather, according to a 50-year-old political compromise, the head of the World Bank is an American and the head of the IMF is a non-American, in practice always a Western European. This process came under scrutiny several

years ago, when Germany's first candidate to head the IMF failed to win support from other major shareholders. Sensitive to adverse press and policy attention to the lack of transparency and accountability, both institutions established committees to improve the appointment process, but the result was pretty much an endorsement of the status quo—a failure on the part of U.S. and European members to put into practice their rhetoric.

Broad Conditionality, Narrow Accountability

All these accountability problems have been magnified by the increase in and transformation of the activities of the IMF and World Bank. Figures compiled in both institutions show a dramatic increase in the number and scope of conditions placed on loans in the past several decades. During the 1980s, loans to a sample of 25 countries were conditional on between 6 and 10 "performance criteria." During the 1990s the number of performance criteria had jumped to about 25. Borrowing countries are now being required to mobilize, redefine, strengthen, or upgrade government processes in an ever wider range of areas.

The conditionality in which both the Fund and the Bank engage affects budgets and policies in areas such as health care and education. Although both organizations now consult with civil society and a broader range of stakeholders, their formal accountability remains unchanged. Their official interlocutors, as set out in their Articles of Agreement, are the Treasury, Finance Ministry, Central Bank, or equivalent agency of a borrowing country. Such transactions erode democratic accountability in several ways. First, domestic policy discussions on health or education are weighted in favor of the Treasury or Central Bank view because these agencies are the conduits of Bank and Fund financing. Second, agreements make the Finance Ministry or Central Bank formally accountable to the Fund and Bank for policies for which a Minister of Education or Health should more appropriately be held to account by the population of the country.

Steps Toward Greater Accountability

Aware of the criticisms they face, and frustrated by their limited effectiveness in implementing wider policy reform, both the IMF and the World Bank have begun more explicitly to recognize a wider range of stakeholders in their work. Both are making themselves more accountable to such stakeholders through more transparency, new mechanisms of horizontal accountability, and closer collaboration with nongovernmental organizations. The implications for accountability are several.

Although transparency has improved, several crucial gaps remain. The Fund has yet to publish its internal rules, guidelines, and operating procedures; the Bank, the full output of its Operations Evaluation Department (OED). The omissions are important. The Fund can be accountable to its members only if they (in the broad sense of they) are privy to the guidelines and conventions governing the internal operations of the staff—especially given the weaknesses of the Board in holding staff and management to account. Making public the agencies' self-evaluations would invite outside scrutiny that could not only strengthen external accountability but also ensure that the IMF and the Bank take seriously their own reviews. But transparency, though necessary, is not sufficient.

Monitoring and enforcement are also vital. National political systems rely on a combination of constitutionalism and democracy. In the U.S. system, for example, democracy is served by Congress; constitutionalism, by the Supreme Court. Electoral accountability is given life through "horizontal" accountability— agencies and processes that monitor and enforce the mandate, obligations, rules, and promises of institutions.

The IMF and the World Bank have both recently created several agencies and processes to enhance horizontal accountability. The IMF's new office for independent evaluation published its first study in 2002. More boldly, in 1993 the World Bank Executive Board created an inspection panel to consider complaints from groups claiming to have been adversely affected by the Bank's failure to

follow its own policies or procedures. In 1999 the Bank created a compliance officer-ombudsman's office to deal with similar complaints about its International Finance Corporation and Multilateral Investment Guarantee Agency.

Both the Fund and the Bank are working in a world in which more people aspire to hold them to account. A wide range of nongovernmental organizations and other groups will continue to demand that both be more open and transparent, not just about what they do but about how they do it. The IMF has yet to rise to that challenge—not just by publishing its internal rules but by subjecting itself to independent review to ensure that its rules and guidelines are being respected. For example, an independent agency could monitor closely how the IMF applies its Guidelines on Conditionality (revised in 2002), thus permitting the Fund's interlocutors to speak frankly about their work with Fund staff to produce programs and to record and measure their experiences against the standards the Fund has set itself.

Finally, both institutions have responded to demands for greater accountability by diversifying their contacts, in particular with nongovernmental organizations. They no longer refer to their interlocutors in member countries exclusively as "national authorities." Rather, the World Bank writes of "development partners"; the IMF, of "authorities and civil society" and of the need for its programs to enjoy "ownership by the societies affected." At regional, country, and local levels, World Bank regional directors and IMF resident representatives are being told to seek out and maintain contacts. At the annual and spring meetings, both institutions have been actively involved in more dialogue and meetings with a select group of transnational NGOs. Both have also begun to demand local participation by nonstate actors, as in the poverty-reduction strategy papers being required of countries seeking enhanced debt relief.

This is not to say that NGOs have taken a place as major stakeholders: they have acquired neither control nor a formal participatory role in decisionmaking except at the behest of their own governments. Transnational NGOs have fos-

tered a global debate about what international organizations are doing and have been crucial in monitoring and demanding accountability from global actors. In turn, their own accountability is often questioned, particularly where Northern NGOs have allied with or used political leverage in major shareholding countries such as the United States. Developing countries complain that such NGOs end up with a stronger voice in the Fund and Bank than many smaller developing countries.

Different concerns are raised about local or grassroots NGOs. No one doubts the value of independent groups of citizens committed to monitoring their own governments. The question is whether the Fund and Bank can facilitate such a process. Critics argue that in fostering wider participation, the IMF and the World Bank become gatekeepers of social organizations and power. Because the institutions must choose which NGOs to recognize and consult, they end up making decisions with deep social and political consequences. They create incentives for government officials to moonlight as NGOs, thus avoiding electoral and constitutional accountability. They also create organizations that siphon away resources and talented officials from government departments.

In sum, improving relations with nongovernmental groups will not erase the accountability deficit in the Fund and Bank. It takes the institutions into a minefield of social and political relations through which they must step with maximum knowledge and caution. And it does not obviate the need for the most powerful members of both intergovernmental organizations to modernize the ways in which each is accountable to its developing and transition country members.

Looking Ahead

Improving the accountability of the Fund and Bank became a mantra for economic policymakers in the G-7 in the late 1990s. As a result, both institutions became more transparent and opened dialogues with new groups of stakeholders. These measures may improve governance, but they do not solve the core accountability deficit.

Each institution needs a structure of representation that better reflects the stakes of all member states; a stronger, more independent role for the Executive Board in overseeing the work of the staff and management; a transparent set of operating rules that enables others to monitor how they do their work; and an open and participatory process for appointing the head of each organization. Perhaps most important, the powerful members of each institution must restrain their urge to have the Fund and Bank engage in conditionality across a wide range of issues. As a rule of thumb, the scope of the activities of the Fund and Bank should not exceed the scope of their accountability. If they represent and negotiate with Central Banks and Treasuries, their work should not stray beyond the mandates of such agencies. Rather than impose strong external accountability on weak national systems, they should whenever possible look for and strengthen local kinds of accountability. In the longer term, people need to hold their own governments to account regardless of IMF and World Bank strictures.

Ngaire Woods is Fellow in Politics and International Relations at University College, Oxford. This article is drawn from an article published in International Affairs *in January 2001.*

THE IMF Strikes Back

Slammed by antiglobalist protesters, developing-country politicians, and Nobel Prize-winning economists, the International Monetary Fund (IMF) has become Global Scapegoat Number One. But IMF economists are not evil, nor are they invariably wrong. It's time to set the record straight and focus on more pressing economic debates, such as how best to promote global growth and financial stability

By Kenneth Rogoff

Vitriol against the IMF, including personal attacks on the competence and integrity of its staff, has transcended into an art form in recent years. One bestselling author labels all new fund recruits as "third-rate," implies that management is on the take, and discusses the IMF's role in the Asian financial crisis of the late 1990s in the same breath as Nazi Germany and the Holocaust. Even more sober and balanced critics of the institution—such as *Washington Post* writer Paul Blustein, whose excellent inside account of the Asian financial crisis, *The Chastening*, should be required reading for prospective fund economists (and their spouses)—find themselves choosing titles that invoke the devil. Really, doesn't *The Chastening* sound like a sequel to 1970s horror flicks such as *The Exorcist* or *The Omen*? Perhaps this race to the bottom is a natural outcome of market forces. After all, in a world of 24-hour business news, there is a huge return to being introduced as "the leading critic of the IMF."

Regrettably, many of the charges frequently leveled against the fund reveal deep confusion regarding its policies and intentions. Other criticisms, however, do hit at potentially fundamental weak spots in current IMF practices. Unfortunately, all the recrimination and finger pointing make it difficult to separate spurious critiques from legitimate concerns. Worse yet, some of the deeper questions that ought to be at the heart of these debates—issues such as poverty, appropriate exchange-rate systems, and whether the global financial system encourages developing countries to take on excessive debt—are too easily ignored.

Consider the four most common criticisms against the fund: First, IMF loan programs impose harsh fiscal austerity on cash-strapped countries. Second, IMF loans encourage financiers to invest recklessly, confident the fund will bail them out (the so-called moral hazard problem). Third, IMF advice to countries suffering debt or currency crises only aggravates economic conditions. And fourth, the fund has irresponsibly pushed countries to open themselves up to volatile and destabilizing flows of foreign capital.

Some of these charges have important merits, even if critics (including myself in my former life as an academic economist) tend to overstate them for emphasis. Others, however, are both polemic and deeply misguided. In addressing them, I hope to clear the air for a more focused and cogent discussion on how the IMF and others can work to improve conditions in the global economy. Surely that should be our common goal.

THE AUSTERITY MYTH

Over the years, no critique of the fund has carried more emotion than the "austerity" charge. Anti-fund diatribes contend that, everywhere the IMF goes, the tight macroeconomic policies it imposes on governments invariably crush the hopes and aspirations of people. (I hesitate to single out individual quotes, but they could easily fill an entire edition of *Bartlett's Quotations*.) Yet, at the risk of seeming heretical, I submit that the reality is nearly the opposite. As a rule, fund programs lighten austerity rather than create it. Yes, really.

Critics must understand that governments from developing countries don't seek IMF financial assistance when the sun is shining; they come when they have already run into deep financial difficulties, generally through some combination of bad management and bad luck. Virtually every country with an IMF program over the past 50 years, from Peru in 1954 to South Korea in 1997 to Argentina today, could be described in this fashion.

Policymakers in distressed economies know the fund will intervene where no private creditor dares tread and will make loans at rates their countries could only dream of even in the best of times. They understand that, in the short term, IMF loans allow a distressed debtor nation to tighten its belt less than it would have to otherwise. The economic policy conditions that the fund attaches to its loans are in lieu of the stricter discipline that market forces would impose in the IMF's absence. Both South Korea and Thailand, for example, were facing either outright default or a prolonged free fall in the value of their currencies in 1997—a far more damaging outcome than what actually took place.

Nevertheless, the institution provides a convenient whipping boy when politicians confront their populations with a less profligate budget. "The IMF forced us to do it!" is the familiar refrain when governments cut spending and subsidies. Never mind that the country's government—whose macroeconomic mismanagement often had more than a little to do with the crisis in the first place—generally retains considerable discretion over its range of policy options, not least in determining where budget cuts must take place.

At its heart, the austerity critique confuses correlation with causation. Blaming the IMF for the reality that every country must confront its budget constraints is like blaming the fund for gravity.

Admittedly, the IMF does insist on being repaid, so eventually borrowing countries must part with foreign exchange resources that otherwise might have gone into domestic programs. Yet repayments to the fund normally spike only after the crisis has passed, making payments more manageable for borrowing governments. The IMF's shareholders—its 184 member countries—could collectively decide to convert all the fund's loans to grants, and then recipient countries would face no costs at all. However, if IMF loans are never repaid, industrialized countries must be willing to replenish continually the organization's lending resources, or eventually no funds would be available to help deal with the next debt crisis in the developing world.

A HAZARDOUS CRITIQUE

Of course, in so many IMF programs, borrowing countries must pay back their private creditors in addition to repaying the fund. Yet wouldn't fiscal austerity be a bit more palatable if troubled debtor nations could compel foreign private lenders to bear part of the burden? Why should taxpayers in developing countries absorb the entire blow?

That is a completely legitimate question, but let's start by getting a few facts straight. First, private investors can hardly breathe a sigh of relief when the fund becomes involved in an emerging-market financial crisis. According to the Institute of International Finance, private investors lost some $225 billion during the Asian financial crisis of the late 1990s and some $100 billion as a result of the 1998 Russian debt default. And what of the Latin American debt crisis of the 1980s, during which the IMF helped jawbone foreign banks into rolling over a substantial fraction of Latin American debts for almost five years and ultimately forced banks to accept large write-downs of 30 percent or more? Certainly, if foreign private lenders consistently lose money on loans to developing countries, flows of new money will cease. Indeed, flows into much of Latin America—again the current locus of debt problems—have been sharply down during the past couple of years.

Private creditors ought to be willing to take large writedowns of their debts in some instances, particularly when a country is so deeply in hock that it is effectively insolvent. In such circumstances, trying to force the debtor to repay in full can often be counterproductive. Not only do citizens of the debtor country suffer, but creditors often receive less than they might have if they had lessened the country's debt burden and thus given the nation the will and means to increase investment and growth. Sometimes debt restructuring does happen, as in Ecuador (1999), Pakistan (1999), and Ukraine (2000). However, such cases are the exception rather than the rule, as current international law makes bankruptcies by sovereign states extraordinarily messy and chaotic. As a result, the official lending community, typically led by the IMF, is often unwilling to force the issue and sometimes finds itself trying to keep a country afloat far beyond the point of no return. In Russia in 1998, for example, the official community threw money behind a fixed exchange-rate regime that was patently doomed. Eventually, the fund cut the cord and allowed a default, proving wrong those many private investors who thought Russia was "too nuclear to fail." But if the fund had allowed the default to take place at an earlier stage, Russia might well have come out of its subsequent downturn at least as quickly and with less official debt.

Developing countries don't seek IMF assistance when the sun is shining; they come when they have already run into deep financial difficulties.

Since restructuring of debt to private creditors is relatively rare, many critics reasonably worry that IMF financing often serves as a blanket insurance policy for private lenders. Moreover, when private creditors believe they will be bailed out by the IMF, they have reason to lend more—and at lower interest rates—than is appropriate. The debtor country, in turn, is seduced into borrowing too much, resulting in more frequent and severe crises, of exactly the sort the IMF was designed to alleviate. I will be the first to admit the "moral hazard" theory of IMF lending is clever (having introduced the theory in the 1980s), and I think it is surely important in some instances. But the empirical evidence is mixed. One strike against the moral hazard argument is that most countries generally do repay the IMF, if not on time, then late but with full interest. If the IMF is consistently paid, then private lenders receive no subsidy, so there is no bailout in any simplistic sense. Of course, despite the IMF's strong repayment record in major emerging-market loan packages, there is no guarantee about the future, and it would certainly be wrong to dismiss moral hazard as unimportant.

FISCAL FOLLIES

Even if IMF policies are not to blame for budget cutbacks in poor economies, might the fund's programs still be so poorly designed that their ill-advised conditions more than cancel out any good the international lender's resources could bring? In particular, critics charge that the IMF pushes countries to increase domestic interest rates when cuts would better serve to stimulate

When Economists Attack

"And the IMF could have offered Argentina guidance on how to escape from its monetary trap, as well as political cover for Argentina's leaders as they did what had to be done. Instead, however, IMF officials—like medieval doctors who insisted on bleeding their patients, and repeated the procedure when the bleeding made them sicker—prescribed austerity and still more austerity, right to the end."

—*Paul Krugman (2002)*

"However useful the IMF may be to the world community, it defies logic to believe that the small group of 1,000 economists on 19th Street in Washington should dictate the economic conditions of life to 75 developing countries with around 1.4bn people."

—*Jeffrey Sachs (1997)*

"In the past, countries with IMF programs were able to recover because financial markets had confidence in the IMF and were willing to follow its lead…. Since the 1997–99 crisis, however, the emperor has no clothes: IMF programs fail to impress the markets."

—*George Soros (2002)*

"[T]he IMF is not particularly interested in hearing the thoughts of its 'client countries' on such topics as development strategy or fiscal austerity. All too often, the Fund's approach to developing countries has had the feel of a colonial ruler."

—*Joseph Stiglitz (2002)*

the economy. The IMF also stands accused of forcing crisis economies to tighten their budgets in the midst of recessions. Like the austerity argument, these critiques of basic IMF policy advice appear rather damning, especially when wrapped in rhetoric about how all economists at the IMF are third-rate thinkers so immune from outside advice that they wouldn't listen if John Maynard Keynes himself dialed them up from heaven.

Of course, it would be wonderful if governments in emerging markets could follow Keynesian "countercyclical policies"—that is, if they could stimulate their economies with lower interest rates, new public spending, or tax cuts during a recession. In its September 2002 "World Economic Outlook" report, the IMF encourages exactly such policies where feasible. (For example, the IMF has strongly urged Germany to be flexible in observing the budget constraints of the European Stability and Growth Pact, lest the government aggravate Germany's already severe economic slowdown.) Unfortunately, most emerging markets have an extremely difficult time borrowing during a downturn, and they often must tighten their belts precisely when a looser fiscal policy might otherwise be desirable. And the IMF, or anyone else for that matter, can only do so much for countries that don't pay attention to the commonsense advice of building up surpluses during boom times—such as Argentina in the 1990s—to leave room for deficits during downturns.

According to some critics, though, a simple solution is staring the IMF in the face: If those stubborn fund economists would only appreciate how successful expansionary fiscal policy can be in boosting output, they would realize countries can simply wave off a debt crisis by borrowing even more. Remember former U.S. President Ronald Reagan's economic guru, Arthur Laffer, who theorized that by cutting tax rates, the United States would enjoy so much extra growth that tax revenues would actually rise? In much the same way, some IMF critics—ranging from Nobel Prize–winning economist Joseph Stiglitz to the relief agency Oxfam—claim that by running a fiscal deficit into a debt storm, a country can grow so much that it will be able to sustain those higher debt levels. Creditors would understand this logic and happily fork over the requisite extra funds. Problem solved, case closed. Indeed, why should austerity ever be necessary?

Needless to say, Reagan's tax cuts during the 1980s did not lead to higher tax revenues but instead resulted in massive deficits. By the same token, there is no magic potion for troubled debtor countries. Lenders simply will not buy into this story.

The notion that countries should reduce interest rates—rather than raise them—to fend off debt and exchange-rate crises is even more absurd. When investors fear a country is increasingly likely to default on its debts, they will demand higher interest rates to compensate for that risk, not lower ones. And when a nation's citizens lose confidence in their own currency, they will require a large premium to accept debt denominated in that currency or to keep their deposits in domestic banks. No surprise that interest rates in virtually all countries that experienced debt crises during the last decade—from Mexico to Turkey—skyrocketed even though their currencies were allowed to float against the dollar.

The debate over how far interest rates should be allowed to rise in defending against a speculative currency attack is a legitimate one. The higher interest rates go, the more stress on the economy and the more bankruptcies and bank failures; classic cases include Mexico in 1995 and South Korea in 1998. On the other hand, since most crisis countries have substantial "liability dollarization"—that is, a lot of borrowing goes on in dollars—an excessively sharp fall in the exchange rate will also cause bankruptcies, with Indonesia in 1998 being but one example among many. Governments must strike a delicate balance in the short and medium term, as they decide how quickly to reduce interest rates from crisis levels. At the very least, critics of IMF tactics must acknowledge these difficult trade-offs. The simplistic view that all can be solved by just adopting softer "employment friendly" policies, such as low interest rates and fiscal expansions, is dangerous as well as naive in the face of financial maelstrom.

CAPITAL CONTROL FREAKS

Although currency crises and financial bailouts dominate media coverage of the IMF, much of the agency's routine work entails ongoing dialogue with the fund's 184 member countries. As part of the fund's surveillance efforts, IMF staffers regularly visit member states and meet with policymakers to discuss how best to achieve sustained economic growth and stable inflation rates. So, rather than judge the fund solely on how it copes with financial crises, critics should consider its ongoing advice in trying to help countries stay out of trouble. In this area, perhaps the most controversial issue is the fund's advice on liberalizing international capital movements—that is, on how fast emerging markets should pry open their often highly protected domestic financial markets.

Blaming the IMF for the reality that every country must confront its budget constraints is like blaming the fund for gravity.

Critics such as Columbia University economist Jagdish Bhagwati have suggested that the IMF's zeal in promoting free capital flows around the world inadvertently planted the seeds of the Asian financial crisis. In principle, had banks and companies in Asia's emerging markets not been allowed to borrow freely in foreign currency, they would not have built up huge foreign currency debts, and international creditors could not have demanded repayment just as liquidity was drying up and foreign currency was becoming very expensive. Although I was not at the IMF during the Asian crisis, my sense from reading archives and speaking with fund old-timers is that although this charge has some currency, the fund was more eclectic in its advice on this matter than most critics acknowledge. For example, in the months leading to Thailand's currency collapse in 1997, IMF reports on the Thai economy portrayed in stark terms the risks of liberalizing capital flows while keeping the domestic currency (the baht) at a fixed level against the U.S. dollar. As Blustein vividly portrays in *The Chastening*, Thai authorities didn't listen, still hoping instead that Bangkok would become a financial center like Singapore. Ultimately, the Thai baht succumbed to a massive speculative attack. Of course, in some cases—most famously South Korea and Mexico—the fund didn't warn countries forcefully enough about the dangers of opening up to international capital markets before domestic financial markets and regulators were prepared to handle the resulting volatility.

However one apportions blame for the financial crises of the past two decades, misconceptions regarding the merits and drawbacks of capital-market liberalization abound. First, it is simply wrong to conclude that countries with closed capital markets are better equipped to weather stormy financial markets. Yes, the relatively closed Chinese and Indian economies did not catch the Asian flu, or at least not a particularly bad case. But neither did Australia nor New Zealand, two countries that boast extremely open capital markets. Why? Because the latter countries' highly developed domestic financial markets were extremely well regulated. The biggest danger lurks in the middle, namely for those economies—many of which are in East Asia and Latin America—that combine weak and underdeveloped financial markets with poor regulation.

Moreover, a country needs export earnings to support foreign debt payments, and export industries do not spring up overnight. That's why the risks of running into external financing problems are higher for countries that fully liberalize their capital markets before significantly opening up to trade flows. Indeed, economies with small trading sectors can run into problems even with seemingly modest debt levels. This problem has repeatedly plagued countries in Latin America, where trade is relatively restricted by a combination of inward-looking policies and remote location.

Perhaps the best evidence in favor of open capital markets is that, despite the international financial turmoil of the last decade, most developing countries still aim to liberalize their capital markets as a long-term goal. Surprisingly few nations have turned back the clock on financial and capital-account liberalization. As domestic economies grow increasingly sophisticated, particularly regarding the depth and breadth of their financial instruments, policymakers are relentlessly seeking ways to live with open capital markets. The lessons from Europe's failed, heavy-handed attempts to regulate international capital flows in the 1970s and 1980s seem to have been increasingly absorbed in the developing world today.

Even China, long the high-growth poster child for capital-control enthusiasts, now views increased openness to capital markets as a central long-term goal. Its economic leaders understand that it's one thing to become a $1,000 per capita economy, as China is today. But to continue such stellar growth performance—and one day to reach the $20,000 to $40,000 per capita incomes of the industrialized countries—China will eventually require a world-class capital market.

Even though a continued move toward greater capital mobility is emerging as a global norm, absolute unfettered global capital mobility is not necessarily the best long-term outcome. Temporary controls on capital outflows may be important in dealing with some modern-day financial crises, while various kinds of light-handed taxes on capital inflows may be useful for countries faced with sudden surges of inflows. Chile is the classic example of a country that appears to have successfully used market-friendly taxes on capital inflows, though a debate continues to rage over their effectiveness. One way or another, the international community must find ways to temper debt flows and at the same time encourage equity investment and foreign direct investment, such as physical investment in plants and equipment. In industrialized countries, the pain of a 20 percent stock market fall is shared automatically and fairly broadly throughout the economy. But in nations that rely on foreign debt, a sudden change in investor sentiment can breed disaster.

Nevertheless, financial authorities in developing economies should remain wary of capital controls as an easy solution. "Temporary" controls can easily become ensconced, as political forces and budget pressures make them hard to remove. Invite capital controls for lunch, and they will try to stay for dinner.

STRIKING A GLOBAL BARGAIN

Should the international community just give up on global capital mobility and encourage countries to shut their doors?

Looking further ahead in the 21st century, does the world really want to adopt greater financial isolationism?

Perhaps poor nations won't need the IMF's macroeconomic expertise in the future—but they will need something awfully similar.

Perhaps the greatest challenge facing industrialized countries in this century is how to deal with the aging bulge in their populations. With that in mind, wouldn't it be more helpful if rich countries could find effective ways to invest in much younger developing nations, and later use the proceeds to support their own increasing number of retirees? And let's face it, the world's developing countries need funds for investment and education now, so such a trade would prove mutually beneficial—a win-win. Yes, recurring debt crises in the developing world have been sobering, but the potential benefits to financial integration are enormous. Full-scale retreat is hardly the answer.

Can the IMF help? Certainly. The fund provides a key forum for exchange of ideas and best practices. Yes, one could go ahead and eliminate the IMF, as some of the more extreme detractors wish, but that is not going to solve any fundamental problems. This increasingly globalized world will still need a global economic forum. Even today, the IMF is providing such a forum for discussion and debate over a new international bankruptcy procedure that could lessen the chaos that results when debtor countries become insolvent.

And there are many other issues where the IMF, or some similar multilateral organization, seems essential to any solution. For example, the current patchwork system of exchange rates seems too unstable to survive into the 22nd century. How will the world make the transition toward a more stable, coherent system? That is a global problem, and dealing with it requires a global perspective the IMF can help provide.

And what of poverty? Here, the IMF's sister organization, the World Bank, with its microeconomic and social focus and commensurately much larger staff, is appropriately charged with the lead role. But poor countries in the developing world still face important macroeconomic challenges. For example, if enhanced aid flows ever materialize, policymakers in emerging markets will still need to find ways to ensure that domestic production grows and thrives. Perhaps poor nations won't need the IMF's specific macroeconomic expertise—but they will need something awfully similar.

Kenneth Rogoff is economic counsellor and director of the research department at the International Monetary Fund.

Want to Know More?

For a look inside the International Monetary Fund (IMF) during the Asian financial crisis of the late 1990s, see Paul Blustein's ***The Chastening: Inside the Crisis That Rocked the Global Financial System and Humbled the IMF*** (New York: PublicAffairs, 2001). For a passionate and comprehensive list of critiques about the IMF from the left, right, center, and outer space, see Joseph E. Stiglitz's *Globalization and Its Discontents* (New York: W.W. Norton & Company, 2002). Kenneth Rogoff evaluates alternative grand plans to redesign the international financial architecture in "**International Institutions for Reducing Global Financial Instability**" (*Journal of Economic Perspectives*, Vol. 13, No. 4, Fall 1999). Devesh Kapur assesses the strengths and limitations of the fund's crisis management in "**The IMF: A Cure or a Curse?**" (FOREIGN POLICY, Summer 1998).

Jeremy Bulow and Rogoff identified the problem of moral hazard and IMF lending in "**Multilateral Negotiations for Rescheduling Developing Country Debt: A Bargaining-Theoretic Framework**" (*International Monetary Fund Staff Papers*, Vol. 35, No. 4, December 1988). The issue was enshrined in the policy debate in the **Meltzer Commission Report** to the U.S. Congress in 1999, available on the Web site of the Joint Economic Committee of the U.S. House of Representatives. For a recent discussion of the empirical importance of moral hazard in IMF lending, see Olivier Jeanne and Jeromin Zettelmeyer's "**International Bailouts, Moral Hazard, and Conditionality**" (*Economic Policy*, Vol 33, October 2001) and Rogoff's "**Moral Hazard in IMF Loans: How Big a Concern?**" (*Finance & Development*, Vol 39, No. 3, September 2002).

FOREIGN POLICY's recent coverage of the international financial system and the IMF includes "**The Coming Fight Over Capital Flows**" (Winter 1998–99) by Robert Wade; "**Think Again: The International Financial System**" (Fall 1999) by Zanny Minton Beddoes; "**Trading in Illusions**" (March/April 2001) by Dani Rodrik; and "**The Cartel of Good Intentions**" (July/August 2002) by William Easterly. See also "**Michel Camdessus Talks With FP**" (September/October 2000).

• For links to relevant Web sites, access to the *FP* Archive, and a comprehensive index of related FOREIGN POLICY articles, go to **www.foreignpolicy.com**.

Ranking the Rich

The first annual CGD/FP Commitment to Development Index ranks 21 rich nations on whether their aid, trade, migration, investment, peacekeeping, and environmental policies help or hurt poor nations. Find out why the world's two largest aid givers—the United States and Japan—finish last, why Norway's trade policies harm developing nations, and why the Netherlands ranks number one.

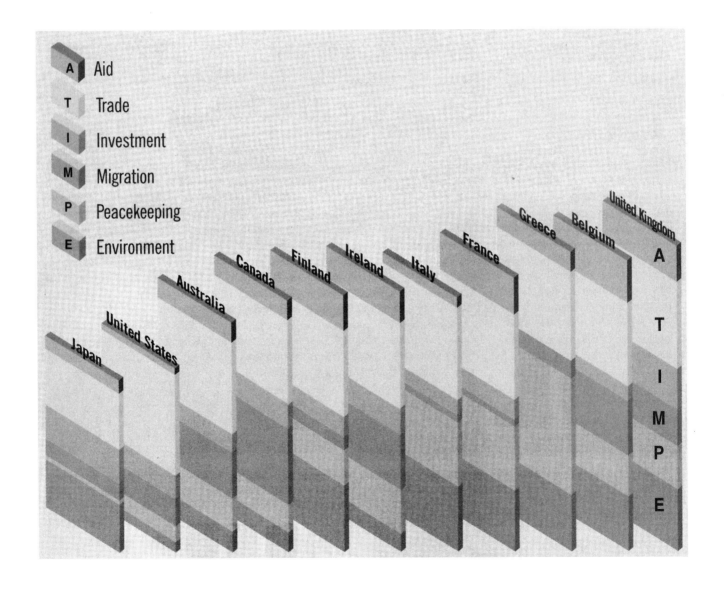

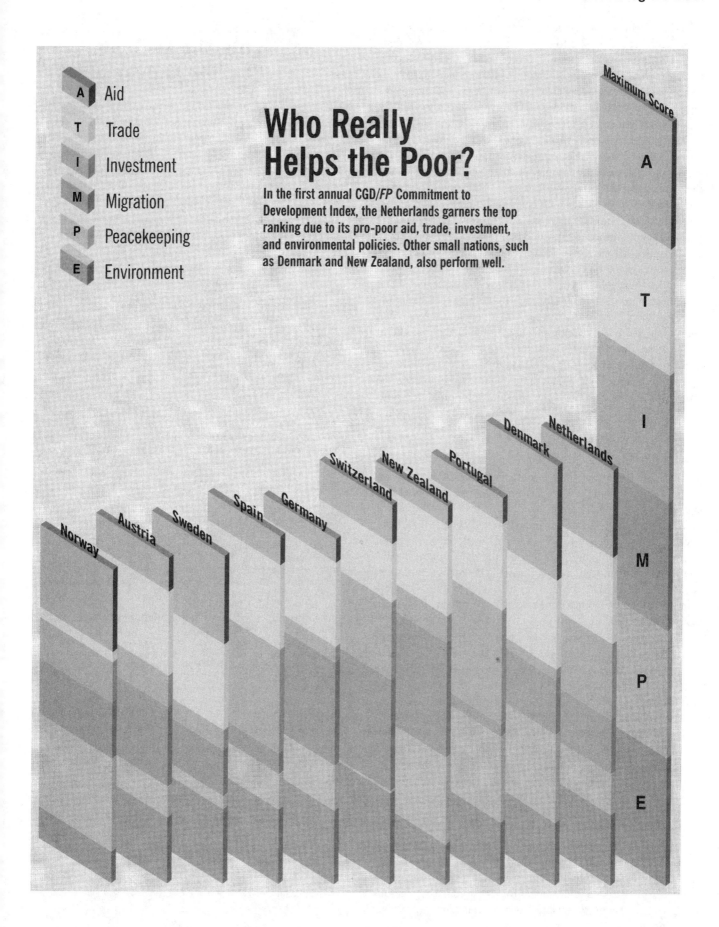

Who Really Helps the Poor?

In the first annual CGD/*FP* Commitment to Development Index, the Netherlands garners the top ranking due to its pro-poor aid, trade, investment, and environmental policies. Other small nations, such as Denmark and New Zealand, also perform well.

A Aid

T Trade

I Investment

M Migration

P Peacekeeping

E Environment

Maximum Score

A

T

I

M

P

E

Norway

Austria

Sweden

Spain

Germany

Switzerland

New Zealand

Portugal

Denmark

Netherlands

Political leaders in the world's richest nations regularly proclaim their fervent desire to end poverty worldwide. At high-profile meetings and summits, politicians push developing countries to tackle corruption, reduce inflation, and slash budget deficits. These leaders also boast of their spending on foreign aid—currently about $58 billion a year—even while they regularly call on each other to spend more.

These objectives and efforts are praiseworthy, no doubt. But cash transfers to poor nations are far from the only or even the most important way rich countries affect poor countries. Indeed, the finger-wagging over foreign aid has actually obscured the critical influence other rich countries' policies have on the development of poor nations. Until now, that is. The first annual CGD/*FP* Commitment to Development Index (CDI), created by the Center for Global Development and FOREIGN POLICY magazine, ranks some of the world's richest nations according to how much their policies help or hinder the economic and social development of poor countries. The CDI looks beyond mere foreign aid flows to encompass trade, environmental, investment, migration, and peacekeeping policies. In this inaugural edition of the index, the CDI ranks 21 nations: Australia, Canada, Japan, New Zealand, the United States, and most of Western Europe.

In ranking these countries' commitment to development, the CDI rewards generous aid giving, hospitable immigration policies, sizable contributions to peacekeeping operations, and hefty foreign direct investment in developing countries. The index penalizes financial assistance to corrupt regimes, obstruction of imports from developing countries, and policies that harm shared environmental resources. Although the governments and leaders of poor nations are themselves ultimately responsible for responding to the many challenges of development, rich countries can and should change their policies to spur economic growth and social development in poorer nations. The CDI highlights and ranks the rich countries' policies themselves, not their final impact. This approach emphasizes what each rich country—regardless of size and reach—can do to improve opportunities for development throughout the world.

The results of the first annual CDI cast traditional assumptions about the most development-friendly countries in a new, unexpected light. For example, the two countries providing the highest absolute amounts of foreign aid to the developing world—Japan and the United States—bring up the rear in the index. Japan ranks last overall, with low marks in migration and aid. The United States ranks high in trade policy but finishes second to last overall due to particularly poor performances in environmental policy and contributions to peacekeeping. By contrast, the Netherlands emerges as the top-ranked nation in the index, thanks to its strong performance in aid, trade, investment, and environmental policies. Two other small countries, Denmark and Portugal, follow in second and third place, respectively. Norway, which is usually regarded as a model global citizen and a force for peace worldwide, comes in a disappointing 10th, mainly due to its poor trade performance. And though New Zealand is not noted for its particularly generous aid giving, that country finishes fourth overall thanks to a strong showing in migration and peacekeeping policies.

How the Index Is Calculated

The final score for each country in the CGD/*FP* Commitment to Development Index (CDI) is an average of scores in six different categories. In the aid, investment, migration, and peacekeeping categories, each country receives a score ranging from 0 to 9 points, with 9 points going to the top performer and 0 to any country that makes no contribution at all in the relevant category. In the trade and environment categories, the CDI measures negative contributions—such as environmental harm and trade barriers—so the countries initially receive a negative score (from 0 to –9); then 10 is added to make the scores positive and comparable to the others.

The **aid** score takes the common measure of "official development assistance"—grants and low-interest loans—as a percentage of the donor country's gross domestic product (GDP), and adjusts that measure to better reflect the quality of the aid. The score discounts "tied aid"—aid that raises costs by compelling recipients to use contractors and consultants form the donor nation—by 20 percent and subtracts administrative costs as well as debt payments that the donor nation receives on past aid. Most aid contributions by the United nations, World Bank, and other multilateral aid programs are not discounted for typing. Aid to poor nations receives greater weight, as does aid to countries with good governance (low corruption, more political voice for citizens, etc.) compared to other countries at similar income levels.

The **trade** score considers rich nations' barriers to exports from developing countries—including tariffs, quotas, and subsidies for domestic producers. A measure of such barriers accounts for three fourths of the score. The remainder of the score directly measures how much rich nations import from developing countries. Imports from the world's poorest nations receive greater weight in this part of the score, as do manufactured imports from all developing countries.

Two thirds of the **investment** score is derived from foreign direct investment flows to developing countries as a share of the investor country's GDP during 1999–2001. These values are discounted according to investor countries' scores on Transparency International's 2003 Bribe Payers Index. The remaining one third of the investment score measures the extent to which governments in rich nations restrict public or private pension funds in their countries from investing in developing countries in particular and, more generally, overseas.

Ninety percent of each nation's **migration** score is derived from the number of legal migrants that each rich country admits from developing nations each year, divided by the receiving country's total population. The remaining 10 percent measures the aid and assistance that rich nations offer to refugees from poor nations.

The **peacekeeping** score counts rich countries' contributions to international peacekeeping operations during 2000–01, as a share of GDP. Included are financial contributions to the U.N. peacekeeping program as well as U.N. and NATO operations, valued at $10,000 per person per month.

The **environment** score measures each rich nation's depletion of the shared commons (two thirds of the score) and its contribution to international environmental initiatives (one third). The depletion measure includes greenhouse gas emissions per capita, consumption of ozone-depleting substances per capita, and fishing subsidies per dollar of GDP. International contributions considered in the index include the ratification of major environmental treaties and protocols, financial contributions to environmental funds, and government support for the development of clean-energy technologies.

A comprehensive and detailed explanation of the methodology—for each policy area and for the index as a whole—is available on the Web site of the Center for Global Development at www.cgdev.org.

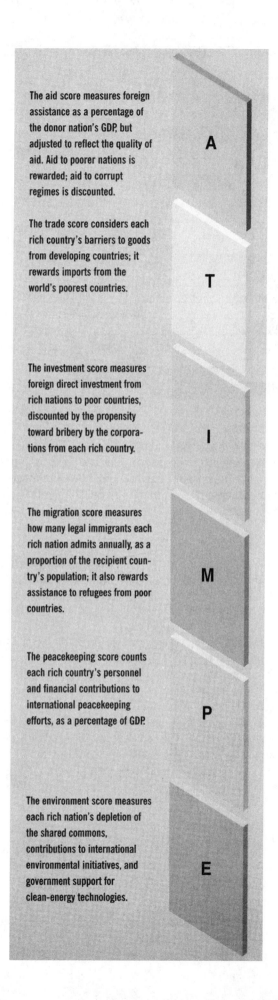

The aid score measures foreign assistance as a percentage of the donor nation's GDP, but adjusted to reflect the quality of aid. Aid to poorer nations is rewarded; aid to corrupt regimes is discounted.

The trade score considers each rich country's barriers to goods from developing countries; it rewards imports from the world's poorest countries.

The investment score measures foreign direct investment from rich nations to poor countries, discounted by the propensity toward bribery by the corporations from each rich country.

The migration score measures how many legal immigrants each rich nation admits annually, as a proportion of the recipient country's population; it also rewards assistance to refugees from poor countries.

The peacekeeping score counts each rich country's personnel and financial contributions to international peacekeeping efforts, as a percentage of GDP.

The environment score measures each rich nation's depletion of the shared commons, contributions to international environmental initiatives, and government support for clean-energy technologies.

In the late 1990s, the U.S. Agency for International Development reassured the U.S. Congress that almost 80 percent of the agency's aid resources went to purchase U.S. goods and services.

The CDI results are critical for two reasons. First, helping impoverished people worldwide build better lives is the right thing to do, and this index can educate policymakers, provoke public discussion, stimulate research, and guide activists seeking that goal. The hard truth is that even the best-performing nations in the CDI have a long way to go to make their policies as helpful as possible for poor families in developing countries. The Netherlands, even though it ranks highest, averages merely 5.6 points on the 10-point scale. Second, what rich countries do to and for the rest of the world comes back to affect them—poverty and instability do not respect borders. Surely the United States would benefit if Mexico were as stable and prosperous as Canada. Surely West European nations would benefit from an economic resurgence in Poland, Hungary, and the Czech Republic. Call it trickle-up economics: When the poor become better off, so do the rich.

STATES OF DEVELOPMENT

Development is a state as well as a process. A society has achieved a state of development to the extent that its citizens live free from want and tyranny and can obtain education and employment. But for the four fifths of the world's population still living in developing countries, the practical question is not what development is, but how to achieve it—and how to speed the process. The CDI measures how well rich countries contribute to the process in six policy areas: aid, trade, environment, investment, migration, and peacekeeping. The countries, scored on a 10-point scale for each policy area, are then ranked by their overall averages. Different scores are calculated in different ways, reflecting the particular issues involved and the availability of data. [For more details on how the index is calculated, see the sidebar on this page.]

Aid Today, rich countries send more than $50 billion a year in grants and low-interest loans to poor nations. Normally, these aid programs are compared via crude sums of dollars disbursed or by total aid as a percentage of gross domestic product (GDP). The CDI improves upon these traditional measures by considering the quality—not just the quantity—of aid. For instance, the index penalizes "tied aid," that is, financial assistance that recipient countries are required to spend on services from the donor nation. (For example, the Canadian or Italian governments may grant loans to a poor nation for highway construction but then require the recipient nation to hire a Canadian or Italian contractor to build the roads, thus preventing the aid re-

Rank	Country	Aid	Trade	Investment	Migration	Peacekeeping	Environment	Final Average
1	Netherlands	6.9	7.0	6.1	4.5	3.5	5.7	5.6
2	Denmark	9.0	6.8	1.0	4.4	7.1	5.0	5.5
3	Portugal	2.2	6.9	9.0	1.0	6.8	5.1	5.2
4	New Zealand	1.7	7.2	2.3	9.0	6.9	3.4	5.1
5	Switzerland	3.3	4.0	6.3	9.0	0.1	7.2	5.0
6	Germany	2.1	6.8	1.4	8.1	3.8	6.0	4.7
6	Spain	2.4	6.8	8.2	1.8	2.9	6.0	4.7
8	Sweden	7.0	6.9	1.8	3.9	1.3	6.1	4.5
9	Austria	2.8	6.8	2.6	6.5	2.6	5.4	4.4
10	Norway	6.6	1.0	3.5	4.6	7.4	2.8	4.3
11	United Kingdom	3.0	6.9	3.4	3.1	3.6	5.0	4.2
12	Belgium	3.5	6.7	1.4	4.5	3.5	4.5	4.0
13	Greece	1.5	6.7	0.0	1.6	9.0	4.6	3.9
14	France	3.1	6.8	1.7	0.8	5.2	4.9	3.8
15	Italy	1.4	7.0	1.5	1.1	5.3	5.3	3.6
15	Ireland	2.6	6.6	2.3	4.5	3.7	1.6	3.6
17	Finland	3.0	6.8	1.7	1.3	2.9	5.4	3.5
18	Canada	1.7	6.6	2.1	6.1	2.4	1.7	3.4
19	Australia	1.7	7.2	1.6	3.7	2.8	1.8	3.2
20	United States	0.8	7.7	2.0	2.3	1.5	1.0	2.6
21	Japan	1.2	4.6	2.8	1.5	0.5	4.0	2.4

The Rankings

The CGD/FP Commitment to Development Index ranks 21 of the world's richest countries according to how much their policies help or hinder the economic and social development of poor nations. The index examines six policy categories: foreign aid, openness to international trade, investment in developing countries, openness to legal immigration, contributions to peacekeeping operations, and responsible environmental practices. In the table, gray cells indicate a particularly favorable score in a given category and black cells indicate poor performance.

cipient from getting the best deal.) In 2001 alone, roughly two fifths of total international aid flows were tied; in the late 1990s, the U.S. Agency for International Development reassured the U.S. Congress that almost 80 percent of the agency's resources went to purchase U.S. goods and services. The CDI aid ranking also subtracts interest payments that donor nations receive on prior loans (equivalent to about $4.7 billion in 2001). Finally, the ranking rewards donors for channeling funds to countries that are relatively poor yet relatively free of corruption compared to other nations at similar income levels.

Denmark tops the CDI aid score, followed by Sweden, the Netherlands, and Norway. These countries are not only among the world's most generous, but only a small proportion of their aid is tied. Japan and the United States rank 20th and 21st, respectively, in the aid category. (The aid scores are based on 2001 data and do not reflect two recent U.S. initiatives: the Millennium Challenge Account and the Emergency Plan for AIDS Relief.) The Japanese aid score suffers because Japan exacts heavy interest payments on old loans. Of course, the United States provides significant private financial contributions to developing countries via churches, foundations, corporations, and private voluntary organizations. Since domestic contributions to such private groups are often tax-exempt, these flows, which would roughly double total U.S. aid, arguably could be credited to U.S. policies. If such private aid flows were included in the

index, the United States' aid ranking would jump to about 14th, assuming no similar contributions from other countries; the United States' overall CDI ranking, however, would remain unaffected.

Trade The CDI trade ranking sides neither with the passionate trade critics who fear a "race to the bottom" in environmental and labor standards nor with the equally passionate advocates who consider international commerce the prime mover of development. The truth about trade is more complicated. On the one hand, Nigeria might be better off without the oil export revenues that have corrupted the state, exacerbated ethnic tensions, and harmed the environment. On the other hand, South Korea, Taiwan, and even China could not have lifted so many from poverty so fast without exporting clothing, shoes, toys, and boom boxes to rich countries.

The CDI measures rich countries' barriers to developing-country exports, as well as the income that poor countries forgo due to internal production subsidies in rich nations. The World Bank estimates that trade barriers in developed economies cost poor nations more than $100 billion per year, roughly twice what rich countries give in aid. [See "How Trade Trumps Aid".] Among the most protected industries in high-income nations are agriculture, textiles, and apparel—not coincidentally the precise areas where poor countries are most competitive, and

Wanted: More Data

The World Bank, the United Nations, and other international agencies devote significant resources to collecting data on social and economic conditions in developing countries. However, the CGD/*FP* Commitment to Development Index (CDI) requires a different kind of information—data on the activities of rich countries that are relevant to development in poor economies. Unfortunately, some of the pieces of information we needed are not available in comparable form from a single source. In some cases, we assembled comparable data one country at a time; in others, we were forced to use outcomes such as foreign direct investment as the best proxy measure for the policies that influence those outcomes. Below is a sampling of the data we wish were available.

(I) Internationally comparable data on international migration and labor movements, including legal inflows by country of origin, skill levels, length of stay, and other agreed categories (such as students and workers on temporary visits)

(2) Data on the impact of rich nations' domestic producer subsidies on trade

(3) Data on the tax treatment of developing-country assets held in rich nations and on the income earned by those assets in rich nations; details on tax-information agreements (if any) between developing countries and each of the 21 countries in the ranking

(4) Internationally comparable data on private aid flows from rich countries to poor ones, including aid from churches, foundations, and other voluntary organizations; data on remittances from migrants back to their home countries.

(5) Data assessing how rich countries affect the security environment of poor countries, from U.S. contributions to keeping major sea-lanes open for trade to French and British subsidies for arms sales to developing countries.

where they could create the most jobs absent such protectionism. Producers in rich nations benefit from a combination of government subsidies and tariffs and quotas on imported goods. Japan, for instance, imposes a 490 percent tariff on foreign rice, while the average cow in Switzerland earns the annual equivalent of more than $1,500 in subsidies. [See "Embarrassment of the Richest".]

In the CDI trade ranking, the United States finishes first, followed by Australia and New Zealand. By contrast, Norway ranks a distant last; it has particularly high tariffs against agricultural imports from poor countries, and its various barriers are equivalent in impact to a flat 61 percent tariff on all goods from developing countries—equivalent, that is, in terms of lost profits for producers in poor nations. Norway's ranking may seem surprising given the country's record as a beneficent provider of aid. However, in Norway, as in much of Western Europe and the United States, agribusiness and other rural interests, though they are no longer competitive abroad, remain politically powerful, and the government has been unable to reconcile domestic politics with its otherwise enlightened approach to the developing world. The problem is less acute in Australia, New Zealand, and the United States, which have more efficient agricultural sectors.

Environment A healthy environment is often dismissed as a luxury for the rich, distinct from and secondary to economic development. Yet poor nations will struggle most with any effects of climate change, such as drought, flooding, and the spread of infectious diseases. The CDI environment ranking reflects the belief that rich nations have special responsibilities for global environmental stewardship. Are such countries reducing their disproportionate exploitation of the global commons? Which nations have signed the Kyoto Protocol on climate change? How much money have they contributed to the Montreal Protocol Fund, which helps developing countries phase out ozone-depleting chemicals? And are developed nations advancing state-of-the-art, environment-friendly energy technologies?

Both Switzerland and Japan have reputations for xenophobia, yet Switzerland finishes near the top and Japan near the bottom of the migration ranking.

According to these measures, Switzerland ranks highest in its environmental policy, thanks to hefty government investment in clean-energy research and development, relatively low emissions of atmosphere-disrupting pollutants, and no fishing subsidies. Sweden finishes second, and Spain third. The Spanish case is particularly interesting since Spain, although it earns only average scores on most environmental indicators, scores high overall due to its heavy government support for wind power technology. Australia, Canada, and the United States rate among the worst environmental performers, due largely to their high per capita greenhouse gas emissions.

Investment International capital flows come in three main varieties. Portfolio investment occurs when foreigners buy securities, such as stocks and bonds, which are traded on open exchanges outside their home country; foreign direct investment (FDI) entails companies from one country buying major stakes in existing companies or building new factories in another nation; and banks lend large sums directly to governments and corporations. For many analysts, the Asian financial crisis of the late 1990s—during which portfolio investors stampeded in and out of various Asian economies—proved the potential dangers of so-called hot-money flows. However, some countries, including the United States in the 19th century and Malaysia during the last 30 years, have benefited greatly from FDI, which is generally more stable than portfolio capital and often brings good management and technology. For example, Singapore could not have raised its per capita income from $2,200 in 1960 to $29,000 in 2000 without ample investment from abroad, which boosted employment and injected new ideas and technologies.

Clearly, foreign investment can bring jobs and foster economic growth in developing countries. The CDI investment

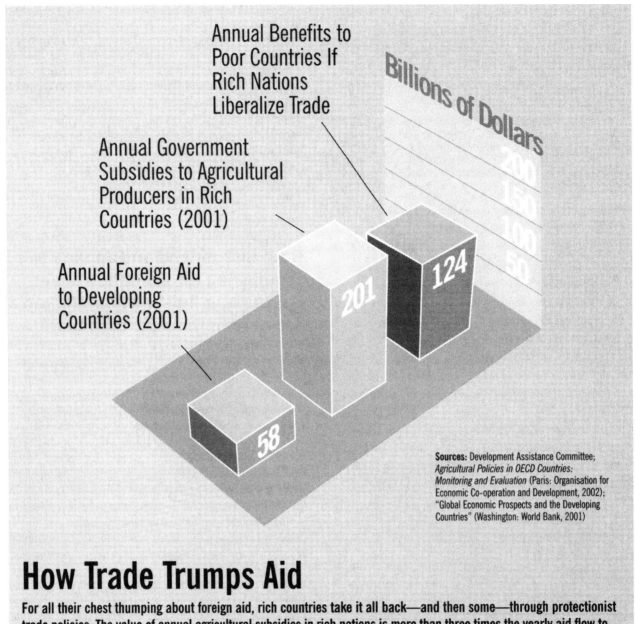

Annual Benefits to Poor Countries If Rich Nations Liberalize Trade

Annual Government Subsidies to Agricultural Producers in Rich Countries (2001)

Annual Foreign Aid to Developing Countries (2001)

Billions of Dollars

200
150
100
50

201

124

58

Sources: Development Assistance Committee; *Agricultural Policies in OECD Countries: Monitoring and Evaluation* (Paris: Organisation for Economic Co-operation and Development, 2002); "Global Economic Prospects and the Developing Countries" (Washington: World Bank, 2001)

How Trade Trumps Aid

For all their chest thumping about foreign aid, rich countries take it all back—and then some—through protectionist trade policies. The value of annual agricultural subsidies in rich nations is more than three times the yearly aid flow to poor countries, causing heavy losses to producers in the developing world. The most protected industries in high-income economies include agriculture and textiles, precisely the sectors where many poor countries tend to be most competitive.

score gives dominant weight to the amount of FDI (as a percent of GDP) flowing from each rich country to all developing countries. However, the CDI "corrects" investment flows by considering the propensity of corporations from rich nations to rely on bribes overseas to conduct their business. Among the countries in the CDI, Italy reportedly has the most corrupt companies, while Australia boasts the least corrupt, according to Transparency International's 2002 Bribe Payers Index. So Australia's FDI, dollar for dollar, counts more than Italy's. Four countries stand out as sources of this "healthy" FDI: the Netherlands, Portugal, Spain, and Switzerland. Although banks and corporations from Japan and the United States often appear to dominate foreign investment in developing countries, U.S. and Japanese in-

vestment scores are relatively low. Indeed, their investment flows are a good deal less impressive when considering the overall size of their economies.

Migration At first glance, it may seem odd to include immigration policy in the CDI. How is the process of development advanced if thousands of Turks exit their native country for Germany or if millions of Mexicans cross the border into the United States? Clearly, migration flows hurt in some ways and help in others. On balance, however, the freer movement of people—like the freer movement of goods—generally enhances development. The easier it is for a Vietnamese laborer to work in Japan, the more Nike will have to pay her to sew clothes

Don't Have a Cash Cow

Annual Subsidy to Cattle Producers, per Head of Cattle (2001)

European Union	$435.76
Australia	$9.14
Canada	$130.53
Japan	$1,296.93
New Zealand	$1.63
Norway	$1,313.16
Switzerland	$1,560.28
United States	$151.63
GDP per capita, Sub-Saharan Africa (2001)	$467.20

Sources: U.N. Food and Agriculture Organization, Organisation for Economic Co-operation and Development

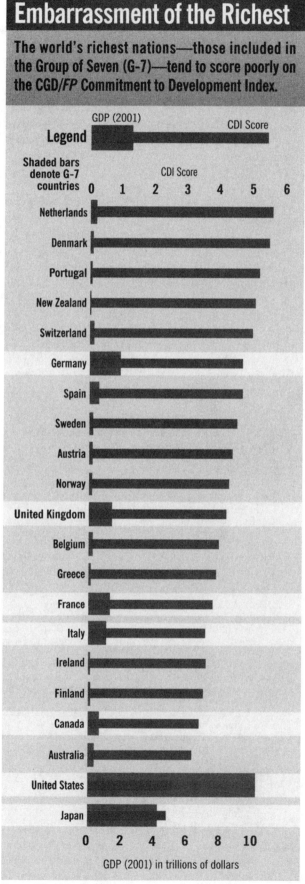

Embarrassment of the Richest

The world's richest nations—those included in the Group of Seven (G-7)—tend to score poorly on the CGD/*FP* Commitment to Development Index.

Source: World Bank's *World Development Indicators* database

in the company's Vietnamese factories. Migrants also send home sums large enough to constitute a major economic force in many developing countries. For example, remittances account for 13 percent of El Salvador's GDP—more than aid, investment, or tourism.

The migration scores in the CDI are surprising. For instance, both Switzerland and Japan have reputations for xenophobia, yet Switzerland finishes near the top and Japan near the bottom of the migration ranking. Why? In Switzerland, noncitizens face great difficulty gaining citizenship; by contrast, everyone from doctors and nurses to nannies and janitors can easily obtain legal entry into Switzerland to work—the indicator the CDI actually measures. Meanwhile, the United States, a nation of immigrants, scores only slightly above Japan. If the United States legalized more of its illegal immigration inflow, as Mexican President Vicente Fox has repeatedly requested, the U.S. score would increase substantially.

Peacekeeping The CDI peacekeeping score rewards financial and personnel contributions to multilateral peacekeeping operations. Greece ranks number one for contributing 2,000 personnel to peacekeeping in nearby Bosnia and Herzegovina and Kosovo—a large number for such a small country. At the other extreme, Switzerland ranks last because it has historically hewed to neutrality and avoided membership in international organizations. (The country only joined the United Nations in September 2002.) Japan scores low as well; it contributed $675 million to U.N. peacekeeping in 2000 and 2001 but provided minimal troops for peacekeeping operations, reflecting the country's constitutional ambivalence regarding the use of military force.

The inclusion of peacekeeping in the CDI reflects the belief that domestic stability and freedom from external attack are prerequisites for economic development. In many cases, rich nations engage daily in military activities that enhance the security of developing countries. These forces keep the peace in places once torn by conflict; their navies protect sea-lanes vital to international trade; occasionally they intervene directly against oppression, as in Kosovo in 1999. In Mozambique, for instance, U.N. peacekeeping paved the way for new elections in 1994 and subsequent economic growth. But one nation's security enhancement may be another's destabilizing intervention—the debate over war in Iraq is a clear example. And the extent to which rich countries encourage arms sales to poor nations or provide aid to repressive regimes may actually undermine security in the developing world. Considering such complexities, the first edition of the CDI focuses solely on peacekeeping contributions rather than on broader aspects of rich nations' security policies.

BETTER POLICIES, FEWER PROMISES

The terrorist attacks on September 11, 2001, challenged those who enjoy the freedom and affluence of life in the world's richest countries to ponder their place and purpose in the larger world. Seven nations—Canada, France, Germany, Italy, Japan, the United Kingdom, and the United States—account for two thirds of the world's economic output. Together, these nations form the Group of Seven (G-7), often characterized in the press

as the "seven leading industrial nations." Yet judging by the results of the first annual CGD/*FP* Commitment to Development Index, the G-7 are not leaders. By virtue of their sheer size, the G-7 engage in more trade, more aid giving, more peacekeeping, and more pollution than any other group of nations. They have the greatest power to help developing countries, but, with the exception of Germany, which ties Spain for sixth place in the index, they generally use their enormous potential the least.

Who currently leads the world in tackling the challenge of development? According to the CDI, the Netherlands, Denmark, and Portugal do. Though they could still perform better, these three nations set an example for other rich nations. But with a combined population smaller than that of Tanzania, these countries can hardly lead alone. The G-7 nations must assume the responsibilities commensurate with their size, power, and economic might. That means reforming all their policies with an eye toward aiding development—as a matter of both morality and enlightened self-interest. These nations' steady progress on the measures included in the CDI could inspire other rich nations to follow suit. If the richest of the rich do not lead, then no one will. But if these countries do step forward, then they will help improve the lives of millions of people who deserve better than they now have—while building a more stable world in the process.

Want to Know More?

For links to background papers and research related to the CGD/*FP* Commitment to Development Index, visit the Web site of the **Center for Global Development** (CGD) at **www.cgdev.org**, as well as **www.foreignpolicy.com**. In particular, see Nancy Birdsall and David Roodman's **The Commitment to Development Index: A Scorecard of Rich-Country Policies"** (Washington: CGD, 2003). Also available are background papers on the different categories of the index: Roodman and William Easterly on foreign aid, William Cline and Josh Catlin on trade, Roodman on the environment, John Williamson on investment, Kim Hamilton and Elizabeth Grieco on migration, and Michael O'Hanlon on peacekeeping.

The British-based group Oxfam calls for further trade liberalization and assails rich countries for their trade barriers against exports from poor nations in **"Rigged Rules and Double Standards: Trade, Globalisation, and the Fight Against Poverty"** (Oxford: Oxfam, 2002), available on the Oxfam Trade Campaign Web site. The Web site of the **Catholic Agency for Overseas Development** (CAFOD) offers reports on international trade; see in particular Duncan Green and Matthew Griffith's **"Dumping on the Poor: The Common Agricultural Policy, the WTO, and International Development"** (London: CAFOD, 2002). Dani Rodrik argues that the conventional recipes of economic integration offer developing nations false hope in **"Trading in Illusions"** (FOREIGN POLICY, March/April 2001).

For works examining the quality of foreign aid, see Eberhard Reusse's *Ills of Aid: An Analysis of Third World Development Policies* (Chicago: University of Chicago Press, 2002) and Graham Hancock's *Lords of Poverty: The Power, Prestige, and Corruption of the International Aid Business* (New York: Atlantic Monthly Press, 1989). Easterly takes a critical look at foreign aid and the problems of economic development in *The Elusive Quest for Growth: Economists' Adventures and Misadventures in the Tropics* (Cambridge: Massachusetts Institute of Technology Press, 2001). Also see Easterly's **"The Cartel of Good Intentions"** (FOREIGN POLICY, JULY/AUGUST 2002) AND MOISÉS NAÍM'S **"WASHINGTON CONSENSUS OR WASHINGTON CONFUSION?"** (FOREIGN POLICY, Spring 2000). Birdsall and Williamson recommend specific aid reforms in *Delivering on Debt Relief: From IMF Gold to a New Aid Architecture* (Washington: Institute for International Economics, 2002). The World Bank tacitly acknowledged the critics but remains optimistic in its report **"Assessing Aid: What Works, What Doesn't, and Why"** (New York: Oxford University Press, 1998)

The Intergovernmental Panel on Climate Change reviews the potential effects of climate change in dispassionate detail in its **"Climate Change 2001: Impacts, Adaptation and Vulnerability"** (New York: Cambridge University Press, 2001). For views on the links between immigration and economic progress, see International Organization for Migration's (IOM) **"The Migration-Development Nexus: Evidence and Policy Options"** (Geneva: IOM Migration Research Series, No. 8, 2002). The World Bank's annual report **"Global Development Finance"** tracks capital flows to developing countries. On the role of peacekeeping operations in helping rebuild developing societies, see Adekeye Adebajo and Chandra Lekha Sriram's, eds., *Managing Armed Conflicts in the 21st Century* (Portland: Frank Cass Publishers, 2001).

• For links to relevant Web sites, access to the *FP* Archive, and a comprehensive index of related FOREIGN POLICY articles, go to **www.foreignpolicy.com**.

Eyes Wide Open

On the Targeted Use of Foreign Aid

DAVID DOLLAR

Conventional wisdom on international development holds that "the rich get richer while the poor get poorer." This saying does not capture exactly what has happened between the rich and poor regions of the world over the past century, but it comes pretty close. In general, poor areas of the world have not become poorer, but their per capita income has grown quite slowly. On the other hand, income in the club of rich countries (Western Europe, the United States, Canada, Japan, Australia, and New Zealand), has increased at a much more rapid pace. As a result, by 1980 an unprecedented level of worldwide inequality had developed. The richest fifth of the world's population—which essentially corresponds to the population of the rich countries—produced and consumed 70 percent of the world's goods and services, while the poorest fifth of the global population, in contrast, held only two percent.

There has been a modest decline in global inequality since 1980 because two large poor countries—China and India—have outperformed the rich countries economically. This shift represents an interesting change that has important lessons for development. However, if one ignores the performance of China and India, much of the rest of the developing world still languishes, and there continues to be an appalling gap between rich countries and poor countries.

Inequality within countries is an important issue as well, but it pales in comparison with inequality between countries across the world. A homeless person pan-handling for two US dollars a day on the streets of Boston would sit in the top half of the world income distribution. Without traveling through rural parts of the developing world, it is difficult to comprehend the magnitude of this gap, which is not just one of income. Life expectancy in the United States has risen to 77 years, whereas in Zambia it has fallen to 38 years. Infant mortality is down to seven deaths per 1,000 live births in the United States, compared to 115 in Zambia. How can these gaps in living standards be understood? And, more importantly, what can be done about it?

Traditionally, one part of the answer to the latter question has been foreign aid. Since the end of the Cold War, aid has been in decline, both in terms of volume (down to about 0.2 percent of the gross national product of the rich countries) and popularity as an effective policy. However, since before September 11, 2001, aid has made something of a comeback, with a number of European countries, notably the United Kingdom, arguing for the importance of addressing global poverty by implementing reforms to make aid more effective. Since September 11, the US government has shown renewed interest as well.

What can came from this renewed interest in foreign aid? Foreign aid bureaucracies have a long history of mistaking symptoms for causes. If this trend continues uncorrected, then it is unlikely that greater volumes of aid will make much of a dent in global poverty and inequality. On the other hand, there is much more evidence about what leads to successful development and how aid can assist in that process. Thus, the potential exists to make aid a much more important tool in the fight against poverty My argument on this matter is comprised of four points.

First, contries are poor primarily because of weak underlying institutions and policies. Features such as lack of capital, poor education, or absence of modern industry are symptoms rather than causes of underdevelopment. Aid focused on these symptoms has not had much lasting impact.

Second, local institutions in developing countries are persistent, and foreign aid donors have little influence over them. Efforts to reform countries through conditionality of aid from the Bretton Woods organizations have generally failed to bring about lasting reform within developing country institutions. It is difficult to predict when serious movements will emerge, but the positive developments in global poverty in the past 20 years have been the result of homegrown reform movements in countries such as China, India, Uganda, and Vietnam.

Third, foreign aid has had a positive effect in these and other cases, and arguably its most useful role has been to support learning at the state and community level. Countries and communities can learn from each other, but there are no simple blueprints of institutional reform that can be transferred from one location to the next. Thus, helping countries analyze, implement, and evaluate options is useful, whereas promoting a "best-practice" approach to each issue through conditionality is not.

Fourth, the financial aspect of foreign aid is also important. In poor countries that have made significant steps toward improving their institutions and policies, financial aid accelerates growth and poverty reduction and helps cement popular support for reform. Hence, large-scale financial assistance needs to be "selective," targeting countries that can put aid to effective use building schools, roads, and other aspects of social infrastructure.

Institutions and Policies

Economists have long underestimated the importance of state institutions in explaining the differences in economic performance between countries. Recent work in economic history and development is beginning to rectify this oversight. In their 2001 study, "Colonial Origins of Comparative Development: An Empirical Investigation," Daron Acemoglu, Simon Johnson, and James Robinson find that much of the variation in per capita income across countries can be explained by differences in institutional quality. They look at a number of different institutional measures, which generally capture the extent to which the state effectively provides a framework in which property is secure and markets can operate. Thus, indicators of institutional quality try to measure people's confidence in their property rights and the government bureaucracy's ability to provide public services relatively free of interest group appropriation and corruption. All countries have some problems with appropriation and corruption, so the practical issue is the extent of these problems. While these differences are inherently hard to measure, some contrasts are obvious; there is, for example, no doubt that Singapore or Finland has a better environment of property rights and clean government than Mobutu's Zaire or many similar locations in the developing world.

Differences in institutional quality explain much of the variation in per capita income across countries, an empirical result that is very intuitive. In a poor institutional environment, households must focus on day-to-day subsistence.

The state fails to provide the complementary infrastructure—such as roads and schools—necessary to encourage long-term investment, while the lack of confidence in property rights further discourages entrepreneurial activity. In this type of setting, any surplus accumulated by individuals is more likely to fund capital flight, investment abroad, or emigration than to be reinvested in the local economy.

[Recent development] efforts failed because they were aimed at symptoms rather than at underlying causes.

In addition, there is evidence that access to markets is also important as well for economic growth. If a region is cut off from larger markets either because of its natural geography or because of manmade trade barriers, then the incentives for entrepreneurial activity and investment are again reduced. In 1999, Jeffrey Frankel and David Romer cautiously concluded that the converse holds as well: better trading opportunities do lead to faster growth. There is still some debate among economists about the relative importance of institutions and trade, but it seems likely that both are important and that in fact they complement each other. Several years ago, Kenneth Sokoloff found that rates of invention were extremely responsive to the expansion of markets during the early industrialization of the United States by examining how patenting activity varied over time and with the extension of navigable waterways. For example, as the construction of the Erie Canal progressed westward across the state of New York, patenting per capita rose sharply county-by-county. The United States had a good system of protecting these intellectual properties, and the development of transport links to broader markets stimulated individuals and firms to invest more in developing new technologies.

Indeed, looking back over the past century, locations with access to markets and good property rights have generally prospered, while locations disconnected from markets and with poor property rights have remained poor. Many of the features that we associate with underdevelopment are therefore results of these underlying weaknesses in institutions and policies. In such environments, there is little incentive to invest in equipment or education and develop modern industry.

But these symptoms have often been mistaken by aid donors as causes of underdevelopment. If low levels of investment are a problem, then give poor countries foreign aid to invest in capital. If a lack of education is a problem, finance broad expansion of schools. If modern industry is absent, erect infant-industry protection to allow firms to develop behind a protected wall. All of these approaches have been pushed by aid donors. In poor countries with weak underlying institutions, however, the results have not been impressive.

Over several decades, Zambia received an amount of foreign aid that would have made every Zambian rich had it achieved the kind of return that is normal in developed economies. If lack of capital was the key problem in Zambia, then that was certainly addressed by massive amounts of aid; but the result was virtually no increase in the country's per capita income. Similarly, large amounts of aid targeted at expanding education in Africa yielded little measurable improvement in achievement or skills. Donors financed power plants, steel mills, and even shoe factories behind high levels of protection, but again there was virtually no return on these investments.

The recent thinking in economic history and development suggests that these efforts failed because they were aimed at symptoms rather than at underlying causes. If a government is very corrupt or dominated by powerful special interests, then giving it money, or schools, or shoe factories will not promote lasting growth and development. These findings suggest that much of the frustration about foreign aid comes from the many failed efforts to develop social infrastructure in weak institutional environments where governments and communities cannot make effective use of these resources—not from the intrinsic inability of aid itself to generate positive results.

There are a number of important caveats about these findings on aid effectiveness. First, humanitarian or food aid is a different story. When there is a famine or humanitarian crisis, international donors have shown that they can bring in short-term relief effectively. Second, there are some health interventions that can be delivered in a weak institutional environment. In much of Africa, donors have collaborated to eradicate river blindness, a disease that can be controlled by taking a single pill each year. That intervention—and certain types of vaccinations—can be carried out in almost any environment. But many other social services require an effective institutional delivery system; other health projects in countries with weak institutions have tended to fail without producing any benefits.

Persistent Institutions

A second important finding from recent work in economic history is that institutions are persistent. Last year, Stanley Engerman and Sokoloff showed how differences in the natural endowments of South and North American colonies centuries ago led to the development of different institutions in the two environments. Furthermore, many of these institutional differences have persisted to this day. If institutions are important and if they typically change slowly over time, then it is easy to understand the pattern of rising global inequality over the past century. Locations with better institutions have consistently grown faster than ones with poor institutions, widening inequalities. Because it is relatively rare for a country to switch from poor institutions and policies to good ones, countries that began at a disadvantage only fell further behind in the years that followed.

The importance of good institutions and policies for development in general and for aid effectiveness in particular is something that donors have gradually realized through experience and research. International donors' first instincts were to make improved institutions and policies a condition of their assistance. In the 1980s in particular, donors loaded assistance packages with large numbers of

conditions concerning specific institutional and policy reforms. Some World Bank loans, for example, had more than 100 specific reform conditions. However, the persistence of institutions and policies hints at the difficulty of changing them. There are always powerful interests who benefit from bad policies, and donor conditionality has proved largely ineffective at overcoming these interest groups. A 2000 study that I co-authored with Jakob Svensson examined a large sample of World Bank structural adjustment programs to find that the success or failure of reform can largely be predicted by underlying institutional features of the country, including whether or not the government is democratically elected and how long the executive has been in power. Governments are often willing to sign aid agreements with large amounts of conditionality, but in many low-income countries the government is either uninterested in implementing reform or politically blocked from doing so. "Aid and Reform in Africa," a set of case studies written by African scholars on 10 African states, reaches similar conclusions: institutional and policy reform is driven primarily by domestic movements and not by outside agents.

Prospects for Reform

The good news is that a number of important developing countries have accomplished considerable reforms in the past two decades. In 1980, about 60 percent of the world's extreme poor—those living on less than one US dollar per day—lived in just two countries: China and India. At that time, neither country seemed a particularly likely candidate for reform. Both had rather poor property rights and government efficiency according to the measures used in cross-country studies, and both were extremely closed to the world market. Over the past two decades, however, China has introduced truly revolutionary reforms, restoring property rights over land, opening the economy to foreign trade and investment, and gradually making the legal and regulatory changes that have permitted the domestic private sector to become the main engine of growth. Reforms in India have not been quite as dramatic, but have still been very suc-

cessful at reducing the government's heavy-handed management of the economy and dismantling the protectionist trade regime. Among low-income countries, there have been a number of other notable reformers as well; Uganda is a good example in Africa, and Vietnam in Southeast Asia.

The general point about all of these low-income reformers is that outside donors were not particularly important at the start of these reform efforts. These movements are homegrown and each has an interesting and distinct political-economy story behind it. Once these reforms began, however, foreign assistance played an important supporting role in each case. Institutional reform involves much social and political experimentation. The way that China has gradually strengthened private property rights is an excellent example, as is the way India reformed its energy sector. Foreign assistance can help governments and communities examine options, implement innovations, and evaluate them. To do this effectively, donor agencies need to have good technical staff, worldwide experience, and an open mind about what might work in different circumstances.

The World Bank is often criticized for giving the same advice everywhere, but this simply is not true. World Bank reports on different countries show that the World Bank typically makes quite different recommendations in different countries. The criticism that comes from government officials in the developing world is a different and more telling one: that the World Bank tends to make a single strong recommendation on each issue, instead of helping clients analyze the pros and cons of different options so that communities can make up their own minds about what to do. We do not know much about institutional change, so it is more useful to promote community learning than to push particular institutional models.

For example, the Education, Health, and Nutrition Program—known by its Spanish acronym, PROGRESA—is a successful program of cash transfers that encourages poor families to keep their children in school that was developed and evaluated in Mexico without any donor support. A number of donors now have helped communities in Central American countries to implement similar programs. In each case, communities need to tailor the program to their particular situation. Systematic re-evaluation is important because the same idea will not necessarily work everywhere. But this is a good example of how donors can promote learning across countries and support institutional change by presenting a variety of reform options for developing countries to follow.

Money Matters

While supporting country and community learning is probably the most useful role for aid, and the one that will have the largest impact, there is a still a role for large-scale financial aid. Studies have shown that there is little relationship between aid amounts and growth rates in developing countries, but there is a rather strong relationship between growth and the interaction of aid and economic policies. This finding, as well as microeconomic evidence about individual projects, suggests that the growth effect of aid is greater in countries with reasonably good institutions and policies. The success of the Marshall Plan is a classic historical example. More recently, states such as Uganda show that the combination of substantial reform and large-scale aid goes together with rapid growth and poverty reduction. In a poor institutional environment, however,

large-scale aid seems to have little lasting economic impact and may even make things worse by sustaining a bad government.

What follows from this is that aid is going to have more impact on poverty reduction if it is targeted to countries that are poor and have favorable institutions and policies. This philosophy underlies a number of new initiatives in foreign aid—European countries, including the United Kingdom and the Netherlands, have reformed and expanded their aid program along these lines. The new US Millennium Challenge Account is based on these principles as well.

Using aid to support learning and being selective in the allocation of large-scale financial resources are linked. When donors tried to push large amounts of money into weak institutional environments, they naturally wanted to have large numbers of conditions dictating how institutions and policies would change. But this neither promoted effective learning nor led to good use of money. The new model argues for much less conditionality—encouraging countries and communities to figure out what works for them—but retaining some form of selectivity in the allocation of financial resources.

Keeping in mind the persistence of institutions and the difficulty of changing them, one should have modest hopes for what foreign aid can accomplish. But as long as there are countries and communi-

ties around the world struggling to change, the international community must support them. Afghanistan today is a good example. The country is trying to develop new institutions at the national and local levels, and the world has a big stake in helping it succeed. The international community does not know for sure what will work, but outsiders dictating a new set of institutions will almost certainly fail. On the other hand, donor agencies can help both national and local governments learn about options, implement policies, evaluate results, and redesign if necessary. As a solid institutional framework develops, there will be increasing scope for large-scale funding of roads, schools, and other social infrastructure. The effort may fail. No doubt the lack of good institutions in Afghanistan reflects extensive historical and political factors that will be hard to overcome. It is important to go in with eyes wide open; trying to reform aid based on what we know is preferable to giving up on aid and closing our eyes to the massive poverty that remains throughout the developing world.

The views expressed are those of the author and do not necessarily reflect official views of the World Bank.

DAVID DOLLAR is Research Manager for the World Bank's Macroeconomics and Growth Team.

The Cartel of Good Intentions

The world's richest governments have pledged to boost financial aid to the developing world. So why won't poor nations reap the benefits? Because in the way stands a bloated, unaccountable foreign aid bureaucracy out of touch with sound economics. The solution: Subject the foreign assistance business to the force of market competition.

By William Easterly

The mere mention of a "cartel" usually strikes fear in the hearts and wallets of consumers and regulators around the globe. Though the term normally evokes images of greedy oil producers or murderous drug lords, a new, more well-intentioned cartel has emerged on the global scene. Its members are the world's leading foreign aid organizations, which constitute a near monopoly relative to the powerless poor.

This state of affairs helps explain why the global foreign aid bureaucracy has run amok in recent years. Consider the steps that beleaguered government officials in low-income countries must take to receive foreign aid. Among other things, they must prepare a participatory Poverty Reduction Strategy Paper (PRSP)—a detailed plan for uplifting the destitute that the World Bank and International Monetary Fund (IMF) require before granting debt forgiveness and new loans. This document in turn must adhere to the World Bank's Comprehensive Development Framework, a 14-point checklist covering everything from lumber policy to labor practices. And the list goes on: Policymakers seeking aid dollars must also prepare a Financial Information Management System report, a Report on Observance of Standards and Codes, a Medium Term Expenditure Framework, and a Debt Sustainability Analysis for the Enhanced Heavily Indebted Poor Countries Initiative. Each document can run to hundreds of pages and consume months of preparation time. For example, Niger's recently completed PRSP is 187 pages long, took 15 months to prepare, and sets out spending for a 2002–05 poverty reduction plan with such detailed line items as $17,600 a year on "sensitizing population to traffic circulation."

Meanwhile, the U.N. International Conference on Financing for Development held in Monterrey, Mexico, in March 2002 produced a document—"the Monterrey Consensus"—that has a welcome emphasis on partnership between rich donor and poor recipient nations. But it's somewhat challenging for poor countries to carry out the 73 actions that the document recommends, including such ambitions as establishing democracy, equality between boys and girls, and peace on Earth.

Visitors to the World Bank Web site will find 31 major development topics listed there, each with multiple subtopics. For example, browsers can explore 13 subcategories under "Social Development," including indigenous peoples, resettlement, and culture in sustainable development. This last item in turn includes the music industry in Africa, the preservation of cultural artifacts, a seven-point framework for action, and—well, you get the idea.

It's not that aid bureaucrats are bad; in fact, many smart, hardworking, dedicated professionals toil away in the world's top aid agencies. But the perverse incentives they face explain the organizations' obtuse behavior. The international aid bureaucracy will never work properly under the conditions that make it operate like a cartel—the cartel of good intentions.

ALL TOGETHER NOW

Cartels thrive when customers have little opportunity to complain or to find alternative suppliers. In its heyday during the 1970s, for example, the Organization of the Petroleum Exporting Countries (OPEC) could dictate severe terms to customers; it was only when more non-OPEC oil exporters emerged that the cartel's power weakened. In the foreign aid business, customers (i.e., poor citizens in developing countries) have few chances to express their needs, yet they cannot exit the system. Meanwhile, rich nations paying the aid bills are clueless about what those customers want. Non-governmental organizations (NGOs) can hold aid institutions to task on only a few high-visibility issues, such as conspicuous environmental

The Aid Cartel's Golden Oldies

Many of the "new" themes that the international aid agencies emphasize today have actually been around for several decades.

Donor Coordination	"[Foreign aid] should be a cooperative enterprise in which all nations work together through the United Nations and its specialized agencies." (U.S. President Harry Truman, 1949)	"Aid coordination… has been recognized as increasingly important." (World Bank, 1981)	"We should improve coherence through better coordination of efforts amongst international institutions and agencies, the donor community, the private sector, and civil society." (World Bank President James Wolfensohn, 2002)
Aid Selectivity	"Objective No. 1: To apply stricter standards of selectivity… in aiding developing countries." (President John F. Kennedy, 1963)	"The relief of poverty depends both on aid and on the policies of the recipient countries." (Cassen Development Committee Task Force, 1985)	"[The International Development Association] should increase its selectivity… by directing more assistance to borrowers with sound policy environments." (International Development Association, 2001)
Focus on Poverty	"[The aid community must] place far greater emphasis on policies and projects which will begin to attack the problems of absolute poverty." (World Bank President Robert McNamara, 1973)	"The Deputies encouraged an even stronger emphasis on poverty reduction in [the International Development Association's] programs." (Former World Bank Managing Director Ernest Stern, 1990)	"The Poverty Reduction Strategy Paper aims at… increasing the focus of… assistance on the overarching objective of poverty reduction." (International Development Association, 2001)
African Reforms	"Many African governments are more clearly aware of the need to take major steps to improve the efficiency… of their economies." (World Bank, 1983)	"African countries have made great strides in improving policies and restoring growth." (World Bank, 1994)	"Africa's leaders… have recognized the need to improve their policies, spelled out in the New Partnership for African Development." (World Bank, 2002)

Sources: William Easterly, "The Cartel of Good Intentions: Bureaucracy vs. Markets in Foreign Aid" (Washington: Center for Global Development, 2002); James Wolfensohn, "Note From the President of the World Bank" (April 12, 2002)

destruction. Under these circumstances, even while foreign aid agencies make good-faith efforts to consult their clients, these agencies remain accountable mainly to themselves.

The typical aid agency forces governments seeking its money to work exclusively with that agency's own bureaucracy—its project appraisal and selection apparatus, its economic and social analysts, its procurement procedures, and its own interests and objectives. Each aid agency constitutes a mini-monopoly, and the collection of all such monopolies forms a cartel. The foreign aid community also resembles a cartel in that the IMF, World Bank, regional development banks, European Union, United Nations, and bilateral aid agencies all agree to "coordinate" their efforts. The customers therefore have even less opportunity to find alternative aid suppliers. And the entry of new suppliers into the foreign assistance business is difficult because large aid agencies must be sponsored either by an individual government (as in the case of national agencies, such as the U.S. Agency for International Development) or by an international agreement (as in the case of multilateral agencies, such as the World Bank). Most NGOs are too small to make much of a difference.

Of course, cartels always display fierce jostling for advantage and even mutual enmity among members. That explains why the aid community concludes that "to realize our increasingly reciprocal ambitions, a lot of hard work, compromises and true goodwill must come into play." Oops, wait, that's a quote from a recent OPEC meeting. The foreign aid community simply maintains that "better coordination among international financial institutions is needed." However, the difficulties of organizing parties with diverse objectives and interests and the inherent tensions in a cartel render such coordination forever elusive. Doomed attempts at coordination create the worst of all worlds—no central planner exists to tell each agency what to do, nor is there any market pressure from customers to reward successful agencies and discipline unsuccessful ones.

As a result, aid organizations mindlessly duplicate services for the world's poor. Some analysts see this duplication as a sign of competition to satisfy the customer—not so. True market competition should eliminate duplication: When you choose where to eat lunch, the restaurant next door usually doesn't force you to sit down for an extra meal. But things are different in the world of foreign aid, where a team from the U.S. Agency for International Development produced a report on corruption in Uganda in 2001, unaware that British analysts had produced a report on the same topic six months earlier. The Tanzanian government churns out more than 2,400 reports annually for its various donors, who send the poor country some 1,000 missions each year. (Borrowing terminology from missionaries who show the locals the one true path to heaven, "missions" are visits of aid agency staff to developing countries to discuss desirable government policy.) No wonder, then, that in the early 1990s, Tanzania was implementing 15 separate stand-alone health-sector projects funded by 15 different donors. Even small bilateral aid agencies plant their flags everywhere. Were the endless meetings and staff hours worth the effort for the Senegalese government to receive $38,957 from the Finnish Ministry for Foreign Affairs Development Cooperation in 2001?

By forming a united front and duplicating efforts, the aid cartel is also able to diffuse blame among its various members when economic conditions in recipient countries don't improve according to plan. Should observers blame the IMF for fiscal austerity that restricts funding for worthy programs, or should they fault the World Bank for failing to preserve high-return areas from public expenditure cuts? Are the IMF and World Bank too tough or too lax in enforcing conditions? Or are the regional development banks too inflexible (or too lenient) in their conditions for aid? Should bilateral aid agencies be criticized for succumbing to national and commercial interests, or

should multilateral agencies be condemned for applying a "one size fits all" reform program to all countries? Like squabbling children, aid organizations find safety in numbers. Take Argentina. From 1980 to 2001, the Argentine government received 33 structural adjustment loans from the IMF and World Bank, all under the watchful eye of the U.S. Treasury. Ultimately, then, is Argentina's ongoing implosion the fault of the World Bank, the IMF, or the Treasury Department? The buck stops nowhere in the world of development assistance. Each party can point fingers at the others, and bewildered observers don't know whom to blame—making each agency less accountable.

THE $3,521 QUANDARY

Like any good monopoly, the cartel of good intentions seeks to maximize net revenues. Indeed, if any single objective has characterized the aid community since its inception, it is an obsession with increasing the total aid money mobilized. Traditionally, aid agencies justify this goal by identifying the aid "requirements" needed to achieve a target rate of economic growth, calculating the difference between existing aid and the requirements, and then advocating a commensurate aid increase. In 1951, the U.N. Group of Experts calculated exactly how much aid poor countries needed to achieve an annual growth rate of 2 percent per capita, coming up with an amount that would equal $20 billion in today's dollars. Similarly, the economist Walt Rostow calculated in 1960 the aid increase (roughly double the aid levels at the time) that would lift Asia, Africa, and Latin America into self-sustaining growth. ("Self-sustaining" meant that aid would no longer be necessary 10 to 15 years after the increase.) Despite the looming expiration of the 15-year aid window, then World Bank President Robert McNamara called for a doubling of aid in 1973. The call for doubling was repeated at the World Bank in its 1990 "World Development Report." Not to be outdone, current World Bank President James Wolfensohn is now advocating a doubling of aid.

By forming a united front and duplicating efforts, the foreign aid community is able to diffuse blame among its members when economic conditions in poor countries fail to improve.

The cartel's efforts have succeeded: Total assistance flows to developing countries have doubled several times since the early days of large-scale foreign aid. (Meanwhile, the World Bank's staff increased from 657 people in 1959–60 to some 10,000 today.) In fact, if all foreign aid

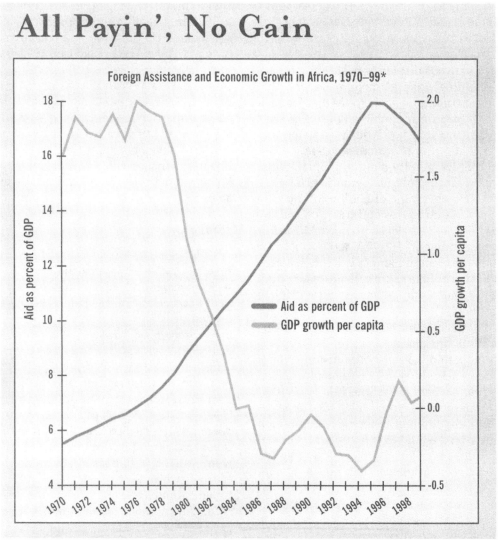

All Payin', No Gain

Foreign Assistance and Economic Growth in Africa, 1970–99*

Aid as percent of GDP

GDP growth per capita

- Aid as percent of GDP
- GDP growth per capita

*The data for each year represents the average per capita GDP growth rate and the average rate of aid (as a percentage of GDP) over the previous 10 years.

Source: *World Development Indicators* (Washington: World Bank, 2001); calculations by author

given since 1950 had been invested in U.S. Treasury bills, the cumulative assets of poor countries by 2001 from foreign aid alone would have amounted to $2.3 trillion. This aid may have helped achieve such important accomplishments as lower infant mortality and rising literacy throughout the developing world. And high growth in aid-intensive countries like Botswana and Uganda is something to which aid agencies can (and do) point. The growth outcome in most aid recipients, however, has been extremely disappointing. For example, on average, aid-intensive African nations saw growth decline despite constant increases in aid as a percentage of their income.

Aid agencies always claim that their main goal is to reduce the number of poor people in the world, with poverty defined as an annual income below $365. To this end, the World Bank's 2002 aid accounting estimates that an extra $1 billion in overseas development assistance would lift more than 284,000 people out of poverty. (This claim has appeared prominently in the press and has been repeated in other government reports on aid effec-

tiveness.) If these figures are correct, however, then the additional annual aid spending per person lifted out of poverty (whose annual income is less than $365) comes to $3,521. Of course, aid agencies don't follow their own logic to this absurd conclusion—common sense says that aid should help everyone and not just target those who can stagger across the minimum poverty threshold. Regrettably, this claim for aid's effect on poverty has more to do with the aid bureaucracy's desperate need for good publicity than with sound economics.

A FRAMEWORK FOR FAILURE

To the extent that anyone monitors the performance of global aid agencies, it is the politicians and the public in rich nations. Aid agencies therefore strive to produce outputs (projects, loans, etc.) that these audiences can easily observe, even if such outputs provide low economic returns for recipient nations. Conversely, aid bureaucrats don't try as hard to produce less visible, high-return out-

"Do Everything" Development

In September 2000, representatives of 189 countries met at the U.N. Millennium General Assembly in New York and adopted the Millennium Declaration concerning peace, security, and development issues. The Millennium Development Goals (MDGs), listed below, emerged from this gathering. Since then, virtually all the leading aid institutions have endorsed the MDGs, including the World Bank, International Monetary Fund, Organisation for Economic Co-operation and Development, and the Inter-American Development Bank.

Goal 1: Eradicate extreme poverty and hunger

Halve, between 1990 and 2015, the proportion of people whose income is less than $1 a day. Halve, between 1990 and 2015, the proportion of people who suffer from hunger.

Goal 2: Achieve universal primary education

Ensure that, by 2015, children everywhere, boys and girls alike, will be able to complete a full course of primary schooling.

Goal 3: Promote gender equality and empower women

Eliminate sender disparity in primary and secondary education preferably by 2005 and in all levels of education no later than 2015.

Goal 4: Reduce child mortality

Reduce by two-thirds, between 1990 and 2015, the under-five mortality rate.

Goal 5: Improve maternal health

Reduce by three-quarters, between 1990 and 2015, the maternal mortality ratio.

Goal 6: Combat HIV/AIDS, malaria, and other diseases

Have halted by 2015 and begun to reverse the spread of HIV/AIDS. Have halted by 2015 and begun to reverse the incidence of malaria and other major diseases.

Goal 7: Ensure environmental sustainability

Integrate the principles of sustainable development into country policies and programmes and reverse the loss of environmental resources. Halve, by 2015, the proportion of people without sustainable access to safe drinking water. Have achieved, by 2020, a significant improvement in the lives of at least 100 million slum dwellers.

Goal 8: Develop a global partnership for development

Develop further an open, rule-based, predictable, non-discriminatory trading and financial system.... Address the special needs of the least developed countries.... Address the special needs of landlocked countries and small island developing states.... Deal comprehensively with the debt problems of developing countries.... In cooperation with developing countries, develop and implement strategies for decent and productive work for youth. In cooperation with pharmaceutical companies, provide access to affordable, essential drugs in developing countries. In cooperation with the private sector, make available the benefits of new technologies, especially information and communications.

Source: United Nations Development Programme and World Bank

puts. This emphasis on visibility results in shiny showcase projects, countless international meetings and summits, glossy reports for public consumption, and the proliferation of "frameworks" and strategy papers. Few are concerned about whether the showcase projects endure beyond the ribbon-cutting ceremony or if all those meetings, frame-works, and strategies produce anything of value.

This quest for visibility explains why donors like to finance new, high-profile capital investment projects yet seem reluctant to fund operating expenses and maintenance after high-profile projects are completed. The resulting problem is a recurrent theme in the World Bank's periodic reports on Africa. In 1981, the bank's Africa study concluded that "vehicles and equipment frequently lie idle for lack of spare parts, repairs, gasoline, or other necessities. Schools lack operating funds for salaries and teaching materials, and agricultural research stations have difficulty keeping up field trials. Roads, public buildings, and processing facilities suffer from lack of maintenance." Five years later, another study of Africa found that "road maintenance crews lack fuel and bitumen... teachers lack books... [and] health workers have no medicines to distribute." In 1986, the Word Bank declared that in Africa, "schools are now short of books, clinics lack medicines, and infrastructure maintenance is avoided." Meanwhile, a recent study for a number of different poor countries estimated that the return on spending on educational instructional materials was up to 14 times higher than the return on spending on physical facilities.

And then there are the frameworks. In 1999, World Bank President James Wolfensohn unveiled his Comprehensive Development Framework, a checklist of 14 items, each with multiple subitems. The framework covers clean government, property rights, finance, social safety nets, education, health, water, the environment, the spoken word and the arts, roads, cities, the countryside, microcredit, tax policy, and motherhood. (Somehow, macroeconomic policy was omitted.) Perhaps this framework explains why the World Bank says management has simultaneously "refocused and broadened the development agenda." Yet even Wolfensohn seems relatively restrained compared with the framework being readied for the forthcoming U.N. World Summit on Sustainable Development in Johannesburg in late August 2002, where 185 "action recommendations"—covering everything from efficient use of cow dung to harmonized labeling of chemicals—await unsuspecting delegates.

Of course, the Millennium Development Goals (MDGs) are the real 800-pound gorilla of foreign aid frameworks. The representatives of planet Earth agreed on these goals at yet another U.N. conference in September 2000. The MDGs call for the simultaneous achievement of multiple targets by 2015, involving poverty, hunger; infant and maternal mortality, primary education, clean water, contraceptive use, HIV/AIDS, gender equality, the environment, and an ill-defined "partnership for development". These are all worthy causes, of course, yet would the real development customers necessarily choose to spend their scarce resources to attain these particular objectives under this particular timetable? Economic principles dictate that greater effort should be devoted to goals with low costs and high benefits, and less effort to goals where the costs are prohibitive relative to the benefits. But the "do everything" approach of the MDGs suggests that the aid bureaucracy feels above such tradeoffs. As a result, government officials in recipient countries and the foreign aid agency's own frontline workers gradually go insane trying to keep up with proliferating objectives—each of which is deemed Priority Number One.

A 2002 World Bank technical study found that a doubling of aid flows is required for the world to meet the U.N. goals. The logic is somewhat circular, however, since a World Bank guidebook also stipulates that increasing aid is undoubtedly "a primary function of targets set by the international donor community such as the [Millennium] Development Goals." Thus increased aid becomes self-perpetuating—both cause and effect.

FOREIGN AID AND ABET

Pity the poor aid bureaucracy that must maintain support for foreign assistance while bad news is breaking out everywhere. Aid agencies have thus perfected the art of smoothing over unpleasant realities with diplomatic language. A war is deemed a "conflict-related reallocation of resources."

Countries run by homicidal warlords like those in Liberia or Somalia are "low-income countries under stress." Nations where presidents loot the treasury experience "governance issues." The meaning of other aid community jargon, like "investment climate," remains elusive. The investment climate will be stormy in the morning, gradually clearing in the afternoon with scattered expropriations.

Another typical spin-control technique is to answer any criticism by acknowledging that, "Indeed, we aid agencies used to make that mistake, but now we have corrected it." This defense is hard to refute, since it is much more difficult to evaluate the present than the past. (One only doubts that the sinner has now found true religion from the knowledge of many previous conversions.) Recent conversions supposedly include improved coordination among donors, a special focus on poverty alleviation, and renewed economic reform efforts in African countries. And among the most popular concepts the aid community has recently discovered is "selectivity"—the principle that aid will only work in countries with good economic policies and efficient, squeaky-clean institutions. The moment of aid donors' conversion on this point supposedly came with the end of the Cold War, but in truth, selectivity (and other "new" ideas) has been a recurrent aid theme over the last 40 years.

Unfortunately, evidence of a true conversion on selectivity remains mixed. Take Kenya, where President Daniel arap Moi has mismanaged the economy since 1978. Moi has consistently failed to keep conditions on the 19 economic reform loans his government obtained from the World Bank and IMF (described by one NGO as "financing corruption and repression") since he took office. How might international aid organizations explain the selectivity guidelines that awarded President Moi yet another reform loan from the World Bank and another from the IMF in 2000, the same year prominent members of Moi's government appeared on a corruption "list of shame" issued by Kenya's parliament? Since then, Moi has again failed to deliver on his economic reform promises, and international rating agencies still rank the Kenyan government among the world's most corrupt and lawless. Ever delicate, a 2002 IMF report conceded that "efforts to bring the program back on track have been only partially successful" in Kenya. More systematically, however, a recent cross-country survey revealed no difference in government ratings on democracy, public service delivery, rule of law, and corruption between those countries that received IMF and World Bank reform loans in 2001 and those that did not. Perhaps the foreign aid community applies the selectivity principle a bit selectively.

DISMANTLING THE CARTEL

How can the cartel of good intentions be reformed so that foreign aid might actually reach and benefit the world's poor? Clearly, a good dose of humility is in order, considering all the bright ideas that have failed in the past. Moreover, those of us in the aid industry should not be so arrogant to think we are the main determinants of whether low-income countries develop—poor nations must accomplish that mainly on their own.

Antiglobalization protesters are largely on target when it comes to the failure of international financial institutions to foment "adjustment with growth" in many poor countries.

Still, if aid is to have some positive effect, the aid community cannot remain stuck in the same old bureaucratic rut. Perhaps using market mechanisms for foreign aid is a better approach. While bureaucratic cartels supply too many goods for which there is little demand and too few goods for which there is much demand, markets are about matching supply and demand. Cartels are all about "coordination," whereas markets are about the decentralized matching of customers and suppliers.

One option is to break the link between aid money and the obligatory use of a particular agency's bureaucracy. Foreign assistance agencies could put part of their resources into a common pool devoted to helping countries with acceptably pro-development governments. Governments would compete for the "pro-development" seal of approval, but donors should compete, too. Recipient nations could take the funds and work with any agency they choose. This scenario would minimize duplication and foster competition among aid agencies.

Another market-oriented step would be for the common pool to issue vouchers to poor individuals or communities, who could exchange them for development services at any aid agency, NGO, or domestic government agency. These service providers would in turn redeem the vouchers for cash out of the common pool. Aid agencies would be forced to compete to attract aid vouchers (and thus money) for their budgets. The vouchers could also trade in a secondary market; how far their price is below par would reflect the inefficiency of this aid scheme and would require remedial action. Most important, vouchers would provide real market power to the impoverished customers to express their true needs and desires.

Intermediaries such as a new Washington-based company called Development Space could help assemble the vouchers into blocks and identify aid suppliers; the intermediaries could even compete with each other to attract funding and find projects that satisfy the customers, much as venture capital firms do. (Development Space is a private Web-based company established last year by former World Bank staff members—kind of an eBay for foreign aid.) Aid agencies could establish their own intermediation units to add to the competition. An information bank could facilitate transparency and communication, posting news on projects searching for funding, donors searching for projects, and the reputation of various intermediaries.

Bureaucratic cartels probably last longer than private cartels, but they need not last forever. President George W. Bush's proposed Millennium Challenge Account (under which, to use Bush's words, "countries that live by these three broad standards—ruling justly, investing in their people, and encouraging economic freedom—will receive more aid from America") and the accompanying increase in U.S. aid dollars will challenge the IMF and World Bank's near monopoly over reform-related lending. Development Space may be the first of many market-oriented endeavors to compete with aid agencies, but private philanthropists such as Bill Gates and George Soros have entered the industry as well. NGOs and independent academic economists are also more aggressively entering the market for advice on aid to poor countries. Globalization protesters are not well informed in all areas, but they seem largely on target when it comes to the failure of international financial institutions to foment "adjustment with growth" in many poor countries. Even within the World Bank itself, a recent board of directors paper suggested experimenting with "output-based aid" in which assistance would compensate service providers only when services are actually delivered to the poor—sadly, a novel concept. Here again, private firms, NGOs, and government agencies could compete to serve as providers.

Now that rich countries again seem interested in foreign aid, pressure is growing to reform a global aid bureaucracy that is increasingly out of touch with good economics. The high-income countries that finance aid and that genuinely want aid to reach the poor should subject the cartel of good intentions to the bracing wind of competition, markets, and accountability to the customers. Donors and recipients alike should not put up with $3,521 in aid to reduce the poverty head count by one, 185-point development frameworks, or an alphabet soup of bureaucratic fads. The poor deserve better.

William Easterly is senior fellow of the Center for Global Development and the Institute for International Economics in Washington, D.C., and former senior advisor of the development research group at the World Bank. He is the author of The Elusive Quest for Growth: Economists' Adventures and Misadventures in the Tropics (*Cambridge: MIT Press, 2001*).

[Want to Know More?]

The Web site of the **World Bank** contains many of the documents cited in this article, as well as relevant works such as **"World Development Report 2002: Building Institutions for Markets"** (Washington: World Bank, 2001) and **"Assessing Aid: What Works, What Doesn't, and Why"** (Washington: World Bank, 1998). Visit the Web site of the **United Nations** to find the **"Monterrey Consensus"** documents from the March 2002 U.N. International Conference on Financing for Development in Mexico. **The Poverty Reduction Strategy Papers** are available on the Web site of the **International Monetary Fund**.

Insightful books on foreign assistance include Judith Tendler's classic *Inside Foreign Aid* (Baltimore: Johns Hopkins University Press, 1975), James Morton's *The Poverty of Nations: The Aid Dilemma at the Heart of Africa* (London: I.B. Tauris Publishers, 1996), and Nicolas van de Walle's *African Economies and the Politics of Permanent Crisis: 1979–99* (Cambridge: Cambridge University Press, 2001). Other useful works include *Tropical Gangsters* (New York: Basic Books, 1991) by Robert Klitgaard and *The World Bank: Its First Half Century, Volume I* (Washington: Brookings Institution Press, 1997) by Devesh Kapur, John P. Lewis, and Richard Webb. In his recent book *On Globalization* (New York: PublicAffairs, 2002), philanthropist George Soros suggests market mechanisms for foreign aid.

Enduring works on bureaucracy include William A. Niskanen Jr.'s *Bureaucracy and Representative Government* (Chicago: Aldine-Atherton, 1971) and James Q. Wilson's *Bureaucracy* (New York: Basic Books, 1989). Lant Pritchett and Michael Woolcock assess the problems of bureaucracy in economic development in **"Solutions When the Solution Is the Problem: Arraying the Disarray in Development"** (Washington: Center for Global Development, 2002).

FOREIGN POLICY has a long history of covering economic development and foreign aid, including Samuel Huntington's **"Foreign Aid: For What and for Whom"** (Winter 1970-71), appearing in *FP*'s inaugural issue. Also see **"Development: The End of Trickle Down?"** (Fall 1973) by James Grant, **"The Third World: Public Debt, Private Profit"** (Spring 1978) by Albert Fishlow, et al., **"Funding Foreign Aid"** (Summer 1988) by David R. Obey and Carol Lancaster, and **"The IMF: A Cure or a Curse?"** (Summer 1998) by Devesh Kapur. More recent *FP* coverage includes Ricardo Hausmann's **"Prisoners of Geography"** (January/February 2001), Dani Rodrik's **"Trading in Illusions"** (March/April 2001), Stephen Fidler's **"Who's Minding the Bank?"** (September/October 2001), and William Easterly's **"Think Again: Debt Relief"** (November/December 2001).

This article is based on a longer research paper by Easterly, **"The Cartel of Good Intentions: Bureaucracy vs. Markets in Foreign Aid"** (Washington: Center for Global Development, 2002). Fox a comprehensive treatment of foreign aid and the problems of economic development, see Easterly's *The Elusive Quest for Growth: Economists' Adventures and Misadventures in the Tropics* (Cambridge: MIT Press, 2001).

For links to relevant Web sites, access to the *FP* Archive, and a comprehensive index of related FOREIGN POLICY articles, go to **www.foreignpolicy.com**.

SPECIAL REPORT THE DOHA ROUND

The WTO under fire

Why did the world trade talks in Mexico fall apart? And who is to blame?

CANCÚN

SOME poor countries' politicians seemed to revel in the collapse of the World Trade Organisation's ministerial meeting on September 14th. The Philippine trade minister, for instance, told Reuters news agency that he was "elated" by it. Tanzania's delegate claimed to be "very happy" that poor countries had stood up to rich-country "manipulation". But others were upset and shocked. According to one observer, the trade minister of Bangladesh had tears in his eyes. "I'm really disappointed," he is reported to have said. "This is the worst thing we poor countries could have done to ourselves."

Disappointment is the right reaction. For the Doha round of trade talks run by the WTO was geared specifically to help poor countries. They will be the biggest victims if the talks cannot be revived, and there seems to be scant prospect of that. The negotiations have not been officially abandoned. Diplomats pledged to continue talking in Geneva, the WTO's headquarters, with a formal meeting to be held no later than December 15th. But momentum has clearly been lost. No one now expects the round to finish by its original deadline of December 31st 2004. Some trade officials privately wonder whether it will ever finish, or whether Cancún's collapse—coming less than four years after the Seattle ministerial meeting broke down in December 1999—marks the end of the WTO as an effective negotiating forum.

The price of posturing

According to the World Bank, a successful Doha round could raise global income by more than $500 billion a year by 2015. Over 60% of that gain would go to poor countries, helping to pull 144m people out of poverty. While most of the poor countries' gains would come from freer trade among themselves, the reduction of rich-country farm subsidies and more open markets in the north would also help. That prize is now forgone.

As the scale of this lost opportunity becomes clear, the postmortems and recriminations are beginning. Three big questions stand out: Why did the talks collapse? Who was to blame? And where does the WTO go from here?

The new alliance [1]

The G21 member countries

Argentina	Ecuador	Paraguay
Bolivia	Egypt	Peru
Brazil	Guatemala	Philippines
Chile	India	South Africa
China	Indonesia	Thailand
Colombia	Mexico	Venezuela
Costa Rica	Nigeria	
Cuba	Pakistan	

Though the speed of collapse caught even seasoned trade negotiators by surprise, the seeds of disaster were sown long before September 14th. The launch of the Doha round in the eponymous capital of Qatar in November 2001 was itself a nail-biting negotiation marked by acrimony between rich and poor. The rhetoric was grand: Doha would reduce trade-distorting farm support, slash tariffs on farm goods and eliminate agricultural-export subsidies; it would cut industrial tariffs, especially in areas that poor countries cared about, such as textiles; it would free up trade in services; and it would negotiate global rules (subject to a framework to be decided at Cancún) in four new areas—in competition; investment; transparency in government procurement; and trade facilitation. These four new areas are referred to as the "Singapore issues" after the trade meeting at which they were first raised.

From the start, countries disowned big parts of the Doha agenda. The European Union, for instance, denied it had ever promised to get rid of export subsidies. Led by India, many poor countries denied that they ever signed up for talks on new rules. Other poor countries spent more time moaning about their grievances over earlier trade rounds than they did in negotiating the new one. Several rich countries too showed little interest in compromise. Japan, for instance, seemed content simply to say no to any cuts in rice tariffs.

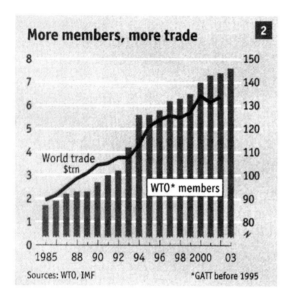

More members, more trade 2

World trade $trn

WTO* members

1985 88 90 92 94 96 98 2000 03

Sources: WTO, IMF   *GATT before 1995

This kind of posturing meant the trade round stagnated for 22 months between the meetings in Doha and Cancún. All self-imposed deadlines were missed; all tough political decisions were put off. That placed a needlessly heavy burden on the Cancún meeting. But it was not an overloaded agenda that killed the talks last weekend. It was that too many countries continued grandstanding at the Mexican resort, rather than seeking the compromises on which trade talks depend.

Focus on farming

Agriculture was the toughest issue dividing negotiators both before and during the Cancún meeting. After months of stalemate, and at the behest of many developing countries, in August America and the EU drew up a framework for freeing farm trade. Though it involved some reform, the plan was much less ambitious than Doha had implied. Export subsidies, for example, were not to be eliminated after all.

Angered by the lack of ambition, a new block of developing countries emerged just before the Cancún meeting to denounce the EU/US framework as far too timid. Led by Brazil, China and India, this so-called G21 (see table) became a powerful voice. It

represented half the world's population and two-thirds of its farmers. It was well organised and professional.

Although it spanned diverse interests—India, for instance, is terrified of lowering tariffs on farm goods, while Brazil, a huge and competitive exporter, wants free trade as fast as possible— the G21 stood together and hammered one message home: rich countries, as the most profligate agricultural subsidisers, should make bigger efforts to cut subsidies and free farm trade. The level of support given to farmers by the rich countries of the OECD has remained more or less unchanged (at over $300 billion) for the past 15 years.

While the fight between Europe, America and the G21 received most attention, another alliance of poor countries, most of them from Africa, was also worried about agriculture, but for different reasons. They feared that freeing farm trade would mean losing their special preferences. (Europe's former colonies, for instance, get special access to the EU's markets for their bananas.) They were even more worried about cutting tariffs than India, fretting that imports would ruin their small farmers. And many, particularly a small group of countries in West Africa, worried most of all about cotton.

Prodded and encouraged by non-governmental organisations (NGOs), especially Oxfam, a group of four West African countries—Benin, Burkina Faso, Chad and Mali—managed to get cotton included as an explicit item on the Cancú agenda. Their grievances were simple, and justified. West African cotton farmers are being crushed by rich-country subsidies, particularly the $3 billion-plus a year that America lavishes on its 25,000 cotton farmers, helping to make it the world's biggest exporter, depressing prices and wrecking the global market.

The West African four wanted a speedy end to these subsidies and compensation for the damage that they had caused. Though small fry compared with the overall size of farm subsidies, the cotton issue (like an earlier struggle over poor-country access to cheap drugs) came to be seen as the test of whether the Doha round was indeed focused on the poor.

But the draft text that emerged halfway through the Cancún meeting was a huge disappointment. The promises on cotton were vague, pledging a WTO review of the textiles sector, but with no mention of eliminating subsidies or of compensation. Worse, it suggested that the West African countries should be encouraged to diversify out of cotton altogether.

This hardline stance had American fingerprints all over it. Political realities in Congress (the chairman of the Senate agriculture committee is a close ally of the cotton farmers) made American negotiators fiercely defensive of their outrageous subsidies. For the Africans, the vague text was a big blow. It caused "anger and bitterness" said one delegate. As a result, the poorest countries dug in their heels when it came to the other big controversial area: that of extending trade negotiations into the four new Singapore issues. Along with many other poor countries, the Africans had long been leery about expanding the remit of the trade talks at all.

Some of their concerns, such as the fear of overloading their few negotiators, were reasonable. Others made less sense, such as the worry (fanned by many NGOs) that rich countries would use these rules to trample on poor countries' sovereignty. Two days into the conference, over 90 countries signed a letter saying that they were not ready to move into these areas.

Unfortunately, the EU and others who cared about the Singapore issues (a group that did not include the Americans) refused to compromise. Only 24 hours before the meeting was due to end, Pascal Lamy, the EU's chief negotiator, reiterated that negotiations had to proceed in all four areas. Only on Cancún's final morning did he budge, offering to give up two of the Singapore issues. There were even hints that Europe could jettison three, leaving only negotiations on trade facilitation on the table.

Rationally, no country should have objected to that, least of all poor countries. The trade blockages that these rules are designed to minimise cost them far more than tariffs. According to the World Bank, the costs of transporting African exports to foreign markets are five times higher, on average, than the tariffs paid on those goods. Complex, inefficient and corrupt customs procedures make up a big share of these transport costs.

By this time, however, reason was playing little role in the progress of the meeting. The group of embittered African countries refused to negotiate on any of the four Singapore issues. South Korea, by contrast, said it could only accept negotiations on all four. At that point, Luis Ernesto Derbez, Mexico's foreign minister and chairman of the Cancún gathering, said he saw no basis for compromise and declared the meeting over. At the centre of all the negotiations between the different factions, Mr Derbez was better placed than anyone to make that judgment.

Within minutes, delegates who had been set to argue all night over agriculture scrambled to catch earlier flights home. Within hours, the Mexican technicians were dismantling the equipment at the conference centre.

The blame game

Who bears responsibility for this? Some delegates, especially from Europe, blamed Mr Derbez for cutting off discussion too hastily. One British politician claimed his action was "utterly unexpected" and "premature". Conspiracy theorists claimed that Mexico ended debate at the behest of the Americans who wanted the meeting to fail all along.

That is nonsense. Given the Mexicans' determination to finish the Cancún meeting on time, the Europeans probably made a tactical mistake in retreating so late on the Singapore issues. But Cancún's failure goes deeper than miscalculations on timing. It happened because of intransigence and brinkmanship by both rich and poor countries; because of irresponsible and inflammatory behaviour by NGOs; and because of the deeply flawed decision-making system of the WTO itself.

The instant post-mortems blamed rich countries most. NGOs accused them of wrecking the talks by pushing poor countries too far on the Singapore issues and giving too little on agriculture. There is much truth to both claims. Europe's ideological attachment to negotiations on investment and competition is hard to fathom, particularly since no European industries were clamouring for them.

On agriculture, moreover, the rich world's concessions were too timid and too grudging. America's bold promises were belied by its actions. Last year's outrageous increase in American farming subsidies, and the cave-in at Cancún by American negotiators to their domestic cotton growers, made far more of an impression on poor countries than Washington's high-minded words about freer farm trade—and rightly so.

Europe was stymied not just by its desire to mollycoddle its own farmers, but by the EU's cumbersome decision-making process. Only after its own internal reforms were agreed to in June could Brussels offer concessions on agriculture, and even then they were meagre. Given the mess that their farm policies create, rich countries should have done far more.

But poor countries, too, bear some responsibility for Cancún's collapse. Although a few emerging economies were tireless negotiators, too many others did no more than posture. Some of the posturing was tactical: for all their public rhetoric, for instance, the G21 group was actively negotiating with both America and Europe. But others, particularly some African countries, could not get beyond their radical public positions. Anti-rich-country rhetoric became more important than efforts to reach agreement.

NGOs, who were at Cancún in force, deserve much of the blame for this radicalisation. Too many of them deluged poor countries with muddle-headed positions and incited them to refuse all compromise with the rich world. The NGO's main mistake, however, was to raise poor countries' expectations implausibly high. Shout loudly and long enough, they seemed to suggest, and you will get your way. That proved a big miscalculation.

Finally, blame belongs to the WTO's own decision-making procedures, or rather the lack of them. Mr Lamy, with reason, called the WTO a "medieval" organisation whose rules could not support the weight of its tasks. Its predecessor, the old GATT system (which was folded into the WTO in 1995), was run by rich countries. Poor countries had little power, but also few responsibilities.

The WTO, by contrast, is a democratic organisation that works by consensus, but with no formal procedures to get there. Any one of the organisation's 148 members can hold up any aspect of any negotiation. Efforts to create smaller information groups are decried as "non-transparent" by those left out. Not surprisingly, this lends itself more to grandstanding than to serious negotiation. The worst problem, though, is that the WTO's requirement for consensus makes it virtually impossible for it to be reformed.

Headed for oblivion?

Can any of these failures be addressed and the Doha round be revived? Some countries are more optimistic than others. The G21, for instance, left Cancún determined to stick together and fight another day. Brazil, in particular, is convinced that sooner or later rich countries will be forced to reform their outrageous farm policies. One weapon it points to is the expiration of the "peace clause."

As part of the trade round before Doha, the Uruguay round, countries pledged not to file formal WTO complaints over the dumping of farm products as long as each country stuck to its (limited) farm-trade commitments. That peace clause runs out at the end of this year. The ensuring flood of disputes, claim some Brazilians, will at least force the Americans and Europeans to negotiate seriously on farm trade.

Optimists also point to the fact that previous trade rounds all took far longer to finish than planned. The Uruguay round, for instance, took eight years rather than three. According to this view, Cancún's collapse is just par for the course. Maybe. But the risk is that trade momentum will simply move elsewhere.

Even before Cancún's failure, the global trade-negotiating process faced unprecedented competition from bilateral and regional trade deals. Last weekend's events can only reinforce that trend. Bob Zoellick, America's top trade negotiator, claimed that countries were approaching him to push for bilateral deals even as the meeting was crumbling. From a global economic perspective, a tangle of such deals is far inferior to freer multilateral trade. For the poorest countries in particular, the chances of getting from a bilateral deal with America what they failed to get from the Doha round are nil.

If the momentum in trade negotiations moves away from the WTO, the consequences for the organisation itself could be grave. There would then be little political impetus to make it more effective, and the WTO's Geneva headquarters could quickly become no more than a court where disputes on existing trade rules are slowly adjudicated. Far from building a stronger multilateral system, the WTO would quietly sink into oblivion as a negotiating forum. Everyone would lose from this but, once again, the biggest losers would be the poor countries.

Cancún's collapse does not make any of these outcomes inevitable, but it does make them much more likely. That is why it is such a tragedy.

Playing dirty
at the WTO

Locked out of some meetings. Not even invited to others. And then all the decisions are made after you've left. It's all in a day's work for 'developing' World delegates at the WTO.

By Mark Lynas

Yash Tandon approached the guard warily and asked permission to pass. Briefly consulting a list of names close to his chest, the security guard shook his head. Permission denied. Just then two other men approached and were waved quickly through. They cast cursory—almost derisory—backward glances at Tandon, who stood aside as they strode by.

A scene outside the VIP lounge of a posh nightclub? No—welcome to the 'green room' of the World Trade Organisation. Professor Tandon, director of an Africa-based academic institute, was an official Ugandan delegate to the fourth WTO ministerial conference in Doha, Qatar. And, as he was discovering, the WTO operates one rule for the rich and another for the poor.

On paper, the WTO is fully open and democratic. Its former director general Mike Moore once said: '[It's] the most democratic international body in existence today. The WTO is not imposed on countries... No country is forced to sign our agreements. Each and every one of the WTO's rules is negotiated by member governments and agreed by consensus.'

But speaking more frankly on an earlier occasion, Moore made a stark admission. 'There is no denying,' he said, 'that some members are more equal than others when it comes to influence.' Moore had probably never spoken truer words. According to NGOs and many delegates from Southern countries, 'democracy' at the WTO has never been anything more than a shallow facade.

One of the most heavily-criticised aspects of the WTO process is the 'green room' system, where powerful members meet informally in closed groups to work out areas of agreement. Attendance, as professor Tandon and others found out, is by invitation only.

'You are representing a country, and it is humiliating and ridiculous for you to hang around in the corridor'

Speaking of one 'green room' meeting at Doha, Zimbabwean ambassador to the WTO Boniface

Chidyausiku said: '[It was] operating like a mafia. [My minister] could not speak since he was not officially invited to the consultations. He could only give notes to his colleagues to intervene.' Several other African delegates simply gave up and went back to their hotels. 'You are representing a country and it is humiliating and ridiculous for you to hang around in the corridor,' said one.

The result of this approach is that by the time the whole conference convened to make its democratic 'consensus decision' everything had already been stitched up behind closed doors. The plenary meetings were little more than sideshows, where government ministers could make speeches and feel involved before rubber-stamping what had already been agreed elsewhere. Any country holding up progress was seen as a 'wrecker' and often simply ignored by the chair.

Green room meetings are not the only ones where attendance is by invitation only. Entire conferences (known as 'mini-ministerials') have been held in which government ministers from a chosen few countries hammer out issues of agreement in advance. When one mini-ministerial was convened in Mexico, developing country delegates seeking invitations from the WTO secretariat were told that the Mexican government was the responsible party. The Mexican government denied having anything to do with the meeting apart from providing facilities. The Southern delegates were left bemused and, yet again, excluded.

Arm-twisting and bullying

Most delegates and government ministers from developing countries are well aware that WTO decisions on opening up trade to multinational corporations are not in their interests. Indeed, almost all low-income African and Asian countries arrived at the Doha conference aligned to negotiating blocs that were determined to stand their ground. Why then did they eventually buckle and sign up to a new trade round? The answer lies partly in the underhand tactics employed by the US and the European Union to forcibly extract the outcomes they wanted.

As Adriano Campolina Soares, head of ActionAid's international food and trade campaign and an observer in Doha, recalls, developing countries strenuously opposed the first draft of the ministerial declaration presented to the conference. 'When the second draft came back, all their concerns had been ignored. In fact, the text was even worse—but now they accepted it. This shows how the WTO works; pressure had been put on these countries behind closed doors.'

This 'pressure' can take many forms. Powerful countries may threaten to cut aid budgets, or they might offer new cash as bribes. Pakistan, for instance, got its $1 billion post-September 11 aid package from the US one day after the Doha conference had concluded. Tanzania, another compliant country, got a massive new World Bank/IMF debt-servicing deal a week after Doha. The threatened cancellation of preferential trade agreements—often crucial to developing country economies—was another favourite US/EU tactic. And once the Southern negotiating blocs were split, individual countries could be picked off one by one.

The Doha conference was held just two months after September 11, and the war on terrorism was a crucial bargaining chip. With the US trade representative Robert Zoellick insisting that a new trade round would help to stamp out terrorism, no countries wanted to be seen as opposing him. The British trade and industry secretary Patricia Hewitt took up the theme, saying the Doha agreement 'signals the determination of the world community to fight terror with trade, as well as arms'.

Another favourite tactic was direct lobbying by rich country representatives against 'difficult' Southern negotiators. Thanks to pressure from the US on his government back home, one ambassador in Doha was sacked before he had even unpacked his luggage. Many others found themselves on a US blacklist and soon lost their jobs.

As Fatoumata Jawara and Aileen Kwa point out in their forthcoming book *Behind the Scenes at the WTO*, the end result is 'to shift the decision-making process away from the formal process, in which all countries are at least notionally equal, into the realm of bilateral horse-trading. Here, only the rich have any real leverage, while most developing countries are so desperate for trade opportunities, aid, debt reduction, etc, that they have little choice but to succumb'.

More basic financial concerns also weigh heavily on developing countries. The WTO holds over 1,000 meetings a year, many of them running simultaneously, and only nations which can afford to support large missions can represent themselves properly. At ministerial conferences the discrepancies can be even more extreme: the World Development Movement discovered that while the EU had 502 people on its delegations to the WTO meeting in Doha, the Maldives had two, and Haiti, the poorest country in the Western hemisphere, had no delegates at all.

One African delegate noted: 'They got the deals they wanted because of sheer fatigue on our part. They have big delegations and can stagger people. We don't. It is very difficult to go on negotiating day and night for several days without sleep.'

Many developing country negotiators also had linguistic problems that made it difficult to keep up with what was going on: green room negotiations were almost exclusively completed in English; even official documents would only be translated into French and Spanish—often much later than the English texts became available. Exhausted Southern delegates found themselves negotiating in the dark.

Dirty tricks

The powerful countries took great pains at the Doha ministerial meeting to limit discussion and set the agenda in advance. The draft declaration was presented at the ceremonial opening session where formal objections from developing countries were impossible to raise. The very next day the chair of the meeting, Qatari trade minister Youssef Hussain Kamal, caused derisive laughter when he mistakenly left his microphone switched on while discussing with Mike Moore how to stop the Indian delegation—one of the staunchest opponents of a new trade round—taking the floor.

The chair at Doha mistakenly left his microphone on while discussing with WTO director general Mike Moore how to stop the Indian delegation—one of the main opponents of a new trade round—taking the floor

At the end of the fifth day at Doha there was still no agreement, and many developing country delegates and ministers—booked on cheaper scheduled flights—began to leave. In their absence a decision was taken to extend the conference (without any discussion among the membership—a serious procedural omission), and the draft declaration was forced through a day after the meeting was originally supposed to have ended.

As Adriano Campolina Soares says: 'If the WTO keeps the same framework of decision-making, developing countries will never be able to defend their interests. The current framework just emphasises the arm-twisting environment. Despite the title 'development agenda', almost all the concerns of developing countries were systematically ignored by the WTO.'

John Hilary, trade policy adviser for Save the Children and another observer at Doha, has reached a similar conclusion. 'Bullying and blackmail have become an integral part of how the WTO works, as we saw all too clearly at the Doha ministerial,' he says. 'Time and again developing countries have been forced to abandon their negotiating positions as a result of economic, political and even personal threats to their delegates. The pretence that the WTO is an equal and democratic negotiating forum lost its credibility a long time ago.'

According to Hilary, things are likely to be just as bad at the upcoming WTO meeting in Cancun, Mexico. The powerful countries have already begun to hint at the pressure they will be putting on the South to sign up to yet more one-sided trade liberalisation. The EU has even stated openly that the WTO is not about philanthropy. Clearly, the desire of the EU and US to extract still further concessions from the world's poor continues unabated.

Mark Lynas is a freelance writer. His book on the human impacts of climate change will be published by Flamingo later this year.

From *The Ecologist,* June 2003, pp. 38-40. © 2003 by The Ecologist, c/o Ecosystems Ltd., London, UK

UNIT 3
Conflict and Instability

Unit Selections

Key Points to Consider

- What are the causes of civil war?

- How do failing states pose an international security threat?

- What accounts for the continuing violence between Palestinians and Israelis?

- What are the dangers of continued tension between India and Pakistan?

- How does the threat of terrorism affect human rights in southeast Asia?

- How have President Mugabe's policies brought Zimbabwe to the brink of instability?

- How might relations between the United States and Uzbekistan contribute to extremism in central Asia?

- How can the North Korean nuclear weapons program be halted?

- What conditions characterize the plight of the world's refugees?

 **Links: www.dushkin.com/online/**
These sites are annotated in the World Wide Web pages.

The Carter Center
   http://www.cartercenter.org
Center for Strategic and International Studies (CSIS)
   http://www.csis.org/
Conflict Research Consortium
   http://www.Colorado.EDU/conflict/
Institute for Security Studies
   http://www.iss.co.za
PeaceNet
   http://www.igc.org/peacenet/
Refugees International
   http://www.refintl.org

Conflict and instability in the developing world remain major threats to international peace and security. Conflict stems from a combination of sources including ethnic and religious diversity, nationalism, the struggle for state control, and competition for resources. In many cases, boundaries that date from the colonial era encompass diverse groups. A state's diversity can increase tension among groups competing for scarce resources and opportunities. When some groups benefit or are perceived as enjoying privileges at the expense of others, ethnicity can offer a convenient vehicle around which to organize and mobilize. Moreover, ethnic politics lends itself to manipulation both by regimes that are seeking to protect privileges, maintain order, or retain power and those that are challenging existing governments. In a politically and ethnically charged atmosphere, conflict and instability often result as groups vie to gain control of a state apparatus that can extract resources and allocate benefits. While ethnicity has certainly played a role in many recent conflicts, conflicts over power and resources may be mistakenly viewed as ethnic in nature. Recent conflicts in Africa and Asia have increasingly centered around valuable resources. Conflicts like the war in the Democratic Republic of Congo generate economic disruption, population migration, environmental degradation, and may also draw other countries into the fighting.

Early literature on modernization and development speculated that as developing societies progressed from traditional to modern, primary attachments such as ethnicity and religious affiliation would fade and be replaced by new forms of identification. Clearly, however, ethnicity remains a potent force, as does religion. Ethnic politics and the emergence of religious radicalism demonstrate that such attachments have survived the drive toward modernization.

Initially inspired and encouraged by the theocratic regime in Iran, radical Muslims have not only pushed for the establishment of governments based on Islamic law but have engaged in a wider struggle with what they regard as the threat of Western cultural dominance. Radical Islamic groups that advocate a more rigid and violent interpretation of Islam have increasingly challenged more mainstream Islamic thought. These radicals are driven by hatred of the West and the United States in particular. They were behind the 1993 New York City World Trade Center bombing, the United States embassy bombings in Kenya and Tanzania in 1998, the devastating attacks that destroyed the World Trade Center and damaged the Pentagon in September 2001, and the bombing of a night club in Bali in 2002 and a hotel in Jakarta in 2003. Although there is a tendency to equate Islam with terrorism, it is a mistake to link the two. A deeper understanding of the various strands of Islamic thought is required to separate legitimate efforts to challenge repressive regimes and forge an alternative to Western forms of political organization from radicals who pervert Islam and use terrorism.

There is no shortage of tension and conflict around the world. Prospects for peace in the Middle East have faded in the midst of Palestinian suicide bombings and Israeli retaliation. Although the Taliban and its Al Qaeda allies have been driven from power in Afghanistan, remnants of these groups continue to attack both the Afghan government and U.S.-led coalition forces, thus ten-

sions remain. President Hamid Karzai's government has limited control outside the capital, and regional warlords continue to compete for power in many areas of the country. Tensions continue between India and Pakistan over the disputed territory of Kashmir, a situation that threatens not only war between two countries armed with nuclear weapons but that also has implications for the war on terrorism. Within India, religious nationalism is behind periodic clashes between Hindus and minority Muslims and Christians. Efforts by lower castes to gain more political power have also resulted in violence. Muslim extremists have been active in southeast Asia raising fears of another front in the war on terrorism. Meanwhile, North Korea's nuclear weapons program has not only heightened tensions with the United States but has also made regional neighbors increasingly wary. Parts of Africa also continue to be conflict-prone. A tenuous peace prevails in much of the Democratic Republic of Congo where the fighting drew neighboring countries into the conflict, resulting in what has been called Africa's "first world war." However, sporadic fighting continues in the eastern part of the country. Angola's long, brutal civil war finally ended with the death of rebel leader Jonas Savimbi in February 2002, but rebel troops must still be reintegrated into society, always a potentially dangerous process. Liberia's president, Charles Taylor, was forced into exile, and a West African peacekeeping force has been deployed to try to restore order after brutal fighting between the government and rebel forces. Zimbabwe continues its slide into economic disaster and instability. In South America, the Colombian government continues to face a serious challenge by leftist rebels, a conflict made more complicated by ties between the rebels and drug traffickers as well as the activities of right-wing paramilitaries.

The threats to peace and stability in the developing world remain complicated and dangerous and clearly have the potential to affect the Western industrialized countries. These circumstances require a greater effort to understand and resolve conflicts, whatever their source.

THE MARKET FOR CIVIL WAR

Ethnic tensions and ancient political feuds are not starting civil wars around the world. A groundbreaking new study of civil conflict over the last 40 years reveals that economic forces—such as entrenched poverty and the trade in natural resources—are the true culprits. The solution? Curb rebel financing, jump-start economic growth in vulnerable regions, and provide a robust military presence in nations emerging from conflict.

By Paul Collier

Every time a civil war breaks out, some historian traces its origin to the 14th century and some anthropologist expounds on its ethnic roots. Don't buy into such explanations too quickly. Certain countries are more prone to civil war than others, but distant history and ethnic tensions are rarely the best explanations for a conflict. Look instead at a nation's recent past and, most important, its economic conditions.

Once a country has reached a per capita income rivaling that of the world's richest nations, its risk of civil war is negligible. Today, about 900 million people live in such societies. Four billion more live in countries that are either already middle income or on track to becoming so, thanks to rapidly growing and diversifying economies. This group, which includes the economic success stories of the post–World War II era, faces fairly low risk of civil war. The potential for conflict is concentrated among the countries inhabited by the world's remaining 1.1 billion people. These countries typically have poor and declining economies and rely on natural resources—such as diamonds or oil—for a large proportion of national income. As the British, French, Portuguese, and Soviet empires successively dissolved during the last century, the number of such countries increased in waves.

Such at-risk countries are engaged in a sort of Russian roulette. Every year that their dismal economic conditions persist increases the odds that their societies will fall into armed conflict. Whether by luck or prudence, many such nations have so far escaped civil war. Others have not. And once civil war has started, the decline in income and the accumulation of arms, fighting skills, and military capabilities greatly increase the risks of further conflict.

To date, academics and policymakers alike have misdiagnosed the nature of the problem; little surprise, then, that their efforts to prevent civil wars have been ineffective. When the world's leaders can identify the real factors most likely to drive such conflicts, they will have a better chance of preventing future wars.

THE MYTH OF ETHNIC STRIFE

Between 1960 and 1999, there were 52 major civil wars for which comprehensive data is available on social, political, historical, economic, and geographic circumstances. Such wars spanned the developing regions, with the typical conflict lasting around seven years and leaving a legacy of persistent poverty and disease in its wake. To understand the causes of these conflicts, economist Anke Hoeffler and I studied each five-year-period from 1960 to 1999 and identified preexisting conditions that helped predict the outbreak of war.

If a country is mountainous and has a large, lightly populated hinterland, it faces an enhanced risk of rebellion… Nepal is therefore more at risk of civil war, geographically speaking, than Singapore.

For example, income inequality and ethnic-religious diversity are frequently cited as causes for conflict. Yet surprisingly, inequality—either of household incomes or of land owner-ship—does not appear to increase systematically the risk of civil war. Brazil got away with its high inequality; Colombia didn't. And, in fact, ethnic and religious diversity actually re-duces the risk of civil conflict. One important exception: Where the largest ethnic group constitutes a majority but lives along-side a substantial minority, such as in Sri Lanka and Rwanda, the risk of civil war roughly doubles. Once wars start, they also tend to last much longer if the nation in question displays two or three dominant ethnic groups.

Conflicts in ethnically diverse countries may be ethnically patterned without being ethnically caused. International media coverage of civil wars often focuses on history and ethnicity be-cause rebel leaders adopt this sort of discourse. Grievances are to a rebel organization what image is to a business. The rebel group needs to stimulate a sense of collective grievance to build cohesion in its army and to attract funding from its diaspora living in rich countries.

Much to the dismay of democratization activists, democracy fails to reduce the risk of civil war, at least in low-income coun-tries. Indeed, politically repressive societies have no greater risk of civil war than full-fledged democracies. Countries falling be-tween the extremes of autocracy and full democracy—where citizens enjoy some limited political rights—are at a greater risk of war. Low-income societies with new democratic institutions are often at enhanced risk: Just consider the current catastrophe in Ivory Coast, where uncertainty over who could stand for the presidential election in 2000 triggered violent clashes and on-going political instability.

Wherever a civil war occurs, observers will invariably find some deep history of conflict. But overwhelmingly, conflicts in the distant past are not generating civil wars in the present. The history that matters is recent history, not that of the 14th cen-tury. If a country recently experienced a civil war, it is much more likely to have another one. The risk fades the longer peace endures.

Civil war is self-perpetuating, partly because it changes the balance of interests within countries. Groups engaged in con-flict invest in armaments, skills, and infrastructure that are only good for violence. These groups' leaders, and indeed all those who gain from lawlessness, prosper during war, even though society as a whole suffers. The part of the elite that prefers peace will have shifted much of its wealth outside the country. Hence, as a result of the conflict, the balance of elite interests shifts toward further conflict.

Geography matters, too. If a country is mountainous and has a large, lightly populated hinterland, it faces an enhanced risk of rebellion. Presumably, rebels are harder to find and defeat in such terrain. Nepal is therefore more at risk of civil war, geo-graphically speaking, than Singapore.

DIAMONDS, A REBEL'S BEST FRIEND

All these factors notwithstanding, economic conditions remain paramount in explaining civil wars. For the average country in our study, the risk of a civil war in each five-year period was around 6 percent, but the risks increased alarmingly if the economy was poor, declining, and dependent on natural re-source exports. For a country with conditions like those in the Democratic Republic of the Congo (formerly Zaire) in the late 1990s—with deep poverty, a collapsing economy, and huge miner exploitation—the risk reaches nearly 80 percent.

When valuable natural resources are discovered in a particular region of a country, the people living in such localities suddenly have an economic incentive to secede, violently if necessary.

Once started, wars last longer in low-income countries, which are prone to rebellion for many reasons: Recruits have less of a stake in the status quo, and central governments are typically weak. Each additional percentage point in the growth rate of per capita income shaves off about 1 percentage point of conflict risk; conversely, wars are more likely to follow periods of economic collapse—such as the conflicts that have surged in Indonesia since the East Asian economic crisis of the late 1990s. If a country's per capita income doubles, its risk of con-flict drops by roughly half. Simply put, economic growth mat-ters because opportunities for youth depend upon a robust economy.

Conflict is also more likely in countries that depend heavily on natural resources for their export earnings, in part because rebel groups can extort the gains from this trade to finance their operations. Diamonds funded the National Union for the Total Independence of Angola (UNITA) rebel group during Angola's long civil war, as well as the Revolutionary United Front (RUF) in Sierra Leone; timber funded the Khmer Rouge in Cambodia. Indeed, methods of extortion abound. For example, multina-tional corporations that extract natural resources must often pay huge sums to ransom kidnapped workers and to protect infra-structure from sabotage at the hands of rebel groups. Laugh-ably, such payments are sometimes charged to the companies' "corporate social responsibility" budgets.

Natural resources also fuel war because they make secession more likely. When valuable natural resources are discovered in a particular region of a country, the people living in such local-ities suddenly have an economic incentive to secede, violently if necessary. Since most countries are ethnically diverse, the lucky, resource-rich locality is likely to be ethnically distinct as well. Often, the weak political force of ethnic romanticism latches on to the stronger force of economic self-interest so that secessionist movements voice ethnic grievances—Biafra (Ni-geria), Cabinda (Angola), and Aceh (Indonesia) come to mind. The incentive to secede is probably compounded by the corrupt, incompetent way to which governments commonly use natural resource wealth: The greed of a resource-rich locality can seem ethically less ugly if a corrupt national elite is already hijacking the resources.

THE WAR DIVIDEND

One striking lesson from these patterns is that the motivations for rebellion generally matter less than the conditions that make a rebellion financially and militarily viable. Civil wars only occur if a rebel organization can build and sustain a private army. These organizations are unlike traditional opposition groups such as political parties or protest movements. They are hierarchical, authoritarian, expensive, and usually small. Where such organizations are financially militarily feasible, rebellions are likely to emerge, promoting whatever political agenda their leaders happen to support.

Global efforts to curb civil war should therefore focus on reducing the viability—rather than just the rationale—of rebellion. Of course, policies should address legitimate grievances, not because addressing them paves a royal road to peace but because they are legitimate.

Botswana and Sierra Leone were similarly poor countries, both sitting on vast diamond deposits… Botswana harnessed this opportunity, becoming the fastest-growing economy in the world. Sierra Leone used the same resources to impoverish itself…

Those nations currently at war and those that have recently emerged from civil war constitute the core of the problem. Many countries have fallen into a conflict trap: a damaging war that sharply increases the risk of further conflict, followed by a fragile peace, and then back to war. The expected duration of a civil war is currently about eight years—double what it was before the 1980s. Wars therefore do more damage now and thus more powerfully provoke further conflict.

No one knows why wars last longer now. Perhaps global markets in both natural resources and arms make rebellion easier to finance and equip. Rebel groups can now sell the future rights to mineral extraction (conditional on rebel victory) to raise funds for weapons purchases. A similar arrangement reputedly helped French oil Giant Elf Aquitaine (now TotalFinaElf) gain its current access to oil in the Republic of the Congo.

So, what specific measures can countries take to reduce the occurrence and likelihood of civil war?

For developing countries already growing rapidly, the most significant risk may be episodes of economic crisis, such as that experienced by Indonesia in the late 1990s. The opportunity to prevent war in such cases only strengthens the justification for international efforts to avert economic crises. In this light, initiatives to reform and rethink the workings of the global financial system are not merely an academic debate or an effort to ease investors' concerns, but rather a much more serious matter with immediate life-and-death consequences.

Poor countries that are not developing but have so far escaped civil war, such as Zambia and Malawi, are also racing against time. If they do not find ways to accelerate their economic growth and development, they will likely stumble into conflict. Recent casualties include Ivory Coast and Nepal. Nations in these conditions should get the message that change is urgent. Often the remedy should go beyond the standard package of market access, debt relief, and aid programs from the developed countries to include credible policy reform and honest governance within vulnerable countries.

Due to their heavy dependence on natural resource revenues, the governments in many at-risk nations face acute problems of corruption and exposure to international price shocks. But natural resources need not be a curse. Twenty-five years ago, Botswana and Sierra Leone were similarly poor countries, both sitting on vast diamond deposits. Over the ensuing quarter century, Botswana harnessed this opportunity, becoming the fastest-growing economy in the world. Sierra Leone used the same resources to impoverish itself, experiencing the most rapid sustained decline of any country; it now ranks at the bottom of the Human Development Index put out by the United Nations Development Programme. These contrasting examples show that good policy and governance are especially vital where natural resources are discovered.

So far, the record has been dismal: There are many more Sierra Leones than Botswanas. But some encouraging signs are emerging. The "Fowler Report" to the U.N. Security Council in 2000 detailed how UNITA evaded U.N. sanctions against arms smuggling and diamond-based financing; as a result, scrutiny of the international diamond trade increased. Such attention may well have contributed to the demise of UNITA in Angola and the RUF in Sierra Leone, two highly durable, diamond-dependent rebel organizations. Moreover, diamonds are now being tracked through the new Kimberley Process certification scheme, making it harder for rebel groups to obtain financing from these goods. Transparency is the first step toward effective national scrutiny.

And if such strategies work with diamonds, why not replicate the process for timber? The Group of Eight industrialized nations flagged the problems posed by natural resources in its 2002 meeting and may take up the issue again in June 2003 at the Evian summit. Specific political leaders have also begun taking action. British Prime Minister Tony Blair has launched an initiative for greater transparency in the reporting of resource revenues by multinational companies in extractive industries; by nationally owned oil companies such as Angola's Sonangol, which are often a state within a state; and by recipient governments. And in a complementary effort, French President Jacques Chirac has recently called for better mechanisms to cushion low-income countries when the prices of their commodity exports plummet. A new compact could emerge: Rich nations take action to cut rebel financial and cushion adverse shocks, while low-income nations adopt better governance of their revenues from natural resources.

ESCAPING THE CONFLICT TRAP

From 1960 to 1999, international interventions—whether economic or miliary—intended to shorten civil wars were disappointing. Some strategies may have succeeded in individual cases (as in the recent destruction of the Taliban regime in Afghanistan), but no type of intervention has worked regularly.

More effective interventions could target the systems that finance and equip rebel organizations, beyond solely focusing on the trade in diamonds and other commodities. Many rebel movements also receive illicit support from neighboring governments. Such support can be exposed and penalized, and the penalties should outweigh the benefits of rebel alliances. Moreover, governments can discourage huge ransom payments by corporations—such as the $20 million reputedly paid in 1984 by the German engineering company Mannesmann-Anlagenbau AG to the Colombian rebel group ELN, or National Liberation Army, for the release of three of the company's staff. Should such payments be tax deductible and so, in effect, subsidized? Governments could ban the insurance arrangements that facilitate and inevitably inflate such payments. National authorities have started to improve scrutiny of national and international banking systems; recent evidence that al Qaeda is shifting its assets into diamonds suggests that this strategy is becoming effective. National and multilateral policymakers should also look again at drug policy. Wouldn't it be easier and more effective to curb the demand for criminally supplied drugs? And certainly, the flow of arms can be curtailed, especially if efforts are made earlier to catch the big operators, like suspected gunrunner Victor Bout, who is though to have supplied arms to rebel groups in several African nations, including Angola.

The best way to break out of the conflict trap is to ensure that countries that have just ended one conflict do not quickly become enmeshed in another. In some nations, the risks of renewed conflicts are so high that an external military peacekeeping force is normally necessary. The operative word is "external" because high military spending by a post-conflict government actually increases the risk of another war. That external military presence must be credible. In Sierra Leone, the RUF took hostage a large U.N. force that it sensed would not fight, yet when confronted by a smaller British force, the rebel group collapsed.

Unfortunately, peacekeeping missions normally do not last long enough to allow economic recovery to take hold and help keep the peace. The peak time for economic recovery is usually during the middle of the first post-conflict decade. That is also when aid is most effective in promoting economic growth. Unfortunately, international aid is frequently mistimed. It pours in during the first year of peace, when the country's institutions are too weak for the money to be used effectively, then tapers out just when it would be most useful.

Governments in countries recovering from civil war also must give greater priority to economic reform: The post-conflict period is a good time to reform because vested interests are loosened up. For example, after the end of civil conflict in Uganda in 1986, the country's economic policies moved from

---

Want to Know More?

The arguments in this article draw on ***Breaking the Conflict Trap: Civil War and Development Policy*** (Washington: World Bank and Oxford University Press, 2003) by Paul Collier. See also the studies and data from **The Economics of Civil War, Crime and Violence** project, available on the World Banks' Web site. Additional research from the project is available in a special issue of the ***Journal of Conflict Resolution*** (Vol. 46, No. 1, February 2002), edited by Collier and Nicholas Sambanis. See also the International Peace Academy's (IPA) program on **Economic Agendas in Civil War,** available on IPA's Web site.

On the links between natural resources and conflict, see Michael T. Klare's ***Natural Resource Wars: The New Landscape of Global Conflict*** (New York: Henry Holt, 2001), Michael Renner's ***The Anatomy of Resource Wars*** (Washington: Worldwatch Institute, 2002), and the Web site of the nongovernmental organization **Global Witness.** Yahya Sadowski contends that ethnic conflict is not as widespread or as "ethnic" as it seems in **"Think Again: Ethnic Conflict"** (FOREIGN POLICY, Summer 1998). For information on the British initiative on transparency in extractive industries, see the Web site of the United Kingdom's **Department of International Development.** And for views on how rebel groups and local movements promote their causes around the globe, see Clifford Bob's **"Merchants of Morality"** (FOREIGN POLICY, March/April 2002).

For political and historical perspectives on the underlying causes of civil wars, see Monty G. Marshall and Ted Robert Gurr's report **"Peace and Conflict 2003: A Global Survey of Armed Conflicts, Self-Determination Movements, and Democracy"** (College Park: Center for International Development and Conflict Management, 2003). See also Chester A. Crocker, Fen Osler Hampson, and Pamela Aall's, eds., ***Turbulent Peace: The Challenges of Managing International Conflict*** (Washington: United States Institute of Peace Press, 2001). For economic views on the sources of conflict, see Herschel Grossman's **"A General Equilibrium Model of Insurrections"** (*American Economic Review,* Vol. 81, No. 4, September 1991), where the author does not distinguish rebels and revolutionaries from "bandits or pirates." And Jack Hirshleifer postulates the Machiavellian theorem that no opportunity for profitable violence will go unexploited in ***The Dark Side of the Force: Economic Foundations of Conflict Theory*** (Cambridge: Cambridge University Press, 2001).

For links to relevant Web sites, access to the *FP* Archive, and a comprehensive index of related FOREIGN POLICY articles, go to www.foreignpolicy.com.

---

among the worst in Africa to among the best in the following decade.

Finally, diasporas in rich countries pose a particular danger in post-conflict situations. They tend to be more extreme than the populations they leave behind, and they finance extremist and violent organizations. For example, the Tamil and Irish diasporas in North America have both been gullible financiers of murder in the past. Diaspora organizations can play an important role in economic recovery; their networks of skills and businesses are potentially valuable. Afghanistan is now trying constructively to deploy its diaspora, for example. But governments of rich nations should help keep the behavior of diaspora organizations in their borders within legitimate bounds.

LOCAL WARS, GLOBAL CASUALTIES

Civil war is not just disastrous for the countries directly affected; it hurts the surrounding regions and often poses risks for even remote, seemingly unaffected nations. Within the country at war, combat-related deaths represent just a small if gruesome part of the costs: War-related economic ruin also intensifies poverty and disease. Throughout the region, economic growth declines and investment flows dry up. Disease spreads across borders through the flow of refuges. And higher military spending induced by real or potential civil war can fuel pointless regional arms races.

Finally, civil war creates territories beyond the control of recognized governments. These no-go areas can be damaging to the international community. Around 95 percent of the global production of hard drugs is located in civil war countries. Witness how sources of supply shift in response to the changing pattern of conflict: As the Shining Path guerrillas were defeated in Peru in the early 1990s, drug production shifted to territory held by the FARC, or the Revolutionary Armed Forces of Colombia. These lawless areas also provide safe havens and training for international terrorists.

Over the last several decades, national, regional, and global organizations seeking to end or prevent civil wars have often focused on the wrong challenges, or on the right challenges but at the wrong time. Certainly, no single, magic policy will fix the problem; a range of initiatives is urgently required across a broad front. But if governments and multilateral organizations can help curb rebel financing and armament, accelerate the economic development of the countries most at risk, and provide an effective military presence in post-conflict settings, the global incidence of civil war will decline dramatically. These are viable objectives, and they are likely much cheaper than the long-term consequences of continued conflict and neglect.

Paul Collier is director of the World Bank's development research group.

Engaging Failing States

Chester A. Crocker

HITTING THE RIGHT TARGETS

TWO YEARS have passed since the September 11, 2001, terrorist strikes aroused the United States from its post–Cold War strategic slumber. The attacks spurred Washington to action and offered an opportunity for fresh thinking in foreign policy. To meet the challenge posed by large-scale terrorism of global reach, the Bush administration has mobilized the country, assembled substantial armed coalitions, overturned two hostile regimes, weakened the leading terrorist network, and adopted a posture of forward defense against future attacks. It has also refocused relations with Russia, China, and Europe to deal with terrorism and the threat of weapons of mass destruction (WMD) in the hands of rogue regimes.

Despite these important achievements, there is something wrong with the big picture. The administration may be hitting its immediate targets, but it is only paying lip service to the broader objective of achieving a safer and better world order. Forcing U.S. global policies into the simplifying framework of a "war on terrorism" creates the illusion that there is one enemy. In reality, no global adversary exists analogous to the Soviet Union during the Cold War. Terrorism is a tool, not an actor, and conflating the menace of terrorism with the threat posed by WMD in the hands of evil regimes further distorts the strategic picture. By concentrating on worst-case scenarios of immediate vulnerability, moreover, the Bush administration overlooks the failed-state crucible in which many threats to U.S. interests are forged and risks alienating the partners and undercutting the credibility required to address them.

Now that the United States has carried out several bold military campaigns to unseat odious rulers, it must face the reality that these are only the first steps in building global security. Acknowledging this truth openly is the only way to mobilize U.S.

and international attention, resources, and staying power. It is time, therefore, for a fresh articulation of Washington's purposes, centered on sustaining regional security, leading coalitions and institutions to help failing and threatened states, and winning the struggle after wars end and regimes change.

In some ways, the situation is ironic. There has never been a better moment to lead a determined United States toward sustained engagement in the international system. In this sense, the attacks of September 11 were both a wake-up call and a golden opportunity to explain to the U.S. public why serious involvement with the outside world is necessary. And the Bush administration's National Security Strategy, released in September 2002, quietly but explicitly identified the importance of dealing with the problem of failed and failing states. American policymakers have been underestimating this challenge for years. State failure directly affects a broad range of U.S. interests, including the promotion of human rights, good governance, the rule of law, religious tolerance, environmental preservation, and opportunities for U.S. investors and exporters. It contributes to regional insecurity, weapons proliferation, narcotics trafficking, and terrorism. Yet since the strategy was sent to Congress a year ago, the administration has made helping failing and failed states a secondary priority.

The reasoning behind this lapse, presumably, is that rogues that seek to acquire WMD constitute a more pressing threat. But putting the problem of state failure on a back burner means that turbulence continues to spread throughout the Middle East, Africa, the Andean nations, South and Central Asia, and parts of Southeast Asia. Concentrating on rogues first and foremost could create the false sense that deeper problems are being resolved. Regime change may be more satisfying than long-term statecraft; it is certainly more telegenic. But unless the United States and its principal partners engage proactively to prevent and contain state failure, rogue regimes may seize power in additional failed or failing states, raising the specter of fresh adversaries that seek WMD and harbor terrorists. Moreover, the United States must learn to rebuild states after overturning their regimes, or the whole enterprise will backfire.

SLIP SLIDING AWAY

STATE FAILURE is a gradual process. Self-interested rulers might progressively corrupt the central organs of government (as in Burma or in Nigeria during Sani Abacha's regime). Corrupt elites might ally themselves with criminal networks to divide the spoils (as in Liberia and parts of the former Yugoslavia). State authority might be undermined and replaced in particular regions, paving the way for illegal trading operations (as in parts of Georgia and Colombia). During transitions away from authoritarianism, state security services might lose their monopoly on the instruments of violence, leading to a downward spiral of lawlessness (as in several Central American states since the early 1990s). The complete collapse of state power in large sections of a country (as in Lebanon during the 1970s and 1980s and Congo for much of the past five years) is only the most extreme version of the phenomenon.

States with shallow domestic legitimacy tend to fail when they lose foreign support, as often happened in the former colonial domains of European empires. Failure is accelerated when the major actors in the international system abandon local regimes no longer deemed acceptable or convenient partners. Afghanistan exemplifies how an already war-torn polity failed after the strategic disengagement of Moscow and Washington in the early 1990s, ending up as a haven for terrorists. States may also fail due to regional contagion exported by rogues and warlords, as has occurred in central and west Africa.

State failure, inextricably linked with internal strife and humanitarian crisis, can spread from localized unrest to national collapse and then regional destabilization. And unattractive entities—some hostile to U.S. security interests, others hostile to Washington's humanitarian and political goals—may rise to fill the political vacuum. Invariably, state failure is accompanied by the victory of guns over normal politics, the rise of corrupt autocrats who thrive on conflict and deny freedom to their people.

It was fashionable for a period in the 1990s to speak of "democratic transitions," as though the final destination were a given. In reality, some countries will make it, some will remain stuck in failed and weak governance, some will become wards of the international system, some will descend into chaotic warlord struggles, and others will revert to authoritarianism. Which outcome emerges in a particular case will depend on many factors—not least on the action (or inaction) of the world's leading powers.

During the 1990s, when faith in globalization was at its peak, it was hoped that ten "big emerging markets" would change the face of global economics and politics and should therefore become the focal points of U.S. involvement. But however pivotal such states may be, the United States cannot confine itself to cherry-picking its way through the "best" of the transitional world while averting its gaze from the world's strategic slums. Much of the contemporary international state system is crumbling beneath the burdens of warfare, stagnant or declining per capita growth, pandemic disease, rampant official corruption, and autocracy. Turkey may be a beacon of hope for the Muslim world, but it has few parallels. For every South Africa, there are dozens of struggling or failing entities such as Zambia and the Central African Republic. India may be heading in an exciting direction, but much of the rest of South and Central Asia is a mess.

Washington needs to look closely at the relationship between conflict, regime change, and state failure. According to a biennial report published by the University of Maryland's Center for International Development and Conflict Management, armed conflict has decreased by 50 percent since its peak in the mid-1980s and is now at the lowest level since the early 1960s. Still, at the end of 2002, there remained 25 countries engaged in ongoing or sporadic violence, the great majority due to civil strife over the control of government, people, or resources. The report identifies 33 societies emerging from recent wars and nearly 50 regimes located somewhere between autocracy and democracy.

New regimes—whether emerging from wars, negotiated transitions and first-ever elections, or foreign- or domestic-led regime change—tend to be fragile. They may require decades to become institutionalized states operating under the rule of law. In the meantime, the societies around them are ripe for exploitation by ambitious and greedy factions. Law and order, justice, accountability, public services, job creation, and investor confidence are all in short supply.

But the challenge is not only to address state failure in the handful of states where regimes get overthrown. It is also to stop and contain the process of failure before it produces worst-case scenarios. In much of the transitional world—those at-risk societies concentrated in Africa, the Middle East, and southwestern Asia—there is a footrace under way between legitimate governmental institutions and legal business enterprises, on the one hand, and criminal networks, often linked to warlords or political factions associated with security agencies, on the other. Frequently, the informal, undocumented economy is caught between these forces, struggling alongside embryonic civil society groups to survive and watching carefully to see which way the winds blow. When state failure sets in, the balance of power shifts ominously against ordinary civilians and in favor of armed entities operating outside the law (or with tacit official approval).

It might appear that globalization would favor those societal actors most closely linked to international networks of commerce, banking, communications, and diplomacy. But that is only true if the legitimate networks are at least as efficient and well organized as those linking corrupt elites, warlords, and mafiosi with the external facilitators who grease the wheels of criminal business enterprise. Those who lead the cells and networks that hollow out failing states focus with a laserlike intensity on exploiting opportunity and creating facts on the ground. Whether loosely arrayed in symbiotic relations or more closely coordinated by a central brain, they find space to operate in the vacuums left by a declining or transitional state—and they eat what they kill.

GETTING THE DIAGNOSIS RIGHT

IN THIS VAST ZONE of transition and turbulence, the greatest problem is not the absence of nations; it is the absence of states with the legitimacy and authority to manage their affairs. Most of these entities are recent creations, the result of European imperial expansion and subsequent withdrawal. As such, they have always derived a major, if not dominant, share of their legitimacy from the international system rather than from domestic society.

These states are in part a legacy of Washington's success in accelerating the breakup of European empires and championing the aspirations of formerly dependent or colonized peoples. Although Americans seldom acknowledge it, the United States has played a central role in driving the international system to expand from 51 sovereign states in 1945 to almost 200 today. How should this legacy be tackled, and what would a serious state-building strategy involve?

Much has been made of the Bush administration's initial disdain for "nation building," a term with unhelpful overtones from the 1960s. Nation building evokes efforts at economic development, political "modernization," and democratization. Although these are worthy and ultimately essential activities, they are not immediately responsive to the challenge posed by state failure.

Of course the United States should play a role in tackling global poverty. State failures (and associated conflicts) overwhelmingly occur in wretchedly poor countries, not in wealthy industrial or post-industrial states. But, given the massive disparity in living standards within and among developing nations, economics alone does little to clarify policy implications: the answer to state failure cannot be reduced to growth and development.

Focusing on political modernization, meanwhile, has been touted as a way to approach the Muslim world. After all, the 56 members of the Organization of the Islamic Conference (plus other countries with large Muslim populations) include a disproportionate number of the world's conflicted and troubled states. A civilization-based strategy, however, would obscure the issue. Islamic societies vary greatly, and many non-Islamic states are also at risk of failure.

Some identify autocracy as the core issue and suggest stepping up the drive for successful democratization as a bulwark against state failure. Still, there are many unknowns, such as how to sequence democratization's key elements, whether it produces stability at first, what its overall relationship to stability is, whether it can be implanted in all soils, and what outsiders can do to help it flourish. Strong democracies lock in political health, but evidence suggests that weak democracies—as well as reforming autocracies—are highly prone to state failure.

Others argue that failed states stem from arbitrarily defined national borders that bear little relation to regional history and ethnic divisions. The implication of such reasoning is that redesigning dysfunctional states should not be discouraged. This would mean allowing nonviable units to be subsumed into larger ones or to break apart through secession or partition.

But redrawing boundaries would inevitably open a Pandora's box. To legitimize secessionism is to invite claimants large and small from throughout the unsettled zones to put their cases forward, consider military options, and seek external backing. If the intended result was to create ethnically homogeneous units, the process could produce thousands of nonviable statelets. And a new standard of greater tolerance for ethnically based territorial adjustments would be opposed by existing governments, which would rightly accuse the major powers of playing with matches.

Moreover, research by World Bank economist Paul Collier suggests that ethnic fragmentation is not itself the problem; conflict and state failure result from a particular type of ethnic imbalance between a dominant majority and a large minority. Contested natural resources and separatist movements supported by well-heeled expatriate communities likewise contribute to state failure.

In specific cases where regional or ethnic factions of a failing state appear interested in working out a full-blown separation, or where partition appears to be the only answer to irreconcilable differences, outside actors may have a crucial role to play in orchestrating a legitimate, negotiated, and supervised separation. The emphasis in such marginal cases should be placed on how the process of change is brought about and who endorses the result. The international system should respect consensual separation, as it did in the former Czechoslovakia and with the 1993 Eritrean split from Ethiopia (even though this move did not solve the problem, and the sides were soon engulfed in a major war).

What about the option of standing back and letting stronger players take the law into their own hands by creating spheres of influence and protectorates or gobbling up weaker entities outright? There are some precedents, such as Morocco's takeover of the former Spanish Sahara in 1979, Indonesia's grab of East Timor in 1976, India's seizure of Goa in 1961, Russia's muscle-flexing in its "near abroad" during the 1990s, and today's regular interventions in Congo by various of its neighbors.

Perhaps the best thing to be said about such a policy is that it avoids involvement and postpones hard decisions. But the costs can be high when conflict returns in a different guise. And the price in terms of regional security and humanitarian losses can be horrendous: millions of people, for example, have perished in the west and central African wars of the past nine years. Equally important, intervention by local powers has not created stronger states in Russia's near abroad, nor in sub-Saharan Africa; rather, it has expanded the zone of state failures. It has become increasingly difficult to accept such passivity under the guise of realism.

Although the Bush administration appears ambitious about coercive regime change, it remains properly cautious on borders, as reflected in its adamant insistence on Iraq's territorial integrity. Comparable caution can be seen in the prudent stance of the UN Security Council's permanent members concerning the nettlesome Western Sahara problem: every conceivable effort has been made to find a formula that preserves some form of Moroccan sovereignty and does not produce yet another weak African statelet. Such prudence is likewise evident in the

Security Council's handling of the Serbia-Kosovo entanglement, in the joint African- and Western-led diplomacy on the Sudanese civil war, in the Norwegian-led (and U.S.-backed) mediation over Tamil separatist aspirations in Sri Lanka, in U.S.-endorsed talks being organized between the Philippine authorities and the Moro Islamic Liberation Front in Mindanao, and in the mediation over autonomy and power-sharing in Indonesian Aceh, led by a nongovernmental organization.

In each of these cases, the focus is on how to strengthen the state politically and preserve its territorial integrity through measures that decentralize power, reform abusive policies, foster responsive politics, and check central-government domination. Fortunately, these are not insurmountable challenges. A wide range of organizations and governments already work to help failing states undertake such measures as power-sharing and wealth-sharing among units and regions, constitutional and electoral engineering to give voice to cultural and ethnic minorities, and community-based projects to foster intercommunal healing and religious reconciliation.

But weak states also need greater administrative and governing capabilities if they are to behave as responsible, sovereign actors, including enhanced legal codes and court systems; upgraded local and regional administrative apparatuses; responsive and well-trained police forces; stronger bank oversight and public financial management; and closer ties between isolated financial, security, and intelligence personnel at home and abroad. Solemn exhortations for states to behave as responsible, sovereign actors are not enough: the United States needs to help them acquire sovereign capacity.

WHAT IS TO BE DONE?

THE UNITED STATES and other leading powers need to plan and coordinate their strategies for dealing with failed states more coherently, fund key programs more generously, and speak more openly and directly about how to strengthen states and why it matters to do so. The most urgent task is to create the political and organizational capacity to swing into action effectively when existing state structures are failing or about to collapse. The concept of military readiness is well understood, but readiness for what happens after the fighting stops is just as important.

Washington does not have the capacity for political follow-through across a broad spectrum of postconflict or post-intervention requirements. As Afghanistan and Iraq illustrate, the U.S. government lacks the inter-agency mechanisms, institutional memory, doctrine, and committed personnel and budget resources necessary for rebuilding failed states and collapsed regimes. Senior executive branch officials have resisted attempts to bring coherence to postconflict reconstruction and state-building efforts. A recent bipartisan commission report sought to address this inexcusable situation and called on the president to designate a senior interagency leader for reconstruction tasks and to create dedicated staffs addressing the issue within the State Department, the U.S. Agency for International Development, and NATO. It also urged creation of a government-wide center for coordination and training in the major dimensions of postconflict policy (security, justice, governance, and economic recovery). These recommendations form the basis for the Winning the Peace Act of 2003 recently introduced in the Senate.

Washington's resistance to early action in failing states is rooted, quite naturally, in the fear of getting U.S. armed forces stuck with the varied tasks of state building. But rather than an excuse for lack of preparation and inaction, the need to limit the burden on the U.S. military should be seen as a reason to make postconflict planning both civilian led and multilateral. The group of eight highly industrialized countries (G-8) and other allies can provide the manpower to help get the job done, and UN institutions can provide expertise and legitimacy. Failed states have become sufficiently common that the leading nations must find a way to authorize and conduct de facto trusteeships.

Decisions on where to invest scarce time, energy, and resources should be based on such factors as the need to avert terrorist buildups and takeovers by WMD-inclined rogues, a country's inherent regional importance and weight, the possibility of regional side effects and contagion, and the potential humanitarian and political price of outright state collapse.

Once target states are selected, the major powers and institutions should focus their resources in four areas: defusing civil conflict, building state institutions, protecting the state from hostile external influences, and managing regional spread. At the June 2003 Evian summit, G-8 officials issued a joint declaration that pertains to the first three requirements. Titled "Fighting Corruption and Improving Transparency," the document proposes a strengthened global effort against bribery, corruption, and financial crime and real transparency in reporting on revenues derived from extractive industries such as energy and mining. Such a measure is directly pertinent to governance in troubled societies such as Sudan, Indonesia, Kazakhstan, Nigeria, and Colombia. What is less clear is whether the G-8 governments will place a high priority on implementing its important provisions. Another initiative, the G-8 Africa Action Plan, outlines a wide range of capacity-building initiatives to support the adherence of African states to the good governance principles of the New Partnership for Africa's Development.

Major powers are beginning to take steps to ward off state failures in Africa and other turbulent regions. Because of the breadth of the problem, a sharper focus on specific countries and regions will be more meaningful than a generalized grab bag of programs. Leading governments need to speak loudly and clearly about just what these joint efforts will involve, connecting the dots between enlightened burden-sharing, the risks of state failure in troubled regions, and their own national interests and budgets. Above all, they need to hammer out joint strategies. When the United States and its allies work together to ward off state failure, it is possible to limit damage and achieve real progress—as in Sierra Leone, Sudan, Macedonia, and Sri Lanka; when they don't, state failure is aggravated—as in Serbia, Congo, Colombia, and Afghanistan before September 11.

Over the last two years, law enforcement and intelligence cooperation on terrorist financing have improved significantly. Interestingly, most of the agencies and regulatory frameworks brought to bear on terrorist financial networks had their origins in earlier campaigns against money laundering and financial crime. One example is the Organization for Economic Cooperation and Development's (OECD) Financial Action Task Force—grouping 29 member countries, two international organizations, and two observer nations—which operates by pooling information, naming and shaming (by means of its published list of noncooperative countries and territories), and imposing financial sanctions.

In the end, the war on terrorism requires many of the same tools and techniques needed to battle the forces causing and thriving off of state failure. If the major powers and international bodies mobilized parallel efforts, perhaps hearkening back to expanded counterterrorist cooperation, it would not take long to disrupt some of the deadly links among narcotics trafficking, arms smuggling, and the illicit trade in precious stones and metals, timber, people, and exotic species. The techniques of forensic accounting could be used to track down ill-gotten gains, helping respectable governments recover looted assets, deterring looters and those who collude with them, and bolstering legal action against transgressors. But to make such achievements possible, the necessary personnel, budgets, and political capital must be dedicated to the effort.

WALKING THE WALK, TALKING THE TALK

A MORE SUBSTANTIAL engagement by the United States and its primary partners is needed to address what is arguably the leading menace on the globe. This effort would begin by rebalancing the focus of public rhetoric in foreign policy toward broader and deeper engagement in state building and toward identifying the magnitude of the challenge posed by state failure. Properly packaged and articulated, a number of Bush administration initiatives—the Millennium Challenge Account, new assistance initiatives for Africa, expanded funding to address the HIV/AIDS pandemic, multilateral trade expansion, upgraded law enforcement and intelligence sharing on terrorist finances—could be part of a comprehensive failed-state strategy.

But it will be difficult to make full use of such instruments and to develop additional tools unless senior administration officials begin to explain what the United States is doing about state failure and why it matters. A broadly defined policy couched in terms of the full range of American interests and values has some chance of becoming sustainable. A narrowly defined foreign policy couched mainly in terms of military confrontation, rogues, and terrorists will not garner the breadth of domestic and international support required for sustainability.

The administration will also need to broaden its international political base and cease treating indispensable allies as little more than a nuisance. American leaders have a wide range of potential institutions and tools at their disposal for a long-term strategy to strengthen the capacities of sovereign states. Rather than serving as rhetorical dart boards for the amusement of domestic audiences, leading multilateral institutions such as the United Nations, NATO, the OECD, the G-8, and the Organization for Security and Cooperation in Europe should be appreciated for what they might, if properly utilized, contribute to the effort.

Often the appropriate role for the globe's superpower will be to push negotiated settlements of intractable regional conflicts, which are often fed by state failure and spawn fresh failures as they burn. But the exact negotiating structure must be tailored to each case. The Quartet (consisting of the United States, the European Union, Russia, and the UN) appears to offer some value in the context of Israeli-Palestinian peace talks. A U.S.-led troika (with the United Kingdom and Norway), plus an African group led by the Kenyans, is dealing with the conflict in Sudan. And in the India-Pakistan case, the United States plays a unique role, supported by like-minded powers. U.S. engagement should be applied to citizens as well as governments. When there are ingredients for internally led regime change, as in Serbia, outside powers should facilitate the transition.

The Bush administration has succeeded recently in achieving a massive increase in funding for the Defense Department, reversing years of deferred investment. The time has arrived to boost the civilian foreign affairs budget as well. Since entering office, Secretary of State Colin Powell, National Security Adviser Condoleezza Rice, and the colleagues have begun to reverse the long-term budget decline in this area. Yet the international affairs budget remains around one percent of the total federal budget. Last December, eight of Rice's predecessors from both parties urged her to support a "substantial increase" in the 2004 international affairs budget so that it might ultimately be restored to its high-water mark of the Reagan years, when it was 30 percent higher than today. But whether the administration will expend the necessary effort to secure such an increase remains to be seen.

Any serious strategy to combat state failure will require more resources and attention than the problem is currently receiving. Polls have repeatedly shown that the public believes the United States spends vastly more on foreign assistance than it actually does and would support levels of aid and effort far higher than those actually in place. But only the president himself can demolish the misperceptions and make the case for doing what is necessary to engage the world's failing states. Unless he does, the administration's efforts to tackle terrorism and rogues and build a safer, better world will falter.

CHESTER A. CROCKER is the James R. Schlesinger Professor of Strategic Studies at Georgetown University's School of Foreign Service.

Collapse of the road map (1)

Progressing to a bloody dead end

An awful inevitability led to the latest outrage. The Palestinians hope for an outside intervention that never comes. The Israelis (see next page) make their own plans

EAST JERUSALEM

FOUR months after it was inaugurated, the American-backed road map has collapsed in blood. On September 6th, the Palestinian prime minister, Mahmoud Abbas, resigned. Three hours later Israel tried to kill Sheikh Ahmed Yassin, the founder and spiritual leader of Hamas, and other leaders of the militant Islamist movement. Three days later Hamas exacted revenge; its suicide bombers killed 15 Israelis, some of them soldiers, in two separate attacks in Tel Aviv and Jerusalem. Israel's first response was to attack the home of Mahmoud Zahar, a Hamas political leader and spokesman, killing his son. The horrors will, without doubt, continue.

Mr Abbas had been America's man. His appointment as prime minister in April was seen as the key to new American "engagement" in the conflict, based on diluting the powers of Yasser Arafat, the Palestinians' elected president, reforming the Palestinian Authority (PA), and implementing the road map. Mr Abbas saw the unilateral Palestinian ceasefire, in which all the militant factions committed themselves not to attack Israeli civilians, as the key to his efforts to steer his people away from violence and back to negotiations.

The ceasefire lasted seven weeks. There was not an absolute end to Palestinian violence, but there was a genuine decline. It finished on August 19th when a Hamas suicide

bomber killed 22 Israelis on a bus in Jerusalem. Since then, Israel has killed 15 Hamas men, plus six innocent bystanders, either by assassination in Gaza or by military assault in the West Bank.

Why has the road diverged so bloodily from the map? In an address to the Palestinian parliament, Mr Abbas put some of the blame on himself and his bad relations with Mr Arafat. Never good, these became venomous. Mr Arafat saw any attempt by the prime minister to gain a degree of independence in decision-making as part of an Israeli-American conspiracy to unseat him. He refused to transfer control of the PA's various police forces to Mr Abbas, as was called for in the road map. And he quietly incited some of his more militant loyalists in Fatah to denounce Mr Abbas's "American" regime.

But, said Mr Abbas, the fundamental cause of his failure was "Israel's unwillingness to implement its commitments under the road map". This is certainly what lost him Palestinian public opinion. During the truce, Israel withdrew its tanks a few score metres from Bethlehem, opened a few roads in Gaza and removed five roadblocks out of 220 in the West Bank. It did little else to loosen the "closure" that isolates each Palestinian village from its neighbour, leaves one out of two Palestinians without a job, and nearly two out of three in poverty.

Nor did Israel freeze settlement construction, as called for in the road

map. Instead the so-called "security fence" that has already swallowed large chunks of Palestinian farmland in the northern West Bank has been extended to include parts of occupied East Jerusalem. "Abu Mazen [Mr Abbas's usual name] can do nothing for us here," shrugged a Palestinian from the East Jerusalem village of Abu Dis, standing next to a new concrete wall that separates him from his lands, family and business on the "Israeli side" of Jerusalem.

Mr Abbas also blamed an American administration that "did not exert sufficient influence on Israel to implement the road map". As a result, support from the various Palestinian factions, including his own Fatah movement, dribbled away. Mr Abbas had sold them the road map in the hope that an end to Palestinian violence would bring American pressure on Israel to end its current occupation policies.

That pressure never came. Mr Abbas, it appeared, carried no more clout with the White House than had the "irrelevant" Mr Arafat. Last week, the prime minister threatened a parliamentary vote of confidence in a last throw to rescue his government. He was told he would lose.

Will his replacement fare better? Mr Arafat moved swiftly to install Ahmed Qurei, the speaker of parliament who is better known as Abu Ala, as his successor. Like Mr Abbas, Mr Qurei is a moderate and one of the architects of the Oslo accords. He is opposed to the "armed *intifada*" and committed to the road map. Unlike

Mr Abbas, he enjoys the trust of this leader, at least for now. He accepted the post, but said he wanted guarantees that America and Europe would intervene to end Israel's violence against the Palestinians, and to see that the road map was implemented. He insisted Israel lift the siege on Mr Arafat since, without him, "no prime minister will succeed". And he proposed a new ceasefire, but one that would include Israel as well as the Palestinian factions.

He is unlikely to get far. Israel views another "temporary Palestinian ceasefire" as just another ruse to evade the PA's responsibility to take on Hamas and Islamic Jihad, root and branch. Stunned by the departure of Mr Abbas and cool to his successor, the Americans, apparently, feel the same. The European Union is warmer to Mr Qurei, but armed with its recent decision to designate Hamas's political wing (as well as its military wing) a "terrorist organisation", it too is placing the onus on him and the PA to act, and act first. But unless Mr Qurei can deliver some substantial return, Palestinian public opinion, including opinion in Fatah and the security services, is unlikely to tolerate a vigorous move against the Islamists.

Collapse of the road map (2)

Apocalypse, soon

The Israelis think of revenge, perhaps against Yasser Arafat

JERUSALEM

IN ISRAELI government circles the debate after the suicide bombs in Jerusalem and Tel Aviv was not about negotiating afresh with the new Palestinian prime minister, Ahmed Qurei. It was rather about removing Yasser Arafat, re-invading the Gaza strip or even reoccupying all of the Palestinian territories and forcibly dismantling the Palestinian Authority.

Ariel Sharon has resisted such apocalyptic advice before. But the proponents, who include some of his senior ministers, apparently concluded, after the bid to kill Sheikh Ahmed Yassin last weekend, that past inhibitions may no longer be operative. When Mahmoud Abbas's brief tenure as Palestinian prime minister imploded, the Americans signalled that Mr Arafat was to be left alone. But aides to Mr Sharon later claimed that they detected a possible softening of the American fiat.

Cabinet members took their views to the public. The foreign minister, Silvan Shalom, said there would never be diplomatic progress so long as Mr Arafat was in the arena. The defence minister, Shaul Mofaz, agreed it would be better if he were gone. Ehud Olmert, the deputy prime minister, suggested that deporting Mr Arafat would be counter-productive, but that Israel could cut all communications to his Ramallah office, turning "him into a prisoner in solitary confinement".

Most of the ministers blame Mr Arafat personally for the terrorist attacks, and, though no hard evidence has been adduced to substantiate that claim, most of the public accepts it. In an opinion poll published in *Haaretz,* an Israeli newspaper, on September 10th, 18% said that Mr Arafat should be assassinated, 28% that he should be deported and 27% that his isolation should be intensified. The same poll, taken before the latest suicide bombings in Jerusalem and Tel Aviv on September 9th, had nearly 70% approving of the attempted killing of Sheikh Yassin, and 60% urging the army to try again.

Israel's attempt, on September 10th, to kill Mahmoud Zahar, the Hamas spokesman, signalled clearly that Mr Sharon's orders to the army, in the wake of the August 19th bombing in Jerusalem, to hunt down the entire Hamas leadership are still in force, despite the costly reprisals. Mr Sharon, on his cut-short visit to Delhi—the Indians have become big customers for Israeli military technology—said that Mr Abbas's downfall showed how cautious Israel must be in its dealings with the Palestinians. As for Mr Qurei, he would meet with him, Mr Sharon said, only if he carried out all of the Palestinians' commitments under the road map.

An Indian 'War on Terrorism' Against Pakistan?

Kanti Bajpai

Since the events of September 11, 2001, many in India have argued that if the United States can justify its wars in Afghanistan and Iraq in the name of combating terrorism, destroying weapons of mass destruction, and changing regimes, then India is justified in attacking Pakistan. Indian external affairs minister Yashwant Sinha is reported to have said, "India has a much better case to go for preemptive action" against Pakistan than the United States had in Iraq. The massacre of more than twenty Hindu men, women, and children in Nadimarg, Kashmir, in late March 2003, renewed calls for sterner action against Pakistan. And yet, Indian anger notwithstanding, military action against Pakistan would be both ineffective and dangerous.

Ever since the Kargil War of 1999, influential Indian strategists have argued that India should be prepared to fight "limited war under nuclear conditions," that is, military operations of a limited conventional nature. This is the only way, they believe, to respond to Pakistan's strategy of sub-conventional warfare—terrorism, and, as in 1999, incursions across the line of control in Kashmir. Pakistan can only be dissuaded from continuing its sub-conventional warfare, in this view, by the threat of military punishment. This new Indian thinking challenges the Pakistani conviction that its nuclear weapons protect it from Indian retaliation.

There are probably two reasons for the new Indian thinking on limited war. The first is the belief that India, with its bigger nuclear forces, has "escalation dominance" and can up the ante at every level of violence. To the extent that India has the whip hand, Pakistan's threat to use nuclear weapons against a punitive Indian strike would be neutralized. Islamabad would be dissuaded from resorting to nuclear weapons by the fear of massive retaliation.

The second reason is a calculation that the nuclear powers, especially the United States, would not allow either country, but Pakistan above all, to use nuclear weapons. Indian thinking on the subject is not available in cold print but quite likely rests on the following kinds of arguments. First, the nuclear powers are determined to preserve the nuclear taboo and in particular to stop anyone outside the nuclear five (the United States, United Kingdom, France, Russia, and China) from using nuclear weapons. Second, a nuclear war between India and Pakistan could draw in other nuclear-weapon states. Imagine that Pakistan appeals to China for help after using nuclear weapons against an Indian conventional attack. Faced with the prospect of Chinese involvement, India turns to Russia. The United States urges both China and Russia to stay out of the fight but finds that Beijing and Moscow are unable to stand on the sidelines. Surely Washington must do everything it can to avoid such a situation, including stopping Pakistan from starting a nuclear crisis in the first place. Third, American troops on the ground in Pakistan since September 11 would be trapped in a nuclear war between India and Pakistan. Clearly, this would be unacceptable to Washington. Finally, in regard to Pakistan, the United States and other nuclear powers do not want a Muslim state using nuclear weapons at a time when other Muslim countries are in the hunt for weapons of mass destruction.

How PERSUASIVE are these Indian suppositions? Does India have escalation dominance, and will this stop Pakistan from using nuclear weapons against an Indian conventional strike? First of all, it is unclear that India can raise the stakes at every level of violence. The most important consideration here is whether India has superiority at the nuclear level—this is essential if one is to play the game of escalation dominance. Given the secrecy of the two nuclear programs, it is hard to say. Some reports suggest that Pakistan has the nuclear lead in terms of deliverable weapons (even though India has more fissile material). Whatever the truth of these reports, Indian nuclear superiority is not an established fact. Furthermore, even if India has nuclear superiority, the logic of escalation dominance may be invalid—Pakistan may not be deterred from raising the level of violence. Everything depends on how Pakistan perceives Indian military goals once the fighting begins. If it concludes that Indian objectives go beyond mere punishment, it might well feel that it has to use nuclear weapons, that it is better to go down fighting than to surrender or be conquered.

Nor can we count on the nuclear powers' restraining Pakistan. Faced with the possibility of nuclear war in South Asia, they might choose to wash their hands of both India and Pakistan and bend their energies to avoid being dragged into the conflict. Even if this is not the case and they do exert themselves on Pakistan, they may not have much influence in a fast-deteriorating military situation. A country, and a government, that thinks it is on the verge of dismemberment and defeat is not likely to be swayed by threats from the international community. After all, the ultimate threat from the nuclear powers would be to do to Pakistan what India was in the process of doing anyway, namely, punishing it militarily. One could argue that international pressure at such a crucial moment would only deepen Pakistan's sense of desperation and push it further toward the suicidal brink.

Leaving aside the contentious matter of escalation dominance and the role of the nuclear powers, there is the crucial question of whether or not any kind of limited war could be effective in reducing the terrorist threat and Pakistan's support of it. A set of thought experiments will show that none of India's military options is very good. There are, roughly, seven options: fomenting terrorism in Pakistan; hot pursuit of terrorists; special-forces attacks against terrorist camps; air and artillery strikes against terrorist facilities; conquering and holding a slice of Pakistani territory; a naval blockade or bombardment of Pakistani ports; and a serious military push into Pakistani

territory. This list is not exhaustive. The Indian air force, as also the Indian navy, could, in addition, mount strikes against other Pakistani targets—dams, cities, and military targets. However, these are even more dangerous and immoral ventures, and most of my critique of the lesser options will apply even more acutely to them.

Let's look at each of the options. First, some strategists argue that India should do to Pakistan what Pakistan has been doing to it for the past fifteen years: it should encourage internal dissent and violence. This sounds like an attractive option at a purely military-strategic level, but the problem with it is twofold. For one thing, going by past experience, Pakistan is not likely to be stopped from interfering in its neighbors' affairs by its own internal instabilities. Ethno-religious violence in Pakistan has been ferocious since at least the 1980s, and no province and city has been left unscathed. But this did not stop Islamabad from mounting anti-Indian rebellions in both Punjab and Kashmir in the 1980s and 1990s. Nor did it stop Pakistan from prosecuting a civil war in Afghanistan and from ousting the Soviet forces. Is there any reason to believe that Pakistani leaders today are more sensitive to internal violence than they were in the past? Ironically, the recent American intervention in Afghanistan has freed Pakistani resources for other theaters, including Kashmir, even as Islamabad deals with domestic disturbances.

Second, and more important, in contemplating the destabilization of Pakistan, India must consider whether such actions are compatible with its public and long-standing complaint that Pakistan is a terrorist state. Indian complaints will carry little moral weight if India itself supports terrorism. Destabilizing Pakistan is not just a moral issue; it is also an issue of statecraft. In the global war against terrorism, India would be part of the problem rather than part of the solution. It could become the object of international sanctions, whereas it has profited—and could profit further—if those sanctions were directed against Pakistan. In short, India would join Pakistan as a terrorist state. This would be a moral and strategic catastrophe for liberal democratic India.

What about the various military actions to which India could resort? These include hot pursuit, special-forces raids, and air and artillery strikes. All these would be directed primarily at destroying terrorist facilities in the Pakistani side of Kashmir, thereby reducing the incidence of terrorism. These options at least have the advantage of being morally justifiable because they will not target innocent civilians. Their military effectiveness, however, is questionable. Hot pursuit strikes would involve Indian troops' crossing the line of control in chase of terrorists. But these attacks probably would not achieve a great deal. Indian troops would probably kill some terrorists and temporarily destroy training and base facilities. The supply of terrorists is large enough, however, and Pakistan could rebuild and relocate the affected facilities fast enough, so that India would have to repeat the attacks, perhaps deeper into Pakistani-held territory, thus making its forces more vulnerable to counter attacks. Indian units that cross the line would also run into Pakistani fire, and casualties might be heavy. India would then be faced with the choice of calling off the pursuit or escalating the engagement.

Most of these limits apply as well to raids by special forces, which also have their own difficulties. They depend, critically, on surprise, stealth, and speed. Indian special forces could attain all three, but we should be cautious. In the history of warfare, the number of special operations that have gone wrong may well outnumber the ones that have succeeded. The capture and destruction of Indian units caught on the Pakistani side would be a public relations disaster if not a serious military reverse. Moreover, and this is crucial in terms of effectiveness, one or two successes will do little to reduce the tempo of terrorism.

Air and artillery fire against the terrorist camps or even against Pakistani military units is another plausible option. The Indian air force and army could launch a series of strikes against terrorist camps and Pakistani military targets. The success of remote fire of this kind, either from the air or from ground artillery, depends on intelligence and accuracy of targeting. If Indian fire is inaccurate, it risks hitting innocent civilians, which would be both morally corrosive and diplomatically counterproductive. Satellite imagery, air reconnaissance, specialized radar, modern locating systems, and precision-guided munitions are just some of the requisites for a bombardment campaign against terrorist facilities and Pakistani military targets. It is unlikely that Indian forces have the requisite capabilities. In any case, the Pakistanis can camouflage, harden, and move these facilities and targets, thereby rendering the Indian campaign ineffective.

Finally, India could opt for a much more ambitious set of punitive actions against Pakistan: conquering and holding a slice of Pakistani territory, a naval blockade or action, and a serious military push into Pakistani Punjab and Sindh. Each of these, too, is fraught with risks. Grabbing and then holding a swath of territory on the Pakistani side of Kashmir or in Pakistan proper is by no means easy. In Kashmir, difficulties of terrain would make it a tough operation to mount and sustain. In Pakistan proper, the prospects are even more daunting, as the Pakistani forces are much stronger there, with considerable armor to back them. India too is strong, but the question is whether or not a potentially expensive operation would serve much purpose. The idea behind it would be, once again, to punish Pakistan for its support of terrorism and to offer it a trade—return of territory for the end of terrorism. What if Islamabad simply shrugs off the loss of territory even as it battles to get it back and continues the terrorism campaign? As we noted earlier, Pakistan is not afraid to fight on several fronts simultaneously. India would then have to decide whether to extend this kind of

salami-slicing to other sectors or to expand the scale of operations.

One way of expanding the conflict is by a naval blockade or other naval actions along the Pakistani coast. India signaled the possibility of naval operations when it sent a flotilla of ships into the Arabian Sea in the summer of 2002. A naval campaign could involve a blockade of Karachi harbor, crippling Pakistan's trade, destruction of Pakistani military and merchant shipping, and destruction of the Gwadar naval base, which the Pakistanis are currently expanding (with Chinese help).

But a naval blockade raises a number of difficulties. For one thing, it could cause great suffering among Pakistani civilians. And it would antagonize many of Pakistan's trading partners, including the major economic and military powers important to India. A large Indian naval presence over a long period of time would get in the way of the American fleet that is positioned in the area as part of the campaign in Afghanistan. This would lead not only to diplomatic friction with the United States, but would also risk the two navies' physically blundering into one another. Finally, a flotilla would be vulnerable to Pakistani air and naval retaliation, in particular by its submarines and the deadly Exocet and Harpoon missiles with which its ships are equipped. An attack against Gwadar is much more attractive, but how would that work to stop Pakistan's terror campaign?

Instead of a naval campaign, India could do what it does best militarily—or at least what it is most predisposed to do—which is to go to war on land, in the plains of Punjab and Sindh. If anything would truly hurt Pakistan, it is the loss of territory in its heartland. The Indian army combined with the air force could invade Punjab or Sindh or both in the kind of "final conflict" that the Indian prime minister frighteningly referred to in one of his speeches during the crisis of 2002.

THIS IS A MILITARY fantasy at best and a nightmare at worst. An invading Indian force would have to get past densely mined Pakistani forward positions. Next, there would be canals and ditches, and, most likely, more mines. Beyond these passive defenses, the Indian army would run into Pakistan's active defenses—its armor and regular infantry units. Contrary to what most Indians and even more foreigners imagine, India does not have a substantial lead in conventional forces on its western border. Indian forces are deployed along the northern border with China and cannot be reduced much further in that sector. Many Indian troops are deployed for internal security duties, particularly in Kashmir and the Northeast, and it would be stretching them dangerously to draw their numbers down significantly. The tide might turn in India's favor if the air force achieved a rapid and decisive victory in the skies. Few people, however, would predict that this could be achieved. In short, the Indian attack would be a very hard slog, and the war might just turn into a stalemate marked by mutual attrition. And that's the good news. The bad news is that if India succeeded in making a breakthrough, Pakistan might unleash its nuclear weapons at Indian troop formations or Indian cities in order to halt Pakistan's military collapse.

The general point that emerges from these thought experiments is that much may be ventured, at great risk, for very small gains. The costs of war would probably be much greater than the costs of the uneasy, flawed, and violent peace that exists today. Some military ventures will gain India little in terms of stopping terrorism even if they succeed. Others could gain it a lot but have little chance of success. Yet others spell catastrophe.

In short, there are great dangers in applying the doctrines associated with the U.S. war on terrorism—retaliation and preemption—in South Asia. While there is no doubt that Pakistan does sponsor terrorism in India, the simple truth is that Pakistan is not as weak as Afghanistan or Iraq, and India is not as strong as the United States. Going after Pakistan militarily will either be ineffective or extremely dangerous. Whatever the United States can or can't, should or shouldn't, do, India's response to terrorism has to be different, and it is mere casuistry to claim that New Delhi should emulate Washington's methods.

KANTI BAJPAI is professor of international politics at Jawaharlal Nehru University, New Delhi.

From *Dissent*, Summer 2003. © 2003 by Dissent Magazine.

The Terror War's Next Offensive

In the aftermath of the Jakarta bombing and the arrest of terrorist suspect Hambali, Thailand and Indonesia are considering new measures to get tough on terrorism—and grappling with how to do so without upsetting delicate political and social balances at home

By Shawn W. Crispin BANGKOK and

Jeremy Wagstaff JAKARTA

IT'S BEING HERALDED as the biggest hit yet against global terrorism in Southeast Asia. But the legal manoeuvres behind the scenes of the recent arrest of Asia's most-wanted terror suspect in Thailand could have an even more profound impact on the region's future.

On August 12, Thai counter-terrorism agents stormed a nondescript apartment block in the historic Thai city of Ayutthaya and apprehended Riduan Isamuddin, or Hambali, suspected of being the operations chief of Jemaah Islamiah, the radical Islamic group with alleged links to the Al Qaeda terrorist organization.

Intelligence officials say Hambali is the mastermind behind some of Al Qaeda's most lethal attacks and played a role in the planning of the October 2002 attack in Bali that killed more than 200 people and the August 5 blast at the JW Marriott Hotel in Jakarta that killed 12 people (*see The Trail of Hambali*).

On the surface, Hambali's arrest marks a major victory for the U.S.-led war on terrorism in Southeast Asia. Thai Prime Minister Thaksin Shinawatra told reporters that catching Hambali, who he says was plotting to bomb the Asia-Pacific Economic Cooperation (Apec) forum meeting in Bangkok in October, means all foreigners suspected of plotting terror acts in Thailand are now under lock and key.

"We have had to take special precautions because our country could be used as a hiding place [for terrorists]," said Thaksin at a military airport in Bangkok on August 15. "Terrorism is now no longer the duty of only one country... we all need to cooperate."

At the same time, the evolving nature of counter-terrorism cooperation between regional governments and the U.S. is raising concerns that the war on terror could also undermine recent democratic gains in Southeast Asia. In Thailand and Indonesia, governments are grappling with how best to react to U.S. pressure to get tough on terrorism, without upsetting delicate political, social and religious balances at home.

"Every government in the region needs to think about how much they should sign onto the international coalition against terror," says Surin Pitsuwan, a former foreign minister of Thailand and an opposition politician. "Without great care, the way the war is fought could have a big impact on social balances that have taken decades to achieve and could take decades to bring back."

It's proving an increasingly difficult balance to strike. One day before Hambali's arrest, Thaksin pushed through new anti-terrorism laws by executive decree, citing unspecified reasons of national security. The move marked the first time Thailand's criminal codes had been amended by the administrative branch, and controversially bypassed the usual parliamentary process for enacting new laws.

NEW ANTI-TERROR LANDSCAPE

Legal experts say Thaksin's decrees will fundamentally alter Thailand's legal landscape, providing tough new penalties for convicted terrorists and including provisions that will allow for

---

GOOD NEWS RAISES CONCERNS

- Hambali's arrest on August 11 is seen as a major victory for the war on terrorism
- One day earlier, the Thai premier decreed new anti-terrorism laws
- Some fear that counter-terror efforts in Southeast Asia will undo democratic gains
- Despite Hambali's arrest, his alleged network is far from being dismantled

---

detention without trial, similar to the controversial internal security acts on the books in Malaysia and Singapore. The new anti-terrorism laws will also give Thai authorities greater legal leeway to track the financial transactions of suspected terrorists.

Opposition politicians, legal experts, and human-rights advocates worry that the broad definition of terrorism contained in the new laws could in the future be mobilized for political rather than legal purposes, similar to how detractors say the governments in Malaysia and Singapore have used their security laws to crack down on political opponents. The opposition Democrat Party has already filed an appeal to the Constitutional Court challenging the legality of the decrees.

"The new laws violate the Thai constitution," asserts Danai Anantiyo, vice-president of the Law Society of Thailand. "They will make it easier for the government to curb people's freedoms," he adds, drawing parallels to past anti-communism laws that miliary dictators used in the 1960s and 1970s to stamp out political opposition to their rule.

Former Foreign Minister Surin says anxieties are growing in Thailand's predominantly Muslim southern provinces that the new laws will be used to clamp down on Muslim activists working for greater local democracy in the region. "People are nervous about the way things are being handled," he says.

Indonesia is also grappling with how to build a legal framework to deal with the problem of terrorism. Government efforts to push anti-terrorism legislation through parliament have been sporadic in the face of public opposition and parliamentary hostility.

Central to the delay has been public criticism that the government, under the influence of the military, is rolling back liberties hard-won by the reform movement that toppled President Suharto in 1998. It was the military that buttressed 32 years of authoritarian rule that limited political activity, and led to the arbitrary arrest of labour leaders, student activists and dissidents. Many Indonesians fear a creeping return of such tactics under the guise of cracking down on terrorists.

So far, it seems unlikely the Indonesian government will adopt internal security laws like those in Singapore and Malaysia. Coordinating Minister for Political and Security Affairs Sisilo Bambang Yudhoyono said on August 14 that the government would revise several articles in the anti-terrorism law, par-

ticularly those concerning pre-emptive and preventive action, as well as the role of intelligence.

"These parts are considered insufficient so that right now the government and security apparatus, including the police, are in a waiting position until the terrorists launch their attack," he said.

The government will review the articles, consult the public and submit the law to parliament as soon as possible, Bambang said. "it will be done openly. The public will know about it," he said.

Unlike Indonesia, it is still unclear just how deep—if at all—international terror groups have taken root in Thailand. Until recently, Thai officials strongly denied any internationally backed terrorist groups were active on Thai soil, with the apparent aim of maintaining foreign confidence in the Thai economy and tourism industry.

THE THAI DECREES INCLUDE PROVISIONS TO THE CONTROVERSIAL INTERNAL SECURITY ACTS ON THE BOOKS IN MALAYSIA AND SINGAPORE

Indeed, Thai counter-terrorism and U.S. Federal Bureau of Investigation agents were on occasion at loggerheads, as senior Thai officials complained that their U.S. counterparts were pushing them to arrest and interrogate terror suspects in ways that violated civil liberties protected under Thai law.

But as Washington calls for tougher action in ferreting out terror suspects, the political winds are shifting in Thailand. For example, on the same day Thaksin was scheduled to meet U.S. President George W. Bush in Washington, Thai counter-terrorism agents arrested three Thai nationals on charges that they were plotting to bomb foreign embassies during the Apec meeting.

Notably, Thaksin said there was "no evidence" that any Thai nationals were involved in Hambali's alleged plot to bomb the Apec meeting.

Meanwhile, senior security officials say a number of other terror suspects have been arrested in recent weeks, including some of Hambali's close aides.

CIVIL LIBERTY CONCERNS

Legal experts and human-rights advocates in Thailand say U.S.-inspired counter-terrorism techniques used in the arrest of Hambali—including military-style abduction, detention without trial and unrestricted wire-tapping—violate Thai constitutional provisions on civil liberties.

Sunai Phasuk, spokesman for Forum Asia, a regional human-rights group, fears that Thaksin's recent acquiescence to Wash-

THE TRAIL OF HAMBALI

The arrest of Riduan Isamuddin in Thailand earlier this month marks another successful manhunt for alleged members of the Al Qaeda leadership. The man known as Hambali was a rarity: According to intelligence reports, he was the only non-Arab member of the Al Qaeda elite. Indonesian by birth, Hambali fled the country along with many other radical Muslims in the midst of a 1985 crackdown. He moved to Malaysia and then joined a flood of young men fighting the Soviet forces in Afghanistan. By the late 1990s he was a senior figure in Al Qaeda, commuting between hotels in Bangkok and Kuala Lumpur, an apartment in Karachi and the southern Afghan city of Kandahar, where he drove a white sedan.

> ### The suspect known as Hambali was a rarity: Born in Indonesia, he was said to be the only non-Arab member of the Al Qaeda elite

It was there, in late 2001, that he lost some of his closest comrades in an American air strike, among them Muhammad Atif or Abu Hafs al-Masri, a former Egyptian policeman and Osama bin Laden lieutenant who was seen as the mastermind of the 1998 bombing of United States embassies in East Africa and the September 11, 2001, attacks. Using a false passport, Hambali fled Afghanistan. Tired, he mistakenly filled in his real name and information on arrival papers in a Southeast Asian country, and narrowly avoided arrest after questioning by immigration officials, according to an FBI report.

It was in Southeast Asia that Hambali was most important to Al Qaeda. According to FBI documents and other sources, he coordinated several attacks in the Philippines and Indonesia, including an aborted plan to bomb targets in Singapore in late 2001. It was Hambali's decision to abandon the Singapore operation after the cell preparing the attack told him it would take 18 months to smuggle the explosives in from the Philippines. Instead, he switched to an earlier plan to bomb the U.S. and Israeli embassies in Manila.

Around that time, in early 2002, things started to go wrong. The Singapore cell was arrested, but Hambali, by then hiding in a Bangkok suburb, allegedly continued to coordinate regional activities of Al Qaeda operatives and cells of the Southeast Asian network, Jemaah Islamiah. Using Internet cafes and a separate, innocuous-sounding e-mail address for each contact—screwdriverhk@yahoo.com was one example—Hambali switched the network's attention away from hard targets such as embassies and military installations to softer ones.

In January 2002 he told a meeting of Jemaah Islamiah operatives in Thailand to target bars, cafés and nightclubs frequented by Westerners in Southeast Asia. Already stashed away in Indonesia, Hambali told a terrorist suspect interrogated by the FBI, was one tonne of explosives.

Ten months later three bombs exploded on the tourist resort island of Bali, killing more than 200 people, most of them foreigners. Indonesian police say he sent $45,000 last month from Thailand to a Malaysian, money which they say may have been used in the attack on the JW Marriott Hotel in Jakarta.

Hambali's arrest leaves a hole in Al Qaeda and Jemaah Islamiah, but it's not likely to have any great impact on their operational capacity, analysts say. Three or four men are qualified to take his place, among them Zulkarnaen, head of military activities for Jl, and Yassin Syawal, son-in-law of the founder of Jl, Abdullah Sungkar. Yassin, while not a member, heads Sulawesi-based Laskar Jundullah, an organization regarded by some analysts to be loosely affiliated with Jl. Other possible replacements are Fathur Rohman al-Ghozi, who escaped from a Manila jail last month, and Malaysian bomb expert Azahari Husin.

With a line-up like that there's little doubt that that Jl has the capability to continue launching attacks. Indonesian intelligence officials have said the organization has about 5,000 members in Indonesia alone. Experts say that beyond that may exist other organizations that are not regarded as Jl, but which share some of its goals, in particular the half-century-old struggle for an Islamic state, a movement known as Danul Islam.

The good news? In the short term, police and intelligence agencies seem to be getting better at knowing what to look for. After floundering in the wake of Bali, Indonesian police have reacted quickly to the August 5 attack on the JW Marriott Hotel, and have already arrested a number of suspects. And with each arrest comes possible new intelligence to help build a better picture of what is going on higher up the chain.

SOUTHEAST ASIA'S MOST WANTED

HAMBALI: Until his recent arrest, Southeast Asia's most wanted terrorist suspect and a leader of the Jemaah Islamiah network

AZAHARI HUSIN: A Malaysian physics lecturer, Azahari is a skilled bomb maker and a suspect in the bombing of the JW Marriott Hotel in Jakarta

NOOR DIN MOHAMED TOP: A Malaysian who used to teach in an Islamic school in Johor. Believed to be a key recruiter for Jl and wanted in the Marriott bombing.

FATHUR ROHMAN AL-GHOZI: An Indonesian bomb maker, he was convicted in Manila for possessing explosives; escaped from jail last month

DUL MATIN: This Indonesian Jemaah Islamiah operative is also considered to be a competent bomb maker

ZULKARNAEN: The head of military activities for Jemaah Islamiah; a possible successor to Hambali in the organization's operational leadership

Jeremy Wagstaff

ington's request to bypass the International Criminal Court and extradite U.S. citizens directly to the U.S. for cases that could be tried by the world body will provide a legal buffer for U.S. agents to employ such techniques in Thailand.

In Thailand the perception is growing that the U.S. is driving Thai policy, with greater emphasis on security and less on democracy.

Washington is providing plenty of economic incentive for Thaksin to take up the counter-terror cause. One example: Thailand's Army Corps of Engineers was recently awarded multi-million-dollar reconstruction contracts in war-torn Iraq. When the Thai parliament declined to allocate enough funds for travel and living expenses for nearly 1,000 troops, Washington agreed to pay the difference.

Another is a possible bilateral free-trade arrangement, where preferential treatment for Thai exports would provide much-needed relief for many beleaguered Thai industries facing pricing pressure from lower-cost producers in China.

"There's a lot of quid pro quo flowing between Bangkok and Washington right now," says a Bangkok-based Western diplomat.

For Thailand and Indonesia, the trick will be in striking the right balance between U.S. security requests and domestic demands for more democracy.

The controversial measures Thaksin recently enacted, and the ones Indonesia is considering, will no doubt go a long way in fighting the war on terrorism. The risk is they may lose the bigger regional battle for democracy in the process.

Sasi-on Kam-on in Bangkok and Puspa Madani in Jakarta contributed to this article

Zimbabwe

MUGABE'S END-GAME

The president says he'll step down, but the country's opposition isn't convinced

BY ALAN MARTIN

DRESSED in a thick wool turtleneck to ward off the winter breeze, Victor is keeping office hours under a large tree just up the road from Zimbabwe's mighty Victoria Falls. The two identifying tools of the 26-year-old's trade—an expensive Nokia cellphone and a duffel bag full of cash—lie casually on a picnic table in front of him. Until two years ago, the well-spoken accountant earned a respectable $US1,200 a month working for a chain of luxury safari camps. Now he makes seven times that amount exchanging Zimbabwe dollars for U.S. dollars on the black market. "This is a country in crisis, a place where economic Darwinism is the order of the day for the average guy," says Victor. "All I'm doing is exploiting a loophole to get by."

The "loophole" is the near collapse of Zimbabwe's financial sector. The banks are running out of money, and Victor's success highlights the growing political and economic paralysis gripping what was once one of Africa's most prosperous countries. Zimbabwe's slide into ruin began three years ago when Zimbabwe President Robert Mugabe, now 79, began targeting white farmers by allowing roving gangs to take over their land. Now with only 600 left on their farms—a far cry from their former ranks of 4,500— food production has slumped, leaving nearly eight million people out of a population of 13 million on the brink of starva-

tion. The collapse is so dramatic that the International Monetary Fund declared Zimbabwe's economy to be shrinking faster than any other in the world. "We are not yet a failed state," says Eddie Cross, an economic adviser for the opposition Movement for Democratic Change. "The lights still switch on. But we are perilously close to the abyss."

Police were out in force in Harare during Tsvangirai's court appearance in June

Unemployment hovers near 80 per cent; inflation, already running at 365 per cent annually, is expected to triple by year-end. The banks would print more money if they could—the expedient fallback of any respectable banana republic—but that's not an option: they have neither ink nor paper. Acute gasoline shortages have helped strangle the country's economy and traffic, leaving motorists at the mercy of profiteers charging over US$2.50 a litre. Diplomats have not been paid in weeks; underpaid judges, doctors and other public servants were left fuming in early July after senior politi-

cians in Mugabe's Zanu-PF party granted themselves a 600-per-cent pay increase.

Only Mugabe's departure will help end Zimbabwe's death spiral. During George W. Bush's week-long tour of Africa in July, South African President Thabo Mbeki, a Mugabe confidant and supporter, told the U.S. leader that Mugabe would leave office this fall. But those opposed to Mugabe doubt it. They say the country is locked in a violent end-game with Mugabe and his supporters in the Zanu-PF stepping up repression in a bid to hold on to power at all costs. The phones of high-profile government opponents are tapped and their movements watched by the Central Intelligence Organisation, the feared secret police. Even e-mail traffic is monitored for subversive content.

Torture is also used to intimidate. Michael Moyo, a young political activist, was jailed in June after taking part in an anti-Mugabe rally. Moments after being released he was cornered by a mob of 20 soldiers and CIO agents as he walked toward his brother waiting in the family car. "They put sugar into my petrol tank and slashed the tires," he says. "Then they told me to push the car while they assaulted me." Badly beaten, he was dumped into the car and left for dead. When he met later with *Maclean's,* he was wearing a cast on his right arm and still had partly healed head wounds. Fear was evident in his eyes. "I have had to remove my children from school," he says, "because they have threatened to kidnap them to get to me."

Stories such as Moyo's are all too familiar to Shari Eppel, a psychologist and human rights consultant to Pius Ncube, the country's outspoken Roman Catholic archbishop. Women are being raped in youth militia camps, she says, and homes belonging to government opponents are being torched. "Mugabe knows that you don't need to kill someone to send a strong message," she says. "He knows that it's easier to get away with torture because it doesn't make international headlines."

This is the mess that Morgan Tsvangirai's Movement for Democratic Change (MDC) could inherit. Tsvangirai was charged with treason and imprisoned for 14 days after being accused of organizing a massive protest in June. He was also charged with treason in the run-up to the 2002 elections for allegedly threatening to have Mugabe killed. But George Bizos, the South African lawyer made famous for his defence of Nelson Mandela in the 1960s, has shredded the state's case against Tsvangirai, leaving only political intervention in the way of his acquittal.

When *Maclean's* visited Tsvangirai's compound in Strathaven, an upscale Harare suburb, he was overseeing the transfer of black-market fuel from drums to jerry cans for use in the upcoming campaign for municipal elections. The former union leader, who was widely believed to have been robbed of the presidency during last year's rigged election, admits the MDC lost a crucial propaganda war before the vote when Mugabe succeeded in painting the opposition as

lackeys of Britain and the U.S. "Africa misunderstood the MDC agenda," he says. "We're not out to reverse the gains of independence, but to democratize Zimbabwe. Our problems don't stem from a colonial legacy, but a crisis of governance. Anyone who looks at Zimbabwe objectively will see that."

The opposition has been buoyed by a recent op-ed piece in the *New York Times* in which Colin Powell, the U.S. secretary of state, slammed the Mugabe regime as "corrupt and repressive." Powell also chastised South Africa for not criticizing Mugabe. Bush reiterated the message during his talk with Mbeki. One of the hallmarks of Mbeki's presidency has been creation of the New Partnership for Africa's Development, a multi-billion-dollar aid program that rewards African countries for promoting fiscal and political accountability. But Bush made it clear that the U.S. would withhold $10 billion in financial support for the program until Mugabe's resigns.

Bush's support is the latest in a series of successes for the MDC. Many foreign observers expected the fledgling opposition party to fizzle after the presidential elections, but instead the MDC has consolidated its political support, says Robert Schrire, a professor of political science at the University of Cape Town. "The MDC has been tested and has proved that it has a lot of talent within the ranks," he says. "Key countries like South Africa are reluctantly coming around to accept Tsvangirai as a key player."

Despite growing support for Tsvangirai, the Zanu-PF may not be willing to step aside even if Mugabe quits. Many observers believe hard-liner Emmerson Mnangagwa, the speaker of parliament, will replace Mugabe. "It would be a du-

plicate of Mugabe's style of governance," said Tsvangirai. "He doesn't tolerate dissent and he's violent in his nature. The country will be worse off under him." Putting Mnangagwa in charge of negotiating with the MDC would be a non-starter, adds Schrire. "The whole regime must go, not just Mugabe," he says. "He's only one of an elite that has robbed the country dry."

Zimbabwe's remaining white farmers can only watch and wait. On several occasions Mugabe has told the country in radio addresses that land seizures and occupations of farms are to stop. But the so-called war veterans—most of them dispossessed peasants born after the war of liberation—continue to do as they please. A farm belonging to Roy Bennett, an opposition MP, was overrun by over 200 Zanu-PF supporters in June. They assaulted workers, killed livestock and vandalized farm equipment. Just days before the attack, the president told supporters at a rally that there "should be no more Bennetts" farming the land.

The MDC says it would give the farmers back their land, but hundreds have fled to neighbouring Zambia, Malawi and Mozambique, where the governments have offered them tax incentives. Tobacco companies have also bankrolled farmers to start up plantations in Zambia. In Zimbabwe, more than 200,000 black peasant farmers have moved onto white-owned land. Many of the new farmers say they have not received seeds or fertilizer promised by the government. Others are using ox plows because tractors promised by the government have not arrived. But none of this seems to matter to Mugabe, who clings to power even as the country collapses around him.

Tashkent Dispatch:
Steppe Back

BY ROBERT TEMPLER

'**I**f you doubt our achievements, look at our buildings." This is the slogan of the moment in Uzbekistan, painted in letters three feet high on banners in the capital, Tashkent. Uzbekistan's president, Islam Karimov, likes to think of himself as a latter-day Tamerlane, the conquering Central Asian emperor of the fourteenth century who uttered the phrase to boast of his extraordinary azure-tiled mosques. But the vast new ministries of the Uzbek capital, all mirrored glass and cheap moldings, hardly recall the grandeur of Central Asia's past. Instead, Tashkent's joyless boulevards, high fences, and bombastic palaces feel like Baghdad in better days.

It's not just the pompous architecture that evokes Iraq. Uzbekistan under Karimov is becoming an increasingly repressive and impoverished place, with a horrible human rights record. Economic power has been grabbed by a tiny corrupt elite who have enriched themselves on the back of an exploitative cotton industry. At least 6,000 people are in prison for their religious beliefs. Men who venture outside their homes wearing skullcaps and beards are arrested for being "wahhabis," the local term for anyone who spends too much time at the mosque. The police extract confessions through torture, and compliant judges sentence dissidents to lengthy terms in Jaslyk, a notorious prison camp where last year two religious detainees were boiled to death. An accident with a kettle, the government says. An example of systematic abuse of prisoners, says the U.N. Special Rapporteur on Torture.

Once, American conservatives allied themselves with Islamic extremists in Central Asia to fight Leonid Brezhnev's Soviet Union. Now, they are eagerly developing a friendship with a ruler little different than Brezhnev to fight Islamic extremists. In exchange for U.S. use of an isolated air base in southern Uzbekistan, Karimov has received hundreds of millions of dollars in assistance and a

free ride on human rights and economic reform. Uzbekistan has agreed to make some minor political and economic adjustments, yet the past year has seen the Karimov government's repression and economic mismanagement worsen. The response from Washington has been silence and more money. This is money badly spent: U.S. support for Karimov will backfire, hurting U.S. interests in the region.

Conservatives have begun touting Karimov. As Stephen Schwartz wrote in a recent issue of The Weekly Standard, the United States "must support the Uzbeks in their internal as well as their external combat, and must repudiate the blandishments of the human rights industry."

But, far from helping in the fight against terrorism, this support is likely to spawn new extremists. Alan Kreuger, a Princeton economics professor, and Jitka Maleckova, a Middle East expert at Charles University in Prague, have found that, while it is difficult to demonstrate links between terrorism and poverty or education, there is a close correlation between countries producing terrorists and having a poor record of political rights and civil liberties. Freedom House ranked Uzbekistan just a little above Saddam Hussein's Iraq. This year, the Heritage Foundation and *The Wall Street Journal* put it one hundred forty-ninth on their joint rankings of economic freedom in 156 countries—worse than Burma. Indeed, Uzbekistan's mix of political and economic repression; underground Islamic movements; and a youthful, disillusioned, and unemployed population could prove fertile ground for terrorist recruiters.

There are some legitimate security concerns in Uzbekistan, but Karimov has blown them out of proportion to justify his hard-line rule. The Islamic Movement of Uzbekistan (IMU), a militant group the White House labeled as terrorist even before September 11, 2001, was decimated fighting alongside the Taliban in fall 2001. An underground group of Islamists known as Hizb-ut-Tahrir have never advocated violence in their call for a Central Asian Islamic caliphate.

These groups have relatively little backing right now, but their popular support is growing—mostly because of Karimov. The concentration of wealth and power in an ever-smaller number of hands close to the president, combined with increasing repression and a weakening economy, is fueling widespread discontent that could turn violent. Visiting officials who lecture Karimov on his economic failures are firmly reminded that his education was in the dismal science. Indeed, during Soviet times, he worked at the Uzbek branch of Gosplan, the central planning agency, where he shuffled goods from one unproductive factory to another while skimming a cut. That's still how he sees economic management. Last year, he effectively closed down Uzbekistan's bazaars, the wholesale markets that are the center of commerce, in an attempt, many believe, to enrich members of the government trying to control wholesale trade. He has had the government buy back, at their original price, businesses that were privatized years ago. Many businesses that are then taken over by families of politicians.

But economic mismanagement is only part of a pattern of ultimately self-destructive behavior by Karimov. He has fomented a rebellion in Tajikistan, armed Abdulrashid Dostum—a particularly vicious warlord in Afghanistan—and even bombed villages in Kyrgyzstan, a country he feels has been too lax in tackling Islamic groups. He has mined the once permeable borders of his country so that farmers visiting their cousins in neighboring countries across the hills have had their feet blown off. He has virtually closed Uzbekistan's borders to trade. Meanwhile, the corrupt and powerful benefit from their complete control over the few flourishing areas of the economy, such as cotton production.

All this has helped the radicals. For most people in Uzbekistan, particularly farmers, there is the same drudgery and abusive state control that existed under the Soviet Union but none of the educational or economic opportunities, medical care, or Black Sea vacations. During Soviet times, Uzbekistan was a well-developed industrial center, because Stalin placed heavy industry far from any frontier that could be overrun during a war. All that has gone, the victim of failed economic policies. Former mining towns, such as Angren at the edge of the Ferghana Valley, appear almost completely dead. There is no visible economy, no shops, no market stalls. Enterprising men go to Russia to work. Unfortunate young women get lured into jobs "waitressing" in Dubai.

As a result, across the country there is an increasing sense of economic failure, political paralysis, and popular discontent. It is now easy to find men in Tashkent who, though fond of their vodka and pork sausages, are drawn to the IMU, which had been a feeble movement, because the IMU is one of the only groups that has ever stood up to the government. Underground mosques are gaining in popularity, since they, like the IMU, are one of the only avenues open to people who wish to express their discontent over Karimov's corruption and mismanagement.

It might make more sense for the United States to tolerate Karimov's misrule if Uzbekistan were delivering important assets. But the long-term risks of uncritical support for Karimov, who has nothing in common with the United States other than a shared fear of Islamic extremism, do not outweigh the limited strategic benefits of a base in Uzbekistan. The Uzbek base is of little help to the Pentagon in the war on terrorism, since it already has bases in Afghanistan from which it can battle Al Qaeda offshoots, as well as in Kyrgyzstan for operations in other parts of Central Asia.

KARIMOV, HOWEVER, is prospering on the back of his new relationship with the United States. He can now ignore diplomats who, several years ago, used to raise concerns about human rights abuses and lack of economic reform. In 2002, the State Department issued a limp statement criticizing a fraudulent referendum Karimov held to extend his term in office. Two days after the statement was released, a senior American official announced a tripling of aid for Uzbekistan. Like Hosni Mubarak and other despots who regard themselves as indispensable to Washington, Karimov only has to make the occasional concession to the United States: A few prisoners may be released ahead of a presidential visit to Washington or a new nongovernmental organization may be allowed to register just as a U.S. aid package is under consideration.

The sort of real changes that are needed—changes that might bring democracy and economic opportunity to Uzbeks—will never occur as long as Karimov is running the country. And so a population that aspires to all things that the United States offers is starting to become sullen and resentful at the unquestioning support Washington gives their dictator. Moderate Muslims who want to worship in peace are finding all forms of religious expression and political opposition closed off to them except the underground mosques. Middle-class families are being squeezed out of their businesses by a rapaciously corrupt elite. Young men with no prospects are turning bitter and disillusioned. We know how this story ends.

Robert Templer is Asia program director of the International Crisis Group.

North Korea: The Sequel

The current crisis with North Korea "has the same solution as the original [in 1994]: get North Korea's nuclear program mothballed and its medium- and long-range missiles decommissioned by buying them out at a set price. That price is American recognition of North Korea, written promises not to target the North with nuclear weapons, and indirect compensation in the form of aid and investment."

BRUCE CUMINGS

The current crisis over North Korea's nuclear program is not simply "déjà vu all over again," in the words of the sage Yogi Berra; it is a virtual rerun of events that transpired a decade ago—played fast-forward. Unfortunately, this replay is more dangerous than the original.

In 1991 the administration of President George H. W. Bush became concerned about activities at North Korea's graphite nuclear reactor complex at Yongbyon. Because the nuclear Non-Proliferation Treaty (NPT) gives nonnuclear countries under nuclear threat the right of self-defense, Bush could do nothing about the North Korean activities until the United States cleared its own nuclear weapons out of South Korea—nuclear artillery and land mines, atomic gravity bombs, and "Honest John" rockets (the last dating to 1958). Bush began this process and then inaugurated, for the first time, high-level talks with Pyongyang. The American nukes disappeared from Korea shortly before Bush, and his new diplomacy, disappeared from the American political scene after Bill Clinton's 1992 electoral victory. Clinton the candidate was focused on the economy and paid no attention to the Democratic People's Republic of Korea.

To grab Clinton's attention, six weeks after the new president's inauguration North Korea declared that International Atomic Energy Agency (IAEA) inspectors were doing the bidding of United States intelligence, announced the North's withdrawal from the NPT, and cranked up the country's formidable propaganda apparatus to make it clear that any sanctions imposed by the UN Security Council would be an "act of war." North Korean President Kim Il Sung thus detonated a crisis that lasted 18 months, reaching fever pitch in May 1994 when the North dumped 8,000 well-cooked fuel rods out of the Yongbyon reactor (containing enough plutonium to make five or six atomic bombs), and again in late June when Clinton nearly went to war over North Korea's nuclear actions. Fortunately former President Jimmy Carter entered the fray, flying to Pyongyang to speak directly with President Kim and extracting a commitment for a total freeze on the Yongbyon complex. After the October 1994 United States–North Korea Agreed Framework codified this deal, the IAEA returned to North Korea, sealed off the reactor, encased the fuel rods in concrete casks, and then watched over the facility for the next eight years.

In the current, rapidly unfolding repeat, North Korea again kicked out the inspectors, began loading new fuel rods, castigated the IAEA for being a tool of Washington, announced (again) its withdrawal from the NPT, and asserted that any Security Council sanctions would be interpreted as a "declaration of war." But thus far it has stopped short of opening the plutonium casks, the clearest "red line" that might again provoke a preemptive American strike at the facility.

President George W. Bush's administration has revived the stuttering, confused, and confounding policies of the early Clinton administration. This rerun began when Assistant Secretary of State for East Asian and Pacific Affairs James Kelly went to Pyongyang last October and offered evidence of renewed nuclear activity involving enriched uranium. According to Kelly, the North Koreans at first denied the activity and then admitted it, but not without a certain belligerent satisfaction. Sometime in 1998, according to leaks from the Bush administration, the North Koreans made a deal with America's longtime ally in Islamabad: their missiles for Pakistan's uranium-enrichment technology. Last summer, the same sources say, evidence that the North was manufacturing enriched uranium came to light. It is a very slow process, but if the North Koreans maximize their efforts, using 1,000 centrifuges that they may or may not have, in four or five years they could, on the model of

Pakistan's nuclear program, manufacture one or two large and unwieldy atomic bombs every year.

Shortly after Kelly's return to Washington, a high-level American official told reporters that the 1994 Agreed Framework that froze the North's Yongbyon reactor was null and void—a self-fulfilling prophecy since Bush's advisers had declared it a dead letter soon after coming to power. (Nothing in the agreement prohibits uranium enrichment—Bush spokesmen to the contrary—but the North certainly violated the spirit of the agreement.)

President Bush's team turned a soluble problem into a major crisis, leaving both sides little room to back away.

Since then, the sequel has quickly emerged: Washington won't negotiate with the North Koreans, which would reward "nuclear blackmail." But wait a minute; we had better talk to them or they'll become a nuclear power—but no, we can't "reward" them. Hold on! North Korea is getting out of line again: we'd better take the problem to the Security Council. Whoa, no we can't, because China won't go along. Let's send a low-level or backchannel envoy to Pyongyang. Nothing doing; Pyongyang wants to talk with someone who actually makes decisions. We can't do that, though, because that would be like recognizing this regime, which the United States has refused to do since February 1946.

Obsessive concentration on the problem with Pyongyang is met by inattention and confusion in Washington, and North Korea keeps winning. As Leon Sigal (one of the exceedingly few reliable American experts on this issue) has noted, "You don't want to get into a pissing match when the other guy has a full bladder."

Push came to shove, and finally Washington enunciated its presumed bottom line—similar to what Clinton's defense secretary, William Perry, articulated in 1994: "We do not want war and will not provoke a war over this or any other issue in Korea"; but if UN sanctions "provoke the North Koreans into unleashing a war… that is a risk that we're taking." This was a careful formulation reflecting Perry's role, along with Assistant Secretary of Defense for International Security Policy Ashton Carter, in studying for months in 1993 the feasibility of a preemptive attack on Yongbyon. As it happened, Clinton could not stomach the risk of war with an important congressional election just four months away; his commander in Korea, General Gary Luck, had told him a new Korean war would last six months and he would need 100,000 body bags on hand.

Today George W. Bush repeatedly asserts that the United States has no intention of "invading" North Korea as hard-liners in the Pentagon revive Clinton's plans for a "surgical strike" against Yongbyon; at the same time, diplomats say Washington is ready to talk to Pyongyang but will not negotiate or reward "nuclear blackmail," and all lament Kim Jong Il's multiple interruptions of their march toward war against Iraq. But the extended dilation of the Iraq crisis, occasioned by Bush's decision

last September to put the problem of Iraq's "weapons of mass destruction" (WMD) in the hands of the UN Security Council and the IAEA, was clearly an ideal time for North Korea to fast-forward the current crisis. Bush had serial plans for the "axis of evil": first Saddam Hussein, then North Korea, and then Iran. Kim Jong Il, however is, understandably, a man in a hurry.

The sequel has the same solution as the original: get North Korea's nuclear program mothballed and its medium- and long-range missiles decommissioned by buying them out at a set price. That price is American recognition of North Korea, written promises not to target the North with nuclear weapons, and indirect compensation in the form of aid and investment. Indeed, William Perry was the point man for getting both jobs done between 1998 and 2000 as Clinton's roving ambassador, moving toward mutual diplomatic recognition and a full buyout of Kim Jong Il's missiles despite intelligence evidence that in 1998 North Korea had begun to import aluminum centrifuge tubes and other technology relevant to a separate nuclear program to enrich uranium.

But George W. Bush cannot yet star in the new sequel because of a host of ostensible foreign policy commitments he has been making since the day of his inauguration. In a display of partisan foreign policy decision-making unlike any previous episode, Bush was initially determined to be the anti-Clinton. Clinton wanted the Kyoto treaty? Bush didn't. Clinton loved multilateral confabs and pressing the flesh with allied leaders? Bush would go unilateral, and consult only with those allies who agreed with him (mainly British Prime Minister Tony Blair). Clinton froze North Korea's reactors and was on the verge of buying out their missiles as well? That was mere appeasement of a reprehensible "rogue state." More deeply, Bush's advisers moved toward a general reversal of previous United States strategy: instead of deterrence, the United States would have what political scientist Thomas C. Schelling once called compellence—marshaling America's overwhelming and unchallenged military might to shape relations with allies and constrain adversaries, which in early 2001 meant Russia and China, both termed potential adversaries by Bush's national security adviser, Condoleezza Rice. Instead of nonproliferation—the overwhelming influence in Clinton's policies toward near-nuclear and "rogue" nations—the United States would have counterproliferation: using the threat or reality of American military force to stop WMD development dead in its tracks.

The cold war doctrine of containment was still in place, however—formally against countries like North Korea, Iraq, and Iran; informally against Chinese expansion or Russian resurgence; and (as always since 1945) through hidden constraints on allies such as Japan and Germany (including keeping a myriad of United States military bases on their soil). Along came Osama bin Laden and friends, a force that could not be deterred or contained, and a new strategy of preemptive attack (the better name is preventive war) was formally announced in September 2002. In the midst of this evolution of strategy, President Bush fatally conflated a group of nations that could easily be contained and deterred, namely Iraq, Iran, and North Korea, with the diabolical and uncontrollable Al Qaeda: thus emerged the "axis of evil." These evil-doers were not suicidal and had return addresses, but no matter: they might give or sell their weapons to terrorists.

George W. Bush, a naïf in world affairs, brought into office with him a highly experienced crew of Republican foreign policy hands: Defense Secretary Donald Rumsfeld often seems to be the main spokesman for the administration's strategies, Dick Cheney has unprecedented weight in foreign policy for a vice president, and Secretary of State Colin Powell tries to carry out diplomacy in an administration that does not believe in it. These three big egos would prefer not to consult with each other—let alone with foreign leaders. The result has been a set of independent kingdoms presided over by a weak and inattentive president, extraordinary divisions and battles over policy, and the most incoherent foreign policy in memory.

Bush himself has compounded matters with continuous if utterly gratuitous outbursts against North Korean leader Kim Jong Il. South Korean President Kim Dae Jung, a Nobel Peace Prize winner, met with Bush in March 2001 only to be informed that the North Korean Kim could not be trusted to keep any agreements (as if the 1994 deal had been based on trust rather than verification); the following October Bush traveled to Shanghai to meet various Asian leaders (including Kim Dae Jung again) and denounced Kim Jong Il as a "pygmy." Most recently, in a discussion with *Washington Post* reporter Bob Woodward, Bush blurted out "I loathe Kim Jong Il!" shouting and "waving his finger in the air." In a less-noticed part of this outburst, Bush declared his preference for "toppling" the North Korean regime.[1] One gets the sense from these impromptu ad hominem eruptions that Bush's resentments might have something to do with the widespread perception that both leaders owe their positions to their fathers.

THE GREATER DANGER

After nearly two years of an amateurish American foreign policy replete with ill-thought-out démarches, incessant internal clashes, and predictable reversals, it was inevitable that one of the "axis" countries threatened with preemptive attack would itself occupy the center stage and call Bush's bluff. Kim Jong Il has done that, but North Korea presents a far more difficult crisis for the Bush administration than does Iraq, not to mention another sharp diversion from what one would assume to be America's primary quarry, Osama bin Laden and Al Qaeda. A harsh and bitter realism and brinkmanship that has been formed in the cauldron of a 50-year war has met a messianic idealism wanting to "rid the world of evil" (in the words of the National Security Council document on the new preemptive doctrine), a conceit that would make Old Testament prophets chortle. But in the post–September 11 atmosphere, the replay of this Korean sequel is considerably more dangerous than the first.

Through its recent provocations, Pyongyang has fueled a fire fanned by an administration that listens to no one but lacks the wherewithal to fight major wars on more than one front. North Korea knows this, and therefore has pushed its advantage while Bush is fixated on Saddam Hussein. Furthermore, Bush completely abandoned Bill Clinton's nearly successful attempt to buy out North Korea's medium- and long-range missiles while keeping its nuclear facilities frozen; how could a devastating new war possibly be justified when that option was left to slide into oblivion? Nonetheless we again hear from William Perry and Ashton Carter, in a January 2003 *New York Times* op-ed, that the United States must "make clear [its] determination to remove the nuclear threat even if it risks war."

Even more damning, insiders say that the outgoing Clinton team fully briefed the Bush newcomers on the intelligence about North Korea's imports of nuclear-enrichment technology from Pakistan, but the Bush administration did nothing about it until July 2002, when it picked up evidence that the North might be beginning to build an enrichment facility. Many knowledgeable experts believe that North Korea clearly cheated on its commitments by importing these technologies, but former officials also believe that whatever the North planned to do with them could have been shut down in the context of completing the missile deal and normalizing United States–North Korean relations. By failing to follow through on the Clinton administration's missile deal and then using the new information garnered in July 2002 to confront the North Koreans in October, President Bush's team turned a soluble problem into a major crisis, leaving both sides little room to back away.

The acute danger today derives from a combination of typical and predictable North Korean cheating and provocation, long-standing United States plans to use nuclear weapons in the earliest stages of a new Korean war, and Bush's new preventive war policy. Bush's doctrine fuses existing plans for nuclear preemption in a crisis initiated by North Korea, which have been standard operating procedure for the United States military in Korea for decades, with the apparent determination to attack states like North Korea simply because they have or would like to have nuclear weapons like those that the United States has amassed in the thousands. As if to make this crystal clear, last September a White House insider leaked Presidential Decision Directive 17, which listed North Korea as a target for preemption.

The September 2002 preventive doctrine came out of Condoleezza Rice's office. She later explained to reporters that preemption is "anticipatory self-defense"—that is, the "right of the United States to attack a country that *it thinks* could attack it first [emphasis added]." In the document we read that other nations "should [not] use preemption as a pretext for aggression." In the Korean theater, however, a new war could erupt over an issue like the recent "June crab wars" that have occurred as North and South Korean fishermen compete for lucrative catches in the Yellow Sea, and a vicious cycle of preemption and counterpreemption could immediately plunge Northeast Asia into general war. Adding to the danger is a new threat to the existing deterrent structure on the peninsula: according to a retired United Sates army general with considerable experience in Korea, American advances in precision-guided munitions now make it possible to take out the 10,000 artillery tubes that the North has embedded in mountains north of Seoul, which were heretofore impregnable and constituted the North's basic guarantee against an attack from the South. To the extent that this is true, in the absence of credible security guarantees, any general sitting in Pyongyang would now move to a more reliable deterrent.

THE GREATER EVIL

All this is truly tragic, given the enormous progress toward reconciliation between North and South Korea, propelled by Kim Dae Jung's leadership after he took office in early 1998. Have we forgotten that Kim John Il welcomed President Kim Dae Jung to Pyongyang in June 2000, the first time that Korean heads of state had shaken hands since the country was divided in 1945? In December 2002 the South Korean people decisively broke with the existing political system and the elites within it that also date to 1945 by electing Roh Moo Hyun, a lawyer with a sterling record of courageous defense of labor leaders and human rights activists during the darkest days of the military dictatorship in the 1980s. A burgeoning movement among younger Koreans against the seemingly endless American military presence in the South has conducted successive, truly massive, and dignified candlelight processions along the grand boulevard in front of the American embassy in Seoul. This movement unites citizens—who were educated on the raucous college campuses of the 1980s while American diplomacy backed the dictatorship and its bloody suppression of the Kwangju uprising in 1980—with the incoming Roh administration and a set of advisers well aware of America's shared responsibility for the current crisis. Thus Bush finds himself managing two difficult relationships on the Korean peninsula amid the building momentum toward war with Iraq and the failed search for Osama bin Laden.

Just as it did a decade ago, a supine American media falls in line with this administration's caricature of the Korea crisis instead of doing serious investigative reporting. The cover story of the January 13 issue of *Newsweek* carried a photo of Kim Jong Il along with the headline "North Korea's Dr. Evil." On the cover of *Newsweek*'s first issue after the death of Kim Il Sung in July 1994 was this racist title: "Korea after Kim: The Headless Beast."

But where is the greater evil? To obtain the requisite votes from nonnuclear states to push the NPT through the United Nations in 1968, the United States, the United Kingdom, and the Soviet Union committed themselves to aid any "victim of an act or an object of a threat of aggression in which nuclear weapons are used." In 1996 the International Court of Justice at The Hague stated that the use or threat of nuclear weapons should be outlawed as the "ultimate evil." The justices could not come to a decision about whether the use of nuclear weapons for self-defense was justified: "The Court cannot conclude definitively whether the threat or use of nuclear weapons would be lawful or unlawful in an extreme circumstance of self-defense, in which the very survival of a state would be at stake." By this standard, North Korea is more justified in developing nuclear weapons than the United States is in threatening a nonnuclear North Korea with annihilation.

Once again North Korea believes that its "very survival" is at stake. Probably it is wrong, but in the current volatile conditions of world affairs, one cannot expect the North to take chances on a matter of such gravity. The only way to unravel this emergent calamity, short of war, is a quick return to the status quo ante 2001—to the compelling and still-feasible dénouement to the original crisis fashioned by Bill Clinton and Kim Jong Il. No one will benefit from the current sequel, except those hard-liners in both capitals who believe that true security lies only in the deployment and brandishing of nuclear weapons.

Note

1. See Bob Woodward, *Bush at War* (New York: Simon & Schuster, 2002), p. 340.

BRUCE CUMINGS, *a Current History contributing editor, teaches at the University of Chicago. His newest book,* Parallax Visions: Making Sense of American–East Asian Relations *(Durham, N.C.: Duke University Press, 2002), has recently appeared in paperback and contains an extended analysis of the first crisis with North Korea.*

From *Current History*, April 2003, pp. 51–57. © 2003 by Current History, Inc. Reprinted by permission.

Blaming the victim:
Refugees and global security

The bulk of the world's refugees remain in the developing world. And the industrialized states, more worried after September 11, are taking new steps to keep them away.

by Gil Loescher

THE ISSUES OF HUMAN SECURITY AND the security of states are intimately linked. For example, a greater respect for human rights, more equitable development, and the spread of democracy in war-torn places like Afghanistan, Kosovo, and the Democratic Republic of Congo would not only prevent and/or resolve the problems of refugee movements, they would also help establish a more stable and secure international order.

As a general rule, individuals and communities do not abandon their homes unless they are confronted with serious threats to their lives or liberty. Flight from one's country is the ultimate survival strategy, the one employed when all other coping mechanisms have been exhausted. Refugees serve both as an index of internal disorder and instability and as *prima facie* evidence of the violation of human rights and humanitarian standards. Perhaps no other issue provides such a clear and unassailable link between humanitarian concerns and legitimate international security issues.

Whether refugees find safe haven in the countries to which they flee depends in part on regional stability. This reality was brought home to me in recent months when I visited Turkey, Syria, and Kenya—countries that serve both as host states to refugees in their respective re-

gions and as transit countries for those seeking to migrate to Europe and North America. All three countries are located in extremely unstable regions from which some of the world's major refugee flows originate. The bulk of the refugees in these regions—Somalis, Sudanese, Iraqis, and Iranians—come from countries where conflict and persecution have persisted for years, making it unlikely that they will be able to return home any time soon.

The refugees I interviewed complained that their greatest concern was the poor security in these countries of first asylum. The human rights records of these countries are poor. Physical harassment, detention, and deportation to other countries where refugees risk greater persecution are commonplace. Police and security forces arbitrarily harass, detain, and arrest refugees. Corruption is rampant, especially among poorly paid border guards and police. Many refugees fear being attacked by agents from their home countries, often with the connivance of the authorities in the countries in which they have sought refuge.

This fear was given real meaning in Kenya, the host country for relatively large numbers of refugees from the Great Lakes region of Africa. In April, an assailant broke into a so-called secure resi-

dence established in Nairobi for refugees at particular risk. The assailant murdered two Rwandan refugee children, aged nine and 10, by slitting their throats. Their mother, a close relative of a former Rwandan president, was also seriously injured with multiple stab wounds. She and the children had been waiting in Kenya for 11 months for their resettlement application to be processed.

Refugees also face severe economic and social insecurity in these countries of first asylum; their freedom of movement is severely restricted; they cannot integrate with local populations; they are given inadequate or no assistance; they are refused permission to work. They live in limbo.

Refugees and local, national, and regional security

Refugees who flee persecution or violence may experience threats to their security on a daily basis, but when the connection between security and migration is highlighted by the media or politicians, it is rarely with reference to threats faced by refugees. Rather, the refugees themselves are usually characterized as posing a security threat to the receiving states and their citizens. This viewpoint is as likely to be expressed in the developing world as it is in the industrialized democracies. Most governments today perceive migration and refugee movements as a threat to their national interests.

For developing countries, displaced populations are both a consequence of conflict and a cause of continuing conflict and instability. Forced displacement can obstruct peace processes, undermine attempts at economic development, and exacerbate intercommunal tensions. Refugee flows also can be a source of regional conflict, causing instability in neighboring countries, triggering external intervention, and sometimes providing armed refugee groups with base camps from which to conduct insurgency, armed resistance, and terrorist activities.

In recent years, both new and long-established refugee populations have come to be viewed by local host governments as a threat to the internal order of the state, as well as a threat to regional, or even global, security. States perceive refugee groups as posing both direct and indirect security threats.

The direct security threat posed by the spillover of conflict and armed exiles is by far the strongest link between forced migration and state security. In many regions, such as the Balkans or the Great Lakes region of Africa, refugee exoduses have been deliberately provoked or engineered by one or more parties to the conflict with the specific objective of furthering their own political, military, or strategic objectives.

The use of refugee camps by combatants draws refugee communities directly into cross-border conflicts and accelerates conflicts and tensions within host countries. In recent years, many refugee camps in Africa, Asia, and Central America have housed not only those fleeing persecution and armed conflict, but also combatants and guerrilla forces who use the relative safety of the camps to launch violent campaigns of destabilization against their countries of origin.

In many host countries, governments prefer refugees to remain in camps and settlement areas close to the border of their home countries. Not only are the physical security and material safety of refugees not guaranteed in these remote areas, but the proximity of camps to countries of refugee origin also makes it easy for combatants to cross borders to engage in guerrilla warfare.

But indirect security threats posed by protracted refugee situations also impose important burdens on receiving states. Developing countries shoulder the social and economic strains of the world's vast majority of refugees and displaced persons. The long-term presence of refugees can exacerbate existing tensions and heighten intercommunal conflict, particularly when a state has ethnic rifts of its own, a vulnerable economic or social infrastructure, or hostile neighbors.

Generally, both the governments and the citizens of host countries view refugees negatively, associating them with problems of security, violence, and crime, and as a threat to social cohesion and employment. They are sometimes seen as posing a threat of insurgency or terrorism. In many regions, these negative perceptions have begun to generate a backlash against refugees—and especially, lately, against Islamic groups. Given the regionalization of conflict and the domestic instability caused by both new and protracted refugee situations, the indirect security threats posed by refugee flows, if left unaddressed, are likely to have serious consequences for regional and global security.

Western government responses

Western governments have failed to recognize these regional refugee situations and have not devoted sufficient resources, either financial or diplomatic, to long-standing refugee problems; long-term refugee needs induce "donor fatigue." For example, the refugees I recently visited have been adversely affected by recent cutbacks in donor funding to the U.N. High Commissioner for Refugees (UNHCR). This means that vital assistance programs in Turkey have been ended, and it also means that UNHCR has not been able to deploy the number of protection officers it needs at the dangerous and insecure refugee camps in Kenya. The deprivations and frustration endured by people living in insecure and precarious camps for long periods of time can generate various kinds of violence and instability. This is particularly the case in the Kenyan camps, where sexual abuse and violence against women and girls are common. In addition, banditry and armed robbery occur regularly within the camps.

Western governments have not developed effective policies to address the often deplorable conditions in regional host countries. The European Union, for example, administers development programs inside Somalia, but spends nearly nothing to secure better protection or improve the environment for the more than 250,000 Somalian refugees in neighboring Kenya. The involvement of the European Community Humanitarian Office in these regions is limited exclusively to projects in countries of origin and to non-refugee emergency projects in countries of first asylum.

International attention and assistance are in large part a reflection of politics, geo-strategic interests, and fickle international donor and media priorities. The U.S. and European governments have not fully examined the impact of their foreign and economic policies on refugee-producing regions. For example, more than a decade of economic and political isolation and sanctions against Iraq have made life so miserable and untenable for large sectors of the Iraqi population that hundreds of thousands have fled their homes to neighboring countries. Western governments have not developed a comprehensive policy to deal with these and other migration and refugee problems.

Ultimately, this is self-defeating. If refugees lack security in one country, they will try to move on to another, safer place. For example, most Iraqi refugees are unable to find either physical protection or material security in their host countries. Thus, insecurity, coupled with a lack of resettlement opportunities, has led large numbers of asylum seekers and migrants to move on to Western countries and to risk their lives by using illegal means of entry, including smuggling and trafficking organizations. Not surprisingly, Iraqi refugees now constitute the largest national group of asylum seekers in Europe.

The "securitization" of refugees

Although the bulk of the world's refugees remain in the developing world, the industrialized states feel increasingly threatened by an influx of refugees. Asylum-seekers are no longer limited to neighboring states; "jet age" refugees appear at the doorstep of distant nations. This comes, of course, on top of a steep rise in the number of undocumented immigrants to the West from the developing world and from Eastern Europe.

Throughout most of North America and Western Europe huge backlogs of asylum applications have built up, exposing the inability of the advanced industrialized states to establish administrative and judicial systems that can cope with the growing numbers of asylum applicants. In 1999, Britain, for example, had a backlog of some 100,000 unprocessed asylum applications.

Many governments find it virtually impossible to apprehend and deport those whose claims to refugee status have been rejected. At a time when governments are seeking to reduce public expenditure, they are simultaneously spending large sums on processing asylum applications and providing social welfare benefits to asylum seekers. There is a widespread belief that Western states have lost control of their borders and that refugees and immigrants pose a threat to national identities and economies of host societies.

Not surprisingly, Western worries about asylum seekers influence electoral politics. Across Western Europe, politicians allege that asylum seekers take away jobs, housing, and school places, dilute national homogeneity and culture, and exacerbate racial and ethnic tensions in local communities. As a result, political parties that invoke a popular xenophobia, campaigning on explicitly anti-refugee and anti-immigrant platforms, are now enjoying success in traditionally liberal democratic states like Denmark, the Netherlands, and France.

More recently, asylum seekers and refugees have come to be seen as direct threats to national security. Some communities of exiles are alleged to support terrorist activities and to be engaged in supporting armed conflicts in their countries of origin. In fact, the vast majority of immigrants are grateful for being given asylum and live by the rules.

> An angry, excluded world outside the West will inevitably turn to forms of extremism that will pose new security threats.

Prompted in part by the perceived threat to security, for two decades Western countries have been initiating a wide range of measures to curb the entry, admission, and entitlements of people claiming refugee status. Restrictive measures include extending border controls through stringent visa requirements, imposing sanctions against airlines and other carriers for transporting undocumented individuals, stationing immigration officials abroad, detaining asylum seekers before they reach national borders, negotiating agreements to send home those refused asylum, and threatening to withdraw financial and development aid if regional host countries will not take back those asylum seekers rejected in the West.

This trend has been exacerbated by the war against terrorism. In an effort to toughen immigration laws to prevent terrorists from entering their countries, North American and European governments have rushed through measures that may threaten the concept of the right of asylum. These include measures that allow for the indefinite detention of noncitizens suspected of terrorism, including asylum seekers and recognized refugees, without adequate rights of appeal. Since September 11, some governments have resorted to detention as a matter of first course, in some cases denying individuals their fundamental right to seek asylum and detaining them indefinitely until a deportation order can be executed. Such measures run foul of the U.N. Refugee Convention and accepted human rights standards.

In Europe, the gulf between the cultural background of contemporary refugee groups and that of Europeans causes special concern. Refugee groups may resist assimilation, and Western publics may be unwilling to tolerate aliens in their midst. These feelings, reinforced by racial and religious prejudices, pose difficult social and political problems for European governments. Xenophobic and racist attitudes are increasingly obvious among some segments of the population, and racist attacks have increased in every country that hosts immigrant minorities. Islamic groups in particular have been targeted, especially since September 11, 2001. The anti-immigrant, anti-refugee backlash is being exploited not only by the extreme right wing, but also by mainstream political parties throughout Europe. As a consequence, ethnic profiling and detention of members of Islamic groups and other minorities, including immigrants and asylum seekers, have increased dramatically.

Greater abuse?

The war on terrorism has given policymakers and law enforcement agencies a ready pretext to abuse the rights of refugees and other immigrants. Newly enacted measures to enhance internal law enforcement mechanisms to protect the state against terrorist threats can lead to an even greater deterioration of the rights of all citizens—and particularly the rights of refugees—leading to their increased vulnerability and exclusion. Indefinite detention, governmental restrictions on disclosure of evidence, the establishment of military tribunals with defined jurisdiction over non-citizens, and an array of possible new interior controls to deter potential terrorist abuse of the asylum systems have resulted in a tightening of visa systems around the world, making it even more difficult for refugees to escape persecution.

In many countries around the world, governments have seized on the rhetoric of anti-terrorism to steamroll domestic opposition. In this highly charged environment, politicians and the media are targeting refugees and immigrants as scapegoats for their countries' economic and social problems.

Politicians often exaggerate the various domestic threats associated with refugees in order to win short-term electoral gain. For example, in the run-up to the upcoming general elections in Kenya, President Daniel Arap-Moi and members of Kenya's parliament have stepped up the rhetoric, blaming refugees for depleting the nation's resources, degrading the environment, and endangering national security.

The way ahead

While the security threat associated with refugees is often exaggerated, history demonstrates that refugee movements are not only a humanitarian problem. They have a strong political and security dimension that can adversely affect domestic and international order. The management of migration, and particularly of refugees and displaced people, is not a side effect of political and economic instability and conflict, but an integral part of regional and international insecurity, and an integral part of conflict settlement and peace building within communities.

Forced displacement is a major factor in national and regional instability. Establishing effective responses to refugee needs should be a vital part of any broad model of security. The real challenges for policy-makers, practitioners, and researchers lie not so much within the humanitarian system itself, but in the wider policy-making world, including security, post-conflict development, the enforcement of human rights, and the development of civil societies.

Reacting to terrorist threats by placing unduly harsh restrictions on the free movement of people will simply lead to greater isolation and deprivation. An angry, excluded world outside the West will inevitably turn to forms of extremism that will pose new security threats. A failure by both the industrialized and developing countries to take action to stem the tide of poverty, violence, persecution, and the other conditions that create refugee flows will be costly in security terms.

In the realms of forced migration and state security, international and regional stability and justice coincide. Policy-makers need to build on this coincidence of factors to achieve the political will necessary to address both these issues more effectively and even-handedly. It is in the self-interest of states and coincides with their search for long-term global stability.

The international upheavals reverberating around the terrorist attacks of September 11 underline the important connections between refugee movements and international security. It is now impossible to overlook the strategic importance of the global refugee problem, and both national governments and international organizations need to make greater efforts at finding solutions to it. Western governments must recognize that the most obvious and logical solution lies in improving conditions in countries that produce refugees as well as in local host or transit countries in the regions of refugee origin.

Gil Loescher, Senior Fellow for Forced Migration and International Security at the International Institute for Strategic Studies in London, is the author of The UNHCR and World Politics: A Perilous Path *(2001).*

UNIT 4

Political Change in the Developing World

Unit Selections

Key Points to Consider

- What are the current trends in democracy throughout the world?

- What are the two views regarding the seriousness of Islam's threat to the West?

- What are the challenges involved in rebuilding Afghanistan?

- What accounts for political tensions in Iran?

- Does the election of a new president in Brazil enhance prospects for prosperity?

- What challenges do Latin American leaders face?

- What shapes the prospects for democracy in Africa?

- How do NGOs affect democracy?

 **Links: www.dushkin.com/online/**
These sites are annotated in the World Wide Web pages.

Greater Horn Information Exchange (GHIE)
http://www.hydrosult.com/niledata/usaid/index.htm

Latin American Network Information Center—LANIC
http://www.lanic.utexas.edu

ReliefWeb
http://www.reliefweb.int/w/rwb.nsf

World Trade Organization (WTO)
http://www.wto.org

Political change in the developing world has not necessarily produced democracy, in part due to the fact that developing countries lack a democratic past. Colonial rule was authoritarian, and the colonial powers failed to prepare their colonies adequately for democracy at independence. Even where there was an attempt to foster parliamentary government, the experiment frequently failed, largely due to the lack of a democratic tradition and political expediency. Independence-era leaders frequently resorted to centralization of power and authoritarianism, either to pursue ambitious development programs or often simply to retain power. In some cases, leaders experimented with socialist development schemes that emphasized ideology and the role of party elites. The promise of rapid, equitable development proved elusive, and the collapse of the Soviet Union discredited this strategy. Other countries had the misfortune to come under the rule of tyrannical leaders who were concerned with enriching themselves and who brutally repressed anyone with the temerity to challenge their rule. Although there are a few notable exceptions, the developing world's experiences with democracy since independence have been very limited.

Democracy's "third wave" brought redemocratization to Latin America during the 1980s, after a period of authoritarian rule. The trend toward democracy also spread to some Asian countries, such as the Philippines and South Korea, and by 1990 it also began to be felt in sub-Saharan Africa and, to a much lesser extent, in the Middle East. The results of this democratization trend have been mixed so far.

Although Latin America has been the developing world's most successful region in establishing democracy, its commitment to democracy has been shaken by widespread dissatisfaction due to corruption, inequitable distribution of wealth, and threats to civil rights. Threats to democracy were evident in the April 2002 coup that briefly ousted Venezuela's president Hugo Chavez from power. Argentina's economic collapse also threatened political instability as the middle class, deeply affected by the crisis, demonstrated little confidence in its political leadership.

Africa's experience with democracy has also been varied since the third wave of democratization swept over the continent beginning in 1990. Although early efforts resulted in the ouster of many leaders, some of whom had held power for decades, and international pressure forced several countries to hold multiparty elections, political systems in Africa range from consolidating democracies to states mired in conflict. South Africa's successful democratic consolidation in the face of major challenges stands in sharp contrast to the circumstances in such countries as the Democratic Republic of Congo, Cote d'Ivoire, Liberia, and Burundi.

Efforts to develop a democratic political system in Afghanistan are complicated by ethnic differences. The assassination of the vice president and an attempt on the life of President Hamid Karzai demonstrate the fragility of the current political arrangement and the persistent challenge of regional warlords. In the Middle East, there has been little progress toward democratic reform. Much of the region remains under the control of monarchies or authoritarian rulers presiding over tightly constrained political systems. In Iran, a popularly elected president is locked in a struggle with Islamic hard-liners who are trying to maintain their authority in the face of eroding revolutionary ideals. Preliminary indications of a push for greater democracy within the Palestinian Authority have given way to political infighting, while Israeli policy has strengthened Yasir Arafat's power. The U.S. invasion of Iraq and the ouster of Saddam Hussein were supposed to pave the way for the establishment of democracy, but persistent attacks against Americans demonstrate resistance to U.S. occupation; it is now clear that rebuilding Iraq and fostering democracy will be a long and costly process.

India's democracy continues to be tested by communal strife resulting from religious and caste differences. Pakistan's president, General Pervez Musharraf, who came to power, in a 1999 coup, clearly intends to remain in power, and many of the former Soviet republics of central Asia appear to be turning away from democracy and toward an authoritarianism rooted in their cultural and political past.

While there has been significant progress toward democratic reform around the world, there is no guarantee that these efforts will be sustained. Although there has been an increase in the percentage of the world's population living under democracy, nondemocratic regimes are still common. Furthermore, some semidemocracies have elections but lack civil and political rights. International efforts to promote democracy have often tended to focus on elections rather than the long-term requirements of democratic consolidation.

POLITICS ABROAD

Democracies: Emerging or Submerging?

Anthony W. Pereira

THE SECOND half of the twentieth century was an age of democracy. The women's movement, anti-colonial struggles, and challenges to what W E. B. DuBois called the "color line" won political inclusion for many people throughout the world. And starting in the mid-1970s, electoral democracies replaced authoritarian regimes in southern Europe, Latin America, and parts of Africa and Asia. This process was extended when the former Soviet Union broke up, and communist regimes collapsed there and in Eastern Europe. According to Freedom House, a nonprofit institution that issues an annual assessment of political and civil rights worldwide, there were only twenty-two democracies with 31 percent of the world population in 1950, out of a total of eighty sovereign states. Today, there are 120 electoral democracies representing 58 percent of the world population, out of 192 sovereign states. The percentage of the world's population living under some form of democracy nearly doubled in the last fifty years. This was a historic shift of epic proportions.

Western triumphalism about the "end of history," however, was never justified. Non-democratic regimes still exist. Morocco, Saudi Arabia, the Gulf States, and Swaziland are traditional monarchies. China, Cuba, North Korea, and Vietnam are nominally communist, one-party regimes whose leaders deeply distrust multiparty democracy. And military regimes of the kind so common in developing countries in the 1960s and 1970s survive. Burma (Myanmar), which the democratic activist Aung San Suu Kyi calls a "Fascist Disneyland," is one of them, as are Pakistan and, until 1999, Nigeria. Although they often declare their intent to supervise a democratic "transition," such regimes do not respect the democratic principles of universal suffrage and elected government.

In addition, some of the world's states have imploded, rendering the form of their regime irrelevant: Sierra Leone, the Congo, Colombia, and Sri Lanka today; Afghanistan, Angola, Burundi, Haiti, Liberia, Rwanda, and Somalia a few years ago. It is likely that there will be more, and that international intervention, when it occurs, will not be entirely successful in promoting peace and rebuilding the state.

Furthermore, many new democracies are illiberal, unable or unwilling to guarantee their citizens important political and civil rights, even if regular elections are held. Leaders with a dictatorial bent manipulate political systems to prevent genuine competition or to steal elections. Observers speak of these polities as "democracies without citizenship" or "delegative democracies"—political systems in which plebiscitary elections create mandates for powerful chief executives who rule virtually unchecked.

Finally, changes in patterns of economic and political organization have shifted authority upward to regional and global institutions (the European Union [EU], the North American Free Trade Agreement [NAFTA], the World Trade Organization [WTO], the International Monetary Fund [IMF], the World Bank, multinational corporations, and financial markets) and downward to local and provincial governments. The scope of the nation-state's power has diminished as its institutions have, more and more, been democratized.

The "Third Wave" and "Transitology"

Given all this, what can be said about the world's new democracies? A survey of the situation suggests that many of them are in trouble, and only sustained transnational pressure to democratize globally will create the political conditions in which they can flourish.

The creation of new democracies in the late twentieth century involved the demise of many different kinds of regimes. In

Portugal and Spain, long-lasting fascist dictatorships ended; in Latin America, brutal military dictatorships ran aground; in Africa and Asia, one-party machines lost their hegemony. Few would question the value of such events for people living in these countries. However, much of the current literature about the emerging democracies reflects assumptions that don't hold universally and agendas that embody the interests of powerful external actors.

U.S. political scientist Samuel Huntington christened the democratic transitions of the mid-1970s and after as the "third wave," a successor to the first wave of democratization in the nineteenth century and the second wave during and after World War II. Another political scientist, Philippe Schmitter, wryly refers to the work of Huntington and the legions of other scholars and practitioners who write about democratic transitions as "transitology." Despite its apparently neutral language, transitology reflects ideological assumptions about new democracies. For example, democracy tends to be defined minimally as competition between elites in competitive elections. It is seen primarily as the product of conscious crafting and deal-making and not, as an earlier generation of scholars believed, of long-term historical conflicts between rulers and ruled.

PERHAPS most important, certain kinds of democratic transition are taken to be inherently better than others. Portugal's 1974 revolution, for example, tends to be downplayed in the transitology literature, because it involved mass insurrection, the collapse of the old regime, and curbs on the prerogatives of business and private property in its early phase. In contrast, the more conservative and negotiated Spanish transition of 1975, even though it was actually more violent than its Portuguese counterpart, is the emblematic transition in the transitology literature.

Transitologists extract certain lessons from the Spanish model that they hope democratizers elsewhere will use: the "masses" threaten to destabilize transitions, and their mobilization should be feared, not encouraged; the prerogatives of the market, as defined by the most powerful actors in the global economy, should not be challenged; and negotiated transitions always produce better democracies than those that result from a regime collapse. Such a perspective is intrinsic to the U.S. government's "democracy promotion" bureaucracy—including the U.S. Agency for International Development (USAID) and the National Endowment for Democracy (NED)—which now spends roughly $700 million per year trying to promote a particular version of democracy overseas.

The third wave involved technology as well as ideas. If the generation of 1968 demanded a revolution, the generation of 1978 actually created one—but it was technological, not political. The information and communications revolution involving satellites, cellular telephones, digital technology, the Internet, fax machines, and the computer chip made data gathering, processing, and transmission cheaper and easier than ever before. It became markedly easier to organize elections in poor countries with limited transportation and communications infrastructures. An entire international regime of election specialists

mushroomed in the 1990s, and the United Nations monitored and sometimes ran elections in many places. New information technology was allied to the political transformations of the third wave to produce a sense among many analysts that the expansion of democracy was limitless, irreversible, and conducive to all sorts of other positive changes, such as economic development and international peace.

Actually Existing Democracy

In fact, transitology is of limited usefulness in understanding the politics of new democracies. Many of these political systems are marked by deep disjunctures between the sophistication of their elections, on the one hand, and the legitimacy of their governments and the actual enjoyment of rights by their citizens, on the other.

Although elected civilian governments are far more common in the world than they were thirty years ago, the continuing or revived popularity of former dictators is remarkable. In South Korea, for example, surveys reveal that the most popular former ruler of the country is the late Park Hung Chee, the autocratic leader of a military regime from 1961 to 1979. In a recent poll of Paraguayans, 70 percent said that they were better off under the dictatorship of Alfredo Stroessner (1954–89) than they are today. Polls of Russians reveal the same sentiments about the Brezhnev era. And in Brazil, only 47 percent of respondents in a recent poll agreed with the statement, "Democracy is preferable to any other form of government." Eighteen percent said that in some circumstances, authoritarianism was preferable, and 29 percent thought that it made no difference. Although many new democracies now exist, therefore, it is striking that large numbers of people in these countries prefer authoritarian leaders and are indifferent or hostile to current democratic governments.

Nor do regular elections necessarily prevent the rule of dictatorial leaders. For example, in Zimbabwe Robert Mugabe succeeded in terrorizing the rural population into at least partially endorsing the continued rule of his Zimbabwe African National Union–Patriotic Front (ZANU–PF) Party. He directly appointed twenty of the hundred and fifty members of parliament, while his loyal followers controlled the electoral system, and liberation war "veterans" (many too young to have actually fought in the war) replaced politically unreliable schoolteachers as polling station managers. Because of these manipulations and (it is widely suspected) ballot rigging, Mugabe was able to maintain his hold on power despite a strong showing by the opposition Movement for Democratic Change (MDC) in the parliamentary elections of June 2000.

Elections are also sometimes ignored when the results do not favor incumbents. In Burma in 1990, the opposition National League for Democracy (NLD) won an election by a large margin but then was prohibited from taking office by the military. Many NLD members, including Nobel Peace Prize winner Aung San Suu Kyi, remain under house arrest or in prison. Similarly, in Algeria in December 1991, a militant Islamic party won the country's first-ever parliamentary elections. In response, senior army officers forced then-president Chadli Benjedid to resign and suspended the electoral process, forming a regime that remains in power today

Political and civil rights are therefore nonexistent for large segments of the population in many new democracies. For this reason, Freedom House classifies thirty-five of the world's hundred and twenty democracies as "electoral" but not "liberal" democracies. The number of people in these countries is 20 percent of the world's population. Why is democracy in such political systems so tenuous?

THE QUESTION is complicated, but one way to answer it is to compare new democracies to what I've already called the emblematic case of third wave transition. Spain in 1975, the year of its democratization, had three advantages that are not shared by many other countries: a mature economy, a functioning state bureaucracy, and a society with a high degree of consensus. Each of these deserves some attention.

In the mid-1970s, Spain had a mature capitalist economy in which low-wage agriculture was no longer the leading sector. It had a relatively egalitarian distribution of income compared to that of most developing countries. Spain today has a more egalitarian distribution of income than the United States and a human development score (calculated by the United Nations Development Program) higher than that of Belgium and the United Kingdom. Furthermore, as Spain weathered its first crises as a new democracy, its economy continued to grow, and it faced the enticing prospect of joining the largest and most successful common market in the world, the European Union (EU)—which it did in fact join in 1986.

As a member of the EU, Spain received substantial funds due to policies aimed at reducing economic inequalities within the union. These funds improved infrastructure and boosted employment, especially in Spain's poorest regions. In the period between 1989 and 1993, Spain received about twelve billion ecus (about ten billion U.S. dollars; the ecu preceded the euro), or almost one-quarter of the EU's total structural-aid spending in that period. This was a significant transfer of resources that most countries in the developing world cannot hope to attain.

In addition, Spanish state institutions perform relatively well. Indeed, the judiciary performs too well from the point of view of former dictators such as Chile's Augusto Pinochet and Guatemala's Efraim Rios Mont, both of whom have been investigated and charged with crimes against humanity in Spanish courts. Once again, the European Union's influence on Spain was strong and significant. As democracy deepened, the EU encouraged Spain to modernize its legal system and incorporate Europe's human rights provisions.

Finally, Spanish society had a fairly high degree of consensus, at both the mass and elite level, at the time of its transition. The conflicts that had created the Franco regime lay in the distant past. The most violent period of repression had ended three-and-a-half decades earlier. There was a strong consensus on both the right and the left that the horrors of the civil war should not be repeated. With the exception of Basque terrorism, there was a relative lack of ethnic tension or separatist sentiment

in the country. And finally, a relatively moderate ruling class was willing to accept, without violence, the re-emergence of the unions and the left that the end of authoritarian rule engendered. What some analysts call the "civic micro foundations" of democracy—trust, tolerance, respect for opposition, and acceptance of equal citizenship—were firmly in place in Spanish society before its transition. All these were strengthened in the years after Franco's death as Spain entered the EU, continued to grow economically, and staged successive elections that resulted in the alternation of ruling parties.

BUT OTHER new democracies frequently lack one or more of the variables that contributed to the success of the Spanish transition. In some cases, a mature and growing economy, a well-functioning state, and civic micro foundations are all missing.

It is in economic matters that the differences are most striking. Certainly, the revolution in information technology has benefited some developing countries. South Korea manufactures silicon computer chips and India produces software. But for many countries, previously huge socio-economic inequalities with rich countries have widened, as have domestic inequalities.

In Latin America, for example, there has been a marked jump in informal-sector employment in recent years in a region that is already the most unequal in the world. In 1980, the International Labor Organization estimated that 40 percent of Latin America's nonagricultural workforce was in the informal sector. (In agriculture, where work tends to be seasonal, that figure would be far higher.) These workers have no pension benefits, no unemployment insurance, no paid leaves of any kind, and no employment contract—they can be fired at will. By 1998, informal-sector employment had climbed to 58 percent, a majority of the workforce. Among women in the nonagricultural labor market, the figure was even higher, 65 percent that same year. With so many people so vulnerable, elections in Latin America have a different character than those in rich countries. The poor seek patrons in the political system, and politicians give away material benefits to win votes. Patronage and clientelism rob elections of much of their deliberative content, and while they serve as a redistributive mechanism in societies without extensive welfare states, their ability to ensure democratic accountability is diminished. This is true in other parts of the world as well.

In addition, the globalized world economy is highly volatile. Short-term, speculative investments in currencies, stocks, and other liquid assets flow in and out of developing countries quickly and erratically, subject to the wild swings of the international investment community. The U.S. Federal Reserve Bank estimates that economic volatility, measured by the variation in the growth rate of the gross domestic product from one year to the next, is almost three times as high in Latin America as it is in the countries that belong to the Organization for Economic Cooperation and Development (OECD). When economies are volatile, so is politics. In short, the prevailing development model excludes large numbers of people who are not property-owning stakeholders, but who live instead with a high degree of economic insecurity and deprivation. This is not a foundation for successful democracy.

Much the same story could be told in the area of state institutions, which were seriously debilitated in many developing countries as governments cut spending to cope with the debt crisis of the 1980s. Courts are often neither legitimate nor competent. Widespread crime and violence vitiate the effective exercise of democratic rights for millions.

In many developing countries, but not in Spain, cold war conflicts were large-scale and recent. They make the McCarthy period in the United States look genteel by comparison. The killing fields of Cambodia are one horrific example of this kind of violence, but there are many others. Much of the violence had an ethnic dimension, making it particularly difficult to ameliorate after a democratic transition. Fear and suspicion remain entrenched, hampering the creation of the civic micro foundations of democracy. Violence continues, especially in rural areas, where landowners, state officials, and other powerful groups use armed forces of various kinds to enforce their rule. Elections certified as "free and fair" by international election observers may be marred by widespread intimidation prior to the vote that is not easily monitored by short-term visitors.

Finally, although new information technology is used by states to organize elections, ordinary citizens often lack access to it. Grassroots movements have on occasion used technology—the Zapatistas in Chiapas used the Internet effectively after 1994, for example, and the Grameenphone program has recently provided cellular phones to 100,000 subscribers in 250 villages in Bangladesh. Information technology has great potential to reduce the cost to grassroots organizations of spreading information and organizing. But all too often ordinary people in developing countries are still excluded from the new technology networks. Jeremy Rifkin estimates that 65 percent of the current world population has never made a telephone call. A *Wall Street Journal* survey estimates that 57 percent of those people who regularly surf the Web live in the United States. Only 1 percent live in the Middle East, 1 percent in Africa, and 5 percent in Latin America. In these regions, only affluent elites have access to the information that the Internet provides. The emergence of what political writer Andrew Sullivan calls "dot.communism," in which universal access to the Internet's plethora of free information, goods, and services creates genuine political equality, at least in cyberspace, is far from a reality.

The transitology literature, then, overestimates the transformative potential of third wave democratization. It exaggerates the degree to which deep historical divisions within countries can be crafted out of existence by the correct decisions of democratizing elites. And it uses as a model Spain's transition, which on closer inspection looks exceptional, with favorable conditions impossible to match in most other new democracies.

Global Economic Trends

In addition to these problems, the transitology literature fails to link democratic transitions with changes taking place in the global economy. While some economic changes facilitate openness, transnational networking, and similar developments supportive of democracy, others severely restrict the choices of governments in developing countries.

Beginning in the 1980s, the so-called "Washington Consensus," or neoliberal orthodoxy, came to dominate official thinking about economic development in the U.S. government, the IMF and World Bank, and private banks and foundations. The Washington Consensus requires sharp attacks on fiscal deficits (and therefore cuts in government spending), financial liberalization (letting the market determine interest rates), trade liberalization (lowering tariffs and other barriers), privatization of state-owned enterprises, deregulation, and more rigorous enforcement of property rights, including intellectual property rights. The supremacy of the market is aggressively asserted over all other values, and the supposedly "statist" policies of the past are characterized *tout court* as failures. In the words of political scientist Benjamin Barber, the triumph of market ideology means "no government intervention (however mild) is safe from criticism, and no market mechanism (however violent and unjust) is subject to rebuke."

Powerful international forces push developing countries to adopt a single prescription for their economic troubles, one that leaves them, especially the smaller ones, with little room for alternative policies. The preeminence of the market was reflected in terminology— developing countries became "emerging markets" in much of the literature, as if their only importance was as an open field for the exports and investments of rich countries. The fact that so many third wave democracies emerged precisely during the rise of the Washington Consensus meant, paradoxically, that their ability to manage their own economies was markedly curtailed. They became, in the words of Malawian political scientist Thandika Mkandawire, "choiceless democracies." They are so indebted to and dependent on international financial institutions that they are not really free to make basic democratic decisions about which of their goods and services are to be allocated through the price mechanism and which are not, how certain markets should be regulated, or how their economic reforms should be carried out.

The U.S. government, or at least parts of it, led the way in the forging of the Washington consensus. What the economist Jagdish Bhagwati calls the "Wall Street–Treasury complex" has been firmly in control of U.S. policy, insisting that any government's attempt to limit the mobility of capital is unacceptable, that developing countries should seek always to privatize their industries and lower tariffs, that the intellectual property rights of U.S. firms must take precedence over local needs, and that top priority must always be given to the wishes of the largest financial conglomerates and multinational corporations.

In its foreign economic policy, the United States has rarely deviated from the Wall Street–Treasury Department line. With a singlemindedness that would make the old *apparatchiks* of the Soviet Union proud, U.S. trade officials, State Department employees, and USAID technicians repeat the mantra in international conferences and in the capitals of developing countries: governments should be minimally involved in the economy; even surplus-generating state enterprises should be privatized;

social goals should not be taken into account when evaluating government performance. The only acceptable criterion of state reform is efficiency, measured by the bottom line of the business world; government should not seek to protect an overarching public interest distinct from the preferences aggregated by the market—indeed, there is no public interest distinct from the market; the U.S. government should aggressively represent the interests of U.S. exporters, investors, and other companies (who after all bankroll its elected representatives) rather than the whole of its population, and changes in developing countries should be evaluated solely in terms of the opportunities they provide to those interests. The relentless repetition of this economic formula does not amount to genuine global leadership. It is a politically unsophisticated form of salesmanship that looks hypocritical, since the U.S. government continues to maintain protectionist barriers against developing country exports in agriculture and other sectors.

The view of politics fostered by the Washington Consensus is a utilitarian one. Democracy is seen as useful because it can generate the marketizing and commercializing changes that are "necessary" for the world economy. But democratic politics is feared as well as desired, because it gives rise to views opposed to the consensus. For this reason, much of the writing about the global economy from Washington these days uses the term "governance" rather than "politics." The difference is telling. Governance refers to the administration of people and things. Politics signifies engagement in political affairs. The first assumes that the challenge facing the world economy is to make the management of the status quo more efficient; it takes the dominance of the current orthodoxy for granted. The second allows for conflict between fundamentally different visions and parties; it is open to the possibility of new political winners and meaningful change.

As THE LAST century began, socialist and other radical movements were attempting to democratize politics at the national level by applying the principles of parliamentary democracy to the management of the "commanding heights" of the economy. Today, the terminology is different, and no one is seriously talking about eliminating markets. But many social movements are engaged in a similar project at a global level, attempting to apply the principles of parliamentary democracy to the institutions that manage the international economy. This is the logical next step in the struggle for democracy, which alone can sustain the third wave of democratization.

Whenever new forms of political authority are consolidated, social movements can open new space for democratic contestation. With the rise of nation-states in Europe after the French Revolution, domestic movements with their tactics of strikes, petitions, demonstrations, and marches were born. When the political form of the nation-state was carried by European powers to Africa and Asia, those colonized people used the rhetoric of European nationalism to win their own political independence. And as globalization proceeds, new social movements are emerging, capable at the very least of stopping the most egregious corporate and governmental manipulations. Global institutions such as the WTO should not be opposed *in toto,* but reformed and used as structures within which democratization can take place. A retreat into fortified, nationalist, autarchic states would probably be the worst thing that could happen to democracy in the world today.

The goal of nineteenth-century radicals still lies visible on the horizon in the twenty-first century. The commanding heights of the global economy are still run by small oligopolies of "Davos men": people, corporations, and states rich enough to have an influence on the rules of the game. On the board of the International Monetary Fund, for example, two-thirds of the voting rights are in the hands of representatives of the United States, Japan, and Europe, who speak for less than one-eighth of the world's population. Global grassroots movements might be able to force this and other institutions to move closer to the principles of universal suffrage and democracy, and thus produce a more humane global economy. The managers of the global institutions of "governance, and the corporate and state interests that prop them up, will not accede to change without a fight. But the fight is well worth the effort.

The market is not sacrosanct. People can socialize it so that human values help to determine the conditions of production and the allocation of goods and services. But they can only do that by questioning and challenging the vision of economic and political change promoted by those who run the world's economy. Advocates of democracy need to transcend the narrow confines of the transitology approach and take account of the global economic environment in which democratic transitions do, or do not, take place. Democracy is increasingly about who makes the global rules, not just about who replaces authoritarian leaders in individual countries.

ANTHONY W. PEREIRA teaches political science at Tulane University in New Orleans. He is the author of *The End of the Peasantry,* which deals with the rural labor movement in northeast Brazil.

Two theories

Is political Islam past its peak or a mounting danger?

BY AND large, and simplifying mightily, analysts of political Islam can be grouped around two opposing theories. An optimistic theory holds that violent Islamism reached a peak in the 1980s and 1990s and has now been defeated. The pessimistic theory holds that the Islamists are gaining strength and continue to pose a grave threat to the political order of Muslim states and possibly to the wider world.

The optimists draw heavily on a magisterial book by a French academic, Gilles Kepel, called "Jihad: the Trail of Political Islam", and on a shorter book by another Frenchman, Olivier Roy, called "The Failure of Political Islam". Mr Kepel's book was first published in 2000, before the attacks of September 11th, but he sees no need to revise his overall conclusion. His argument is that almost wherever the Islamists have tried to capture majority support—from Algeria to Iran to Afghanistan—they have failed. Like communism, the blueprint has lost its appeal. "Muslims no longer view Islamism as the source of Utopia," he says, "and this more pragmatic vision augurs well for the future."

Part of this theory cannot be argued with. Whereas September 11th jolted people in the West into noticing the perils of Islamism, the Muslim world itself had been convulsed by the challenge of political Islam throughout the preceding decades. And by the end of the 1990s it really was beginning to look as if the secularists, not the Islamists, or at least not the violent ones, were beginning to prevail.

Consider how things had changed. In three countries—Iran after its 1979 revolution, Afghanistan after the eviction of the Soviet Union and Sudan after an Islamist coup in 1989—Muslim governments had tried to create something resembling an authentic Islamic state. In Iran and Afghanistan Muslim revolutionaries defied one superpower and humbled another, briefly electrifying Muslims everywhere. And yet by the end of the 1990s the experiment in Sudan had collapsed, the Taliban had turned Afghanistan into a reviled dystopia and the Iranian theocracy was losing its hold not only on its foreign admirers but also on its own people. As these failures sank in, the simple slogan that "Islam is the solution" began to look somewhat less plausible.

During the same decade, Islamists suffered another sort of setback. In countries where armed *jihadis* decided to do battle with the state, they were defeated. Egypt crushed the Gemaa Islamiya. After more than 100,000 deaths, the Algerian civil war of the 1990s petered out with a victory, if a Pyrrhic one, for the regime. The mujahideen guerrillas who routed the Red Army in Afghanistan failed to turn Bosnia into another *jihad*. Hence Mr Kepel's conclusion that by the end of the 1990s violence had become a dead-end for the Islamists. It did not lead to power, and it terrified the middle classes. If political Islam has a future, says theory one, it will henceforth consist of working within the rules of democracy and making the appropriate compromises.

The advocates of theory two, the pessimists, agree with some of this analysis. They accept that the *jihadis* failed when they clashed head on with the state. But they are less sanguine about the present state of affairs. They worry that although the states have crushed the violent groups, the influence of political Islam has not only survived but grown stronger. And they do not altogether, or at all, buy the claim of the non-violent Islamists to have adopted democratic values. Of course the Islamists say they have: they calculate that they would do well in free elections. But what if these parties turn out to be Trojan horses? What if they used elections to capture power but had no intention of letting anyone vote them out afterwards?

Again, part of this theory cannot be argued against. Despite the setbacks of the 1990s, the influence of political Islam—the idea, that is, that

Islam should have something important to say about the way society is governed—has spread. Sometimes this growing influence can be measured in electoral victories, such as the landslide victory in Turkey in November 2002 of the Justice and Development Party led by Recep Tayyip Erdogan, or the capture a month earlier by Pakistan's Muttahida Majlis-i-Amal, an alliance of six Islamist parties, of 60 out of 342 seats in the National Assembly plus control of the provincial government of North-West Frontier. But success in elections is only one bit of the evidence.

The power of opposition

Even where Islamists do not win elections, or are not allowed to, they often form the only opposition that governments really worry about. There is no longer much of a "left" left in large parts of the Muslim world. Whereas the secular parties are often shells, the Islamists tend to have large and active memberships, often linked to mosques, who combine impressive welfare work with what is known as *dawa* (spreading the faith), designed to Islamise society from the grassroots. And whereas "civil society" in most poor countries is starved of money, cash pours into the Islamists. This makes them powerful even when they are not in power. "It would not be an exaggeration", says Graham Fuller, a former vice-chairman of the CIA's National Intelligence Council, "to state that Islamists are probably more focused on civil society and the creation of institutions within it than any other political force in the Muslim world."

Besides, the Islamists do not need always to win power in order to get their way. Secular regimes that feel Islam snapping at their heels have long responded by throwing tidbits its way just in case. Indonesia's former president, Suharto, tried to present himself as a devout Muslim in a vain attempt to save his regime. So did Pakistan's Zia ul Haq, a military dictator, and Zulfikar Ali Bhutto, a secular socialist. Other regimes have bowed to Islamist demands by putting bits and pieces of *sharia* into the law of the state, or, like Egypt and Syria, saying in their constitutions that Islamic law is one source for national law. In Malaysia, Mahathir Mohamed, under pressure from the Islamists of the PAS opposition, broke with his country's secular tradition in 2001 and declared Malaysia "an Islamic state". Morocco's king styles himself "defender of the faith".

Indeed, such have been their gains in opposition, and in cultivating their grassroots networks on the ground, that some Islamists think they might be better off opposing and cultivating than being in power. Rashid al-Ghannushi, an influential Tunisian exile in London, has said that if the Islamists' social-missionary work comes into conflict with their political interests, the former must be put first, because Islam's social achievements are liable to be more permanent than its political ones. "The most dangerous thing", he says, "is for the Islamists to be loved by the people before they get to power and then hated afterward." (Iran take note.)

The pessimists say the "successes" notched up this way by political Islam in opposition shift the centre of gravity in Muslim societies towards religion, bigotry and censorship and away from secular liberalism. The optimists demur. Do not think that the aim of Islamist parties is only to give religion a bigger say, they argue. These groups are trying to reform society in numerous ways, using Islam to make their message more appealing. They are not lying: it really is Islam that shapes their view of the world. But they are pragmatists inspired by the progressive values they discern in their faith, not ideologues scouring scripture in search of some ready-made blueprint.

One extreme optimist of this kind is Mr Fuller, that former CIA analyst. He calls political Islam (except in places where the Islamists have grabbed power by force) "the single largest, most vibrant, growing, widespread and active movement in the Muslim world in seeking to strengthen democracy, human rights, civil society and, generally, liberal economies." But even in the Muslim world, this rosy view is controversial, to say the least.

Emad Shahin, a political scientist at the American University of Cairo, is a believer: political Islam, he says, is "the wave of the future". But just down the corridor, Bahgat Korany voices the misgivings of many Muslim intellectuals. "On the one hand, the Islamists are a vibrant force of civil society, an oppositional force with some semblance of a conceptual alternative. On the other, I have lots of questions about what sort of political system they are trying to establish. Plus I wonder whether this is the road to permanent democracy." Some of the Islamists, he says, are too apt to cite religious text as all the authority they need, and not ready enough to admit the possibility of error.

A lot of non-Muslim scholars are a good deal more sceptical than this. Emmanuel Sivan, a professor at the Hebrew University of Jerusalem, argues that people like Mr Fuller, who think the Islamists are the progressives in Muslim societies, pay too much heed to what such groups say to outsiders and too little to what they say among themselves. Though there are liberal voices among the Islamists, says Mr Sivan, these are drowned out by the extremists. The liberals publish learned papers; the extremists have mastered the art of spreading their message far and wide by means of sermons and debates distributed on audiocassettes. Mr Sivan reckons that of the 400 or so tape-cassette preachers, none is a liberal.

It may not matter whether western scholars believe that Islamists are genuine democrats. It does matter what fellow Muslims think. This is because most Muslims say they want democracy. The same Pew survey in June that reported growing

Muslim rage at America also found a strong appetite for democratic freedoms. A poll of some 50 countries—from kingdoms such as Jordan and Kuwait as well as authoritarian states such as Uzbekistan and Pakistan—found strong support for freedom of expression, freedom of the press, multi-party systems and equal treatment under the law. In some parts of the Muslim world, these values scored higher than in parts of central and eastern Europe.

But can you trust them?

Encouraging. But will Muslims who support democracy in theory support it in practice if they cannot trust the Islamists to abide by the rules of the game? Mr Sivan argues that by monopolising the opposition in Muslim countries the Islamists have, as it were, put the fear of God into the very middle classes which might otherwise be democracy's champions. Like Latin America's middle classes when Marxism was strong, they hesitate about taking a chance on fair elections, lest that should hand power permanently to the extremists. "The reason democracy has not penetrated Islam has nothing to do with some essential opposition from the religion," says Mr Sivan. "It's not long since Catholics were looked upon in this light. Democratisation has failed to a large extent because of the radical danger: the fear that it will be one vote, one man, one time." Will it?

AFGHANISTAN

Not a dress rehearsal

Without new thinking, Afghanistan will fail again

CHAGHCHARAN AND KABUL

AFGHANISTAN was not supposed to be simply a dress rehearsal for the invasion of Iraq. It was meant to be a premiere, the blueprint for how to rescue a failed state without colonising it. Apart from America's starring role, it does not much resemble Iraq. It is poor; it has no oil. It became a danger only because it was allowed to be weak, not because it was allowed to be strong. Unlike in Iraq, the locals, almost two years after the fall of the Taliban, want more American engagement, not less. And not just American: the rebuilding of Afghanistan is a shared undertaking of many nations, with a clear United Nations mandate.

In one way, though, it does resemble Iraq: the stakes are high. Failure would be dreadful for Afghanistan, which will not get a better chance to make something of itself any time soon, but also for America, which has staked much credibility on building a terror-free Islamic democracy. But failure would also be bad news for the outside world, including France and Germany, which can hardly wag fingers at America if they are complicit in Afghanistan's undoing. Will the world rise to the challenge?

Perhaps it is beginning to. On August 11th the international peacekeeping force in Kabul, known as ISAF, was placed under the strategic command of NATO. British, Turks, Dutch and Germans have all headed the force of 5,000 or so peacekeepers, which operates separately from the 12,000-strong American-led coalition force still in the country. Their troops and those of 26 other nations have done much to stabilise Kabul; the city is seamy, but also safe and booming. The cost has been high, not least in lives. In June, a suspected al-Qaeda suicide-bomb attack killed four German soldiers; in May, a plane flying Spanish peacekeepers home went down, killing 75. Canada will now assume tactical, or day-to-day, command. Its soldiers will play an important role in deterring terrorism and reducing crime in the capital. But the real peace must be won where ISAF has so far had no interest in going, outside Kabul.

Expansion of ISAF to the provinces appears to be a political non-starter. Instead will come Provincial Reconstruction Teams (PRTs)—clusters of lightly armed soldiers meant to assist rebuilding. The Americans already have several PRTs, Britain and New Zealand each have one, Germany will likely take over another in Kunduz. If all goes well, there might soon be a dozen such units around the country. PRTs have a vague mandate. They are not coalition forces, but receive air support from the coalition. They are not ISAF either, but draw on ISAF's experience. They have the advantage of being mobile and malleable—key assets in Afghanistan. "We're reinventing the century of the Roman empire, but why not?" says one general. Still, PRTs will at best be only a catalyst. They are too small to effect much change by themselves: the 72-man British one in Mazar-i-Sharif is charged with an area the size of Scotland.

Separately from ISAF, the American-led coalition—"Operation Enduring Freedom" to initiates—has been working hard to stabilise the southern bit of Afghanistan, though it is hard to quantify its success. Neo-Taliban caught taking pot-shots are certainly mown down by Apache helicopters. But Osama bin Laden and Mullah Omar, the Taliban leader, remain unaccounted for, and the south of the country is getting more dangerous, not less. Major operations—at present including Italian alpine troops and the first recruits of the new Afghan national army—often result in coralling a few men, who may be terrorists or, just as easily, shepherds. Detainees are spirited off to the Bagram base, north of Kabul, and interrogated. Several have died there. But as major combat dies down, questions are sure to be asked about the cost-effectiveness of what the coalition is up to, given that the Americans are spending $10 billion or so a year on it, versus the half-billion being spent by ISAF.

The good news

Whether or not Afghanistan can survive comes down, in part, to one simple question: are the forces of national integration there greater than the forces of local disintegration?

Optimists say yes. They think Afghanistan is more stable than at any time in the past 24 years. Many, perhaps all, of the terrorist training camps in the country have been destroyed. A soon-to-be-released study by foreign and local aid agencies suggests most Afghans think security has improved. Incrementally, optimists say, a semblance of national government is returning. On August 13th, the government in Kabul removed the warlord Gul Agha Sherzai from his post as governor of Kandahar and stripped Ismail Khan, the powerful governor of Herat, of his other role as regional military commander. A paper decree may not worry Mr Khan much, but it, and the removal of Mr Sherzai, shows the determination of the government.

On the optimists go. Afghanistan enjoys a legitimate government, confirmed by a representative *loya jirga*. There is considerable progress in writing a constitution and organising elections for next year, with suffrage for women. The new national currency, the afghani, is widely accepted and stable. The economy grew by 28% last year, according to preliminary IMF estimates. Two million or more refugees have returned home to rebuild their lives. That—together with the remarkable absence of any ethnic separatist movements—underlines Afghans' belief in their own country. There has been no major humanitarian crisis. Donors remain committed to their promises. America has tripled its aid to $1 billion this year; it will pressure others to do the same.

And the bad

Pessimists scoff at much of this. Taking a marker pen, they score off on a map the third of the country—the south and

southeast—that donors now think is too dangerous to visit. Aid workers have not recovered from the brutal execution of a Red Cross worker by neo-Taliban earlier this year. Even old hands are uneasy; for the first time they are the targets. This week, the UN suspended its road missions in the south of the country. On August 13th, a bomb in the southern province of Helmand killed at least 15 people, and some 45 more, including two local aid workers, died in an array of other bloody incidents. Taliban leaders operate in Quetta and other Pakistani cities, openly distributing weapons and propaganda. Pakistan itself is destabilising its neighbour. Its army has made incursions into Afghanistan. There have been heavy exchanges of mortar fire.

Every plan for Afghanistan made tangible reconstruction a priority. Unfortunately, there has been very little of it. No major roadbuilding projects have been completed. The Kabul to Kandahar road, engineers whisper, is being laid with pencil-thin tarmac to meet the year-end deadline personally set by George Bush; a couple of winters might wreck it.

The pessimists—they would call themselves realists—have little time for the Afghan president, Hamid Karzai. America's boy, they say, isolated and ineffectual. The real power in the provinces is with the warlords. Governors may pay lip service to Kabul but they will not give up their militias. Even if they wanted to, the disarmament process does not have the money to make it possible. The constitution will be fudged; the election delayed. Women will remain powerless.

The economy has grown, yes, but any growth looks good when starting from zero; it is still less than half the size it was in 1978. Returning refugees are struggling to make a go of it; many are drifting to slums in Kabul. To say that there is no humanitarian crisis is to miss the point. There is crisis by attrition.

Guns, butter and poppies

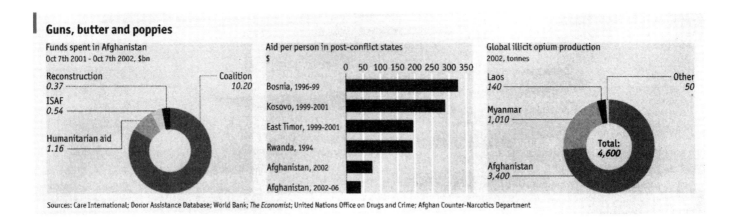

Funds spent in Afghanistan
Oct 7th 2001 - Oct 7th 2002, $bn

Reconstruction
0.37

Coalition
10.20

ISAF
0.54

Humanitarian aid
1.16

Aid per person in post-conflict states
$

0   50 100 150 200 250 300 350

Bosnia, 1996-99

Kosovo, 1999-2001

East Timor, 1999-2001

Rwanda, 1994

Afghanistan, 2002

Afghanistan, 2002-06

Global illicit opium production
2002, tonnes

Laos
140

Other
50

Myanmar
1,010

Total:
4,600

Afghanistan
3,400

Sources: Care International; Donor Assistance Database; World Bank; *The Economist*; United Nations Office on Drugs and Crime; Afghan Counter-Narcotics Department

Most Afghans still have no access to health care. Rates of maternal and infant mortality remain among the highest in the world. Cholera and other diseases are in the ascendant. As to donors, best not to ask. Well under $1 billion of the $4.5 billion promised at the Tokyo conference last year has shown up (though, to be fair, that sum was to be spread over the next few years). Too much of what does arrive ends up going on fat salaries and snazzy cars for foreigners.

Reasons to be doubtful

It is possible to debate these points endlessly, and many in Kabul do little else. But such arguments may be irrelevant in the face of two factors which, without a much more aggressive commitment of resources, could devastate Afghanistan, no matter what the fine-print of the new constitution or the efforts of economists.

The first is water. Afghanistan is a marginal country. About 80% of Afghans depend on what they can grow. But Afghanistan lacks water and cultivable land. Even in the halcyon 1970s, less than 5% of the land was irrigated. The war halved that. Then came the drought, seven years long in some places. Most of the livestock died, staple crops failed. In the south and south-west of the country, water-tables are dangerously low. Even with the best possible governance, that part of Afghanistan is a poor proposition. Drought was an ally of the Taliban. They could not have pushed north without picking up farmers along the way who, having lost their wheat and goats to drought, thought to earn something by shouldering a gun. Rebuilding the irrigation system would help a bit: creating reservoirs in the mountains of the central Hazarajat region could do more: both would be hugely costly.

The second worry is drugs. Three-quarters of the world's opium, and nearly all of Europe's heroin, originates in Afghanistan. Drought is now an ally of opium traders as it was of the Taliban. Pushtun herders used to move livestock around the country. Now they move opium; buying, selling, trading know-how. They use the economics of the old North Sea herring trade, locking in farmers by paying advances on next year's harvest. Afghanistan and Britain—the lead donor for counter-narcotics—say they have evidence drug money is funding terrorism in southern Afghanistan.

The view from rock bottom

The worry is that drought, drugs and insecurity could start to feed off each other. A glance at the map suggests they already do. Three of the country's five big drug producing provinces—Helmand, Uruzgan, and Kandahar—are unsafe and parched. Poppy cultivation is spreading to new areas, and with it insecurity. The nightmare is a new Colombia: a place where drug lords capture and wreck government and the economy alike.

Take Ghowr, one of the poorest and most isolated provinces in Afghanistan. It lies four days' drive west of Kabul on often impassable dirt roads. Poverty here, as in much of Afghanistan, is a constant: Ghowr has been failing since seaborne trade superseded the Silk Road.

Most of Ghowr's 3,000 mullahs are illiterate; they make a living passing out magic charms and spells. It has only five doctors and ten foreign-aid workers, mostly from France's Médicins du Monde. Many of its 750,000 people are newly returned refugees.

They have almost nothing. The pale green steppe, folding immensely away to the horizon, is drought-ravaged. The livestock have died or been sold. Farmers—that is, almost everyone in Ghowr—have fallen deeply in debt. The most desperate have sold their daughters. The rest have planted poppies. But the hoped-for opium bonanza has not arrived. The soil was poor. Mistakes were made. A late frost—Ghowr lies at high altitude—dried the bulbs before the opium gum could be scraped off. Nearly all of the harvest was lost. So the farmers' debts are still mounting, with much of the money owed to opium traders who will insist on a new planting next year.

A recent visit to Chaghcharan coincided with the planned arrival of Mr Karzai; a welcome sign of national integration. The bazaar was festooned with Afghan flags run up by traders themselves from strips of red, green and black cloth. Ghowr's leading mullahs gathered in the mosque to prepare a list of problems Mr Karzai might like to address. Illiteracy and a lack of water topped the list. Opium was not to be mentioned. It was poison, the mullahs said, but perhaps Mr Karzai would understand people had no choice. "In other countries they use opium for enjoyment. Here they grow that their children not die," said one sadly.

A group of women spoke secretly, too fearful to put their complaints directly to Mr Karzai. A few women now have jobs and women can walk freely across the bridge (in Taliban times women had to wade the river downstream so as not to arouse the men, they said). But not much else has changed. Perhaps only 50 women in the province can read and write. Girls still have trouble getting an education. Daughters are still regarded and spoken of as chattels, to be traded for debt or profit, or as settlement in blood feuds.

Mr Karzai was delayed. An illness, said the Americans. No available plane, said the president's office. Disappointed elders drifted back to their villages on the backs of motorbikes. The president did make it the following week. The militia paraded for him. But it was too late to see Mehman Gul, a 15-year-old girl who had been carried from a village by her father. She had been set down on the hospital floor on the day of Mr Karzai's earlier planned visit but, sadly, she died an hour or so later, of typhoid and hunger.

Shooting up

Ghowr is too marginal ever to become a big opium producer, but even so it could become dependent upon poppy cultivation and be shaped by it, like the country as a whole. Opium now accounts for perhaps a third of the Afghan economy. Last year's opium harvest of 3,400 tonnes was worth $1.2 billion in the country. By the time it is injected into veins in Edinburgh or Prague as heroin, profitably bulked out with brick dust and glucose, it is worth $25 billion.

The drug trade is becoming institutionalised. Opium is now being processed into morphine and heroin inside Afghanistan. That means a lot more money for commanders on the ground, something made apparent by the switch to ever more expensive jeeps. Democracy plays into the hands of more sophisticated narco-enriched commanders. They are already thinking about ways to buy, or muscle, a vote which will protect their opium

interests in next year's election. "There's no way we can compete with the cash these guys have," says a morose UN insider.

The Afghan government has made some progress. Poppygrowing has been declared illegal. A new policy body, the Counter-Narcotics Department, or CND has been instituted to direct drug policy in key ministries. Its goal is a 70% eradication by 2008 and 100% by 2013. The CND is being bankrolled by the British government. But it remains woefully ill-equipped. Almost none of its 28 staff officers has any relevant experience. There is little money for communications or vehicles and nothing at all for intelligence gathering. "They're expecting results for nothing," says Mirwais Yasini, the director. If traffickers get hauled in, trying them is another matter. The country even lacks a laboratory which can identify whether shipments seized are contraband.

If the stick is small, the carrot is smaller still. There is little money for alternatives which would allow farmers to grow viable money-earning crops such as saffron. Mr Yasini reckons a basic scheme could cost $300m over three years. Again, the outside world could be doing much more to help.

An attempt to buy out farmers last year only encouraged more areas to plant poppies, so something more radical and innovative is needed: the insertion of several hundred counter-narcotics police officers about the country, a kind of policeman's ISAF. The narco-cops would need to be ceded powers by the Afghan government to disrupt the flow of drugs and assist in crop eradication. They would have to be supported with EU-funded initiatives such as the purchase of wheat at above market prices and money for irrigation, husbandry and rural credit schemes.

America could also play a more useful role by acknowledging and supporting the efforts of Iran—whose policy on drugs is in many ways more intelligent—and by cracking down on (rather than supporting) warlords and commanders its special forces know to be trafficking opium.

And those who played a part in the wrecking of Afghanistan have a responsibility to help put it back together. Few expect Russia to cough up for the carnage unleashed by the Soviet Union, but it could supply survey maps and geologists to help Afghanistan exploit its own natural resources. If Afghanistan could discover a legal export—gold and gemstones are possibilities—to match opium, it might yet prove the pessimists wrong.

ONE COUNTRY, TWO PLANS

BY MARINA S. OTTAWAY

The United States cannot turn to a ready-made model of occupation and reconstruction in postwar Iraq because none fits the country's condition. Turning Iraq into a politically and economically stable nation is as complex a task as planning for war. U.S. military superiority, which made success of the war a foregone conclusion, does not ensure successful reconstruction.

Making an intrinsically difficult task even more complex, the United States is currently guided by two conflicting models of political reconstruction, each subject to a different logic and different imperatives. Under the first, the United States would help Iraq create a decentralized, participatory democracy; under the second, the United States would swiftly give control to an interim Iraqi government.

Iraq has all the characteristics that have impeded democratic transitions elsewhere: a large, impoverished population deeply divided along ethnic and religious lines; no previous experience with democracy; and a track record of maintaining stability only under the grip of a strongly autocratic government. The United States enjoys no clear advantage in trying to develop a new political system for Iraq. It has no historical ties to the country and little understanding of Iraqi culture and society. Many Iraqis resent the United States as an occupying power. And the factor that made the war so successful—reliance on a relatively small, mobile force whose strength lay in technological superiority rather than manpower—is a serious liability to reconstruction. Stabilizing a country requires a large, visible presence on the ground, not sophisticated weapons.

Before the war, when U.S. President George W. Bush was trying to win domestic and international support for intervention, his administration committed itself to the democratic reconstruction of Iraq and the region. "America's interests in security, and America's belief in liberty, both lead in the same direction: to a free and peaceful Iraq," declared Bush in February 2003. In the following weeks, administration officials outlined how they intended to achieve that goal.

The plans presented during this period exuded confidence that the United States had the capacity not only to replace Iraqi President Saddam Hussein's regime but to alter the character of the state and the very social fabric of Iraq. Under U.S. military occupation, U.S. officials and contractors would vet the Iraqi civil service. They would exclude hard-line members of the regime and the Baath party and rehabilitate those not overly tainted by association with the former government. The United States would also create and train a Baath-free military and police force.

The U.S. Agency for International Development (USAID), which is charged with implementing reconstruction, outlined an extraordinarily ambitious program in its "Vision for Post-Conflict Iraq." U.S. contractors would oversee the rehabilitation of physical infrastructure and government services, restoring health services to 25 percent of the population in 60 days and to 50 percent (but to 100 percent of women and children) in six months; they would implement a new educational curriculum for the schools within a year; and they would restore the country's roads and electrical grids with equal speed.

The plans for political reconstruction were even more remarkable. USAID stated that "the national government will be limited to essential national functions, such as defense and security, monetary and fiscal matters, justice, foreign affairs, and strategic interests such as oil and gas." Local government would be responsible for everything else and would be "required to operate in an open, transparent, and accountable matter." Citizens would participate in planning the future of their communities and would control the civil administration through elected local assemblies. At a sweep of the U.S. pen, Iraq would turn from a

A Moving Target: The Cost of Iraq's Reconstruction

Source of Estimate	Annual Cost	Total Cost
The Bush Administration "Supporting Our Troops Abroad and Increasing Safety at Home" (White House Office of the Press Secretary, March 25, 2003)	$3.5 billion for Iraqi relief and reconstruction	N/A
Center for Strategic and International Studies "A Wiser Peace: An Action Strategy for a Post-Conflict Iraq" (Washington: Center for Strategic and International Studies, 2003)	$6.2 billion to $7.9 billion over two years for security and police forces, transitional administration, national dialogue process, justice team, debt restructuring, and employment and education programs*	$6.2 billion to $7.9 billion*
Congressional Budget Office (CBO) "An Analysis of the President's Budgetary Proposals for Fiscal Year 2004" (Washington: Congressional Budget Office, 2003)	Between $1 billion and $4 billion per month for occupation. The CBO has provided no estimates of reconstruction costs.	Unknown
United Nations Development Programme "U.N. Estimates Rebuilding Iraq Will Cost $30 Billion" (*New York Times*, January 31, 2003)	At least $30 billion over the first three years	At least $30 billion
Council on Foreign Relations "Iraq: The Day After" (New York: Council on Foreign Relations, 2003)	$20 billion a year for reconstruction, humanitarian aid, and post-conflict peace stabilization	Unknown
American Academy of Arts and Sciences Committee on International Security Studies *War with Iraq: Costs, Consequences, and Alternatives* (Cambridge: American Academy of Arts and Sciences, Committee on International Security Studies, 2002)	A "Marshall Plan" for Iraq would cost $75 billion over six years.	$31 billion to $115 billion
Center for Strategic and Budgetary Assessments "Potential Cost of a War With Iraq and Its Post-War Occupation" (Washington: Center for Strategic and Budgetary Assessments, 2003)	$20.6 billion to $118.6 billion over the next five years for occupation force, humanitarian aid, governance training, reconstruction and recovery, and debt relief	$103 billion to $593 billion

*This is a recommendation, not a projection.
For links to these documents, please visit www.foreignpolicy.com.

centralized, hierarchical country into a model of participatory democracy.

But this vision for Iraq was completely uninformed by the situation on the ground. As soon as U.S. and British troops entered Iraqi cities, it became clear that the coalition did not have complete control and could only establish it with a much larger U.S. presence and the use of repression. Initial conditions could not be ignored. With resentment toward the occupation mounting even before the war ended, the United States could either stick to the original reconstruction plan and pacify the country by force or take a new approach. Unwilling to increase the size of the occupation force, the U.S. government opted for a new policy.

Within days of the first rumblings of opposition to the U.S. presence, Bush administration officials began discussing a short, light-handed occupation and the swift transfer of power to an Iraqi interim authority (as if that was what they had envisaged all along). Forgetting the detailed plans for a decentralized, participatory system, U.S. officials declared that the United States would not impose a particular political order on Iraq.

Quietly, however, the original, highly interventionist plans for political reconstruction proceeded. In mid-April, USAID awarded the Research Triangle Institute, a North Carolina con-

tractor that often implements USAID projects, a $7.9 million contract, expected to grow to as much as $167.9 million over 12 months, to strengthen "management skills and capacity of local administrations and civil institutions to improve delivery of essential municipal services such as water, health, public sanitation and economic governance; includes training programs in communications, conflict resolution, leadership skills and political analysis." Huge by the standards of political reconstruction programs, such a contract shows that the administration has not abandoned the technocratic project of remaking Iraq into a decentralized, participatory system—despite the United States' lack of full control.

AT A SWEEP OF THE U.S. PEN, IRAQ WOULD TURN FROM A CENTRALIZED, HIERARCHICAL COUNTRY INTO A MODEL OF PARTICIPATORY DEMOCRACY.

Also in early April, USAID issued an invitation to contractors to bid on a 12-month, $70 million Iraq Community Action Program, which "will create community committees responsible for identifying and prioritizing community needs, mobilizing community and other resources, and monitoring project implementation." This agenda is not the program of a light-handed occupation.

Thus, consciously or not, the United States is simultaneously applying two contradictory reconstruction policies in Iraq. Each has a separate logic and coherence, but combining the two renders both illogical and incoherent. Where the United States has little control, invasive projects of social and political transformation cannot succeed. U.S. contractors cannot channel political participation through the new structures they are supposed to create unless there is an occupying force large enough to curb the influence of religious and tribal leaders. Hoping that a light occupation and a quick transfer of power will result in the democracy the United States promised the world is either deeply cynical or excessively optimistic. Hiding a heavy-handed American occupation behind the facade of a quickly formed Iraqi interim authority could theoretically reconcile the two approaches. Yet Iraqis are likely to notice the strings and turn on both the puppets and puppeteer.

In the coming months, developments on the ground will reveal whether, after some initial confusion, the Bush administration is making a serious attempt to turn Iraq into a more democratic country, rather than simply one friendly to the United States. One strong indicator will be whether the military finally takes responsibility for establishing and maintaining law and order in the country, thus giving civilian administrators, U.S. contractors, and Iraqi citizens a chance to undertake the physical and political reconstruction process. If the United States insists that what is left of the Iraqi police can and must do the job, it does not matter whether a civilian or a military administrator is in charge of Iraq, or how many contracts USAID signs. The reconstruction of Iraq will de facto be undertaken, as in many other countries, by the strong and the ruthless.

Marina Ottaway is the author of Democracy Challenged: The Rise of Semi-Authoritarianism *(Washington: Carnegie Endowment, 2003) and a senior associate at the Carnegie Endowment.*

Iran's Crumbling Revolution

Jahangir Amuzegar

WESTERN REPORTERS tend to describe the current situation in Iran in alarmist terms, suggesting that the people are near revolt, the regime faces collapse, and the country is prone to political upheaval. Even if these assessments are premature or extreme, the relentless confrontations between the "reformist" *Majles* (national assembly) and the "conservative" Council of Guardians (which has veto power over *Majles* legislation and vets all candidates for elective office) augur a turbulent political future. The 1979 revolution faces a profound challenge from a new and disenchanted generation, widely known in Iran as "the Third Force." For this broad swath of society born after 1979, Ayatollah Ruhollah Khomeini's promise of a just and free Islamic society has proven a sham. After nearly a quarter-century of theocratic rule, Iran is now by all accounts politically repressed, economically troubled, and socially restless. And the ruling clerical oligarchy lacks any effective solutions for these ills.

The changes wrought by this turmoil call for a new and nuanced U.S. policy toward the Islamic Republic—particularly if the United States goes to war against Iraq. Since the high-profile inclusion of Iran in President George W. Bush's "axis of evil," proposals to deal with that "rogue" state have run the gamut from a preemptive military strike to the pursuit of diplomatic engagement. Between these two extremes, suggestions have included covert action to destabilize the ruling regime, assistance to internal and external opposition groups, financial aid for foreign-based Iranian media, and a call for international condemnation of the ayatollahs. To know what shape U.S. policy should take, however, it is necessary to understand how Iran arrived at its current parlous state.

THE BEST OF ENEMIES

AFTER THE SEIZURE of the U.S. embassy in Tehran in November 1979 by a radical group calling themselves "Students Following the Imam Line," the United States suspended diplomatic relations with Iran. But this "absence" of diplomatic ties has always been somewhat unreal. Mutual demonization has gone hand in hand with participation by the two countries in venues such as the claims tribunal set up in The Hague to arbitrate U.S.-Iran financial disputes. Formal encounters and even cooperation have taken place in the context of multilateral conferences on the future of Afghanistan and on antidrug efforts. The United States had maintained unilateral sanctions on trade and investment with Iran but also carried out clandestine arms-for-hostages deals. There are even reports of joint efforts to combat al Qaeda terrorists and Iraqi oil smugglers, as well as other forms of bilateral cooperation. But neither party has been willing to publicize these supposed contacts.

In fact, despite occasional signs of rapprochement in the last quarter-century, the relationship has remained stalled. Informal polls in both countries have shown no strong domestic opposition to resuming ties, but the influence of powerful hard-line minorities in each country and a number of outstanding disputes that push domestic political buttons have held back all efforts at conciliation. At the same time, Tehran's perception of Washington's eagerness to improve ties has encouraged Iranian foreign-policy makers to increase their demands.

> ## U.S. policy apparently calls for regime change in Iran but trusts the Iranian people to do it themselves.

Washington initially pointed to five major obstacles to the resumption of relations: Iran's state sponsorship of international terrorism, its pursuit of weapons of mass destruction, its opposition to the Arab-Israeli peace process, its threats to neighbors in the Persian Gulf, and its regime's violations of human rights at home. In recent years, the last two issues seem to have lost some of their potency and are now only infrequently raised. On the other hand, a new accusation of Iran's harboring of al Qaeda operatives has recently been added to the list.

The Islamic Republic, for its part, originally demanded that the United States accept the legitimacy of the 1979 revolution, not interfere in Iran's internal affairs, and deal with the Iranian regime on the basis of "respect and equality," As Tehran became more secure domestically and reduced its international isolation, further conditions were added: lifting U.S. economic sanctions, releasing frozen Iranian assets in the United States, and removing the U.S. Navy from the Persian Gulf. The Clinton administration's mildly conciliatory gestures emboldened the ruling clerics to demand even more: an end to one-sided support for Israel and a formal apology for Washington's past misdeeds. Some of these demands have been emphasized more than others, depending on domestic polities within Iran.

After the election of President Muhammad Khatami in 1997 and his subsequent suggestion that "the wall of mistrust" between the two countries be torn down, reconciliation once more began to seem possible, particularly toward the end of Bill Clinton's administration. But mutual hostility was suddenly raised to a new high under George W. Bush. Lumping Iran with Iraq and North Korea in an "axis of evil." Bush castigated Iran for a series of wrongdoings. In particular the pursuit of weapons of mass destruction. This announcement was presumably designed to please hawks at home, garner international support for the war on terror, and warn Tehran of the consequences of any new mischief.

The message, not surprisingly, backfired. The implied threat in downgrading Iran from "rogue" to "evil" status pleased the Iranian opposition in exile but invoked a fiery response from the regime. Government officials dismissed the accusation as another brutish manifestation of the United States' "global arrogance," and many Iranians took it as a deep insult to their national dignity. Significantly, the conflict also caused considerable unease among U.S. allies in Europe and Japan. Ironically enough, the U.S. move also led to increased official contacts between Iran and Iraq.

In an apparent attempt to control the damage and separate good from evil, Washington subsequently drew a distinction between the powerless "elected" reformers in the *Majles* and the relatively weak local councils, and the "unelected" clerics who hold the levers of power. A White House press release on July 12, 2002, assured all Iranians who sought freedom and human rights that they had no better friend than the United States. This communiqué, however, which was apparently timed to coincide with the third anniversary of the prodemocracy student uprisings at Tehran University, again caused a backlash.

The unelected rulers whom Bush sought to condemn used the message to arouse public anger against the United States. Supreme Leader Ali Khamenei (who is backed by hard-line fundamentalists), Khatami (who spearheads the elected reformists), and former president Hashemi Rafsanjani (who leads the modern technocrats) joined together to denounce Bush's statement as interference in Iran's domestic politics. Even Ayatollah Jalaleddin Taheri—a leading dissident who had chastised the theocracy days earlier and was expected to welcome U.S. support—joined the three in asking for anti-American demonstration. Indeed, many reformers, eager to prove their patriotic bona fides, were more vehement than were conservatives in repudiating the White House's message. As a result of this united front, government-sponsored

demonstrations in Tehran and other major cities became a forum for a brand of virulent anti-Americanism rarely witnessed during Khatami's presidency.

Meanwhile, rumors circulated in Tehran that in response to the U.S. pressure, conservative hard-liners were planning to declare a state of emergency, dissolve the *Majles*, and dismiss the Khatami government. Although this crackdown did not happen, the conservative-led judiciary did close newspapers, harass and jail dissidents, forbid the teaching of Western music, insist that shops and restaurants close at midnight, and in general suppress domestic opposition—all in the name of social order.

Noting the failure of its proreform effort, the Bush administration articulated a new "dual track" approach based on "moral clarity." Zalmay Khalilzd a senior National Security Council staff member, unveiled the latest U.S. policy at an influential Washington think tank. The United States now would not seek to impose change in Iran but would instead support the Iranian people in their own quest for democracy. A literal interpretation of this dual-track policy is that the United States finds the Islamic Republic's behavior destructive and unacceptable and is thus calling for the regime change—but that Washington trusts the Iranian people to do it themselves. Subsequent calls in mid-November 2002 by State Department officials and Voice of America broadcasts for the Islamic Republic to "listen to its people" who were demanding "a change in the way they are being governed" were again strongly rejected by Khatami's government as interference in Iran's internal affairs.

THE ROAD TO TEHRAN

TO FIND a truly effective policy within this new dual-track posture will require an understanding of what makes Iran tick—a combination of a fierce sense of national independence, the Byzantine dynamics of Iran's domestic politics, the freewheeling character of Shi'a theology and the emergence of the Third Force. A look at the interplay of these four factors reveal why past U.S. strategies (such as sanctions, containment, and diplomatic pressure) have not been able to change the Islamic Republic's behavior.

Iranian's fierce nationalism is characterized by intense suspicion and outright resentment of outside influences. For example, Ayatollah Khomeini climbed to the Peacock Throne not on the wings of Koranic angels but mainly by championing freedom from U.S. interference. Khomeini's portrayal of the shah as Washington's stooge drew wide appeal because it channeled resentment about British and Russian influence during the previous 200 years of Iranian history. Indeed. The well-publicized chants of "Death to America" by government-organized demonstrators resonate much less with the vast majority of Iranians than does Khomeini's famous comment in the wake of the revolution that "America cannot do a damn thing." Thus any U.S. strategy that even remotely raises the specter of foreign interference in Iran is doomed to fail.

Beyond looking at this yearning for independence, outside observers must also take into account the state of Iran's domestic politics. Contrary to the popular caricature, both the "reformers" and the "conservatives" in Iran are cut from the same cloth. Both camps are byproducts of the same revolution, and both are sworn to abide by the basically undemocratic and even subtly xenophobic 1979 constitution. Few of the current reformers are Jeffersonian democrats; in fact, they are firmly committed to the union of mosque and state. Neither they nor their conservative opponents share America's human rights culture. Consequently, any U.S. policy that favors one group over the other will have little chance of success. Overt support for the opposition abroad or dissidents at home will enable the conservative power-holders to brand the reformers as American lackeys. Similarly, any compromise with the conservatives will likewise be interpreted by the reformers as a sellout by Washington.

Third, it would be a major strategic mistake to treat the Shi'a clerics in Iran as a unified force that rejects modernization or even Westernization. The highly unstructured hierarchy of the Shi'a sect allows Iran's ten or so grand ayatollahs to have not only their own disciples and private financing but also to independently issue religious edicts, or *fatwas*. Khatami's "politics of inclusion" and his plea for a "dialogue among civilization," both rooted in his view of Islamic scripture, can be understood in the same terms. Any successful U.S. strategy, therefore, should harness the positive influence of Iran's internal religious diversity and seek to direct it toward political change.

Finally, the government-encouraged baby boom of the early 1980s has now spawned a new generation, the Third Force, which sees neither the fundamentalists' concept of *velayat-e faqih* (the supremacy of Shi'a jurists) nor Khatami's "Islamic democracy" as the answer to Iran's current predicament. This highly politicized generation has no recollection of the 1979 revolution and no particular reverence for the eight-year "holy war" between Iran and Iraq. Rather, they focus on their frustrated ambitions for a better future. This group includes almost everyone who is not in power and a few who are, representing a wide swath of Iranian society. The common bond among these disparate groups is their disenchantment with the revolution and its aftermath and their distrust of the clerics' ability to cope with Iran's many problems. The Third Force, although still lacking resolute leadership and a specific platform, is united by a common goal of an independent, free, and prosperous Iran blessed by the rule of law. Indeed, some members have proposed a new constitution separating mosque and state, to be established by an internationally observed referendum.

RELIGION AND RECESSION

THE UNITED STATES should seize this moment to plan for Iran's political endgame because the regime's particular brand of politics and religion is in a state of ferment. At the same time, the government has been further weakened because it has failed to deliver on its promises of economic development.

The Iranian public and the press openly question the role of Islam in the country.

Ayatollah Khomeini built a governing ideology on concepts of independence, freedom, and the *velayat-e faqih*. This fusion of statecraft with piety through the absolute power of a supreme leader (the *rahbar*) is now beginning to crumble. The first crack appeared in 1997 when the philosophy's principal architect, Grand Ayatollah Hossein Ali Montazeri, rejected the unquestioned power of the *rahbar* on the grounds that Islam forbids the supremacy of fallible humans. Emboldened by this attack on Khomeini's orthodoxy, a number of mid-ranking clerics and seminarians have subsequently denounced theocratic intrusion into daily life, refusing to accept the inviolability of the *rahbar*'s religious edicts and even allowing fresh interpretations of the Koran itself. The new generation of clerics, taking their cue from older theologians such as Montazeri, now openly questions the legitimacy of absolutist religious power and even speaks of the need for an Islamic reformation. Some young seminarians in the holy city of Qom are now even questioning whether the unity of mosque and state is in their interest, since the unpopularity of the Islamic regime has reduced the number of clerics in the *Majles* and local councils and has also shrunk sources of private funding.

The latest condemnation of the regime came from Ayatollah Taheri, who in July 2002, while resigning from his post as the leader of Friday prayers in Isfahan, lambasted the religious hard-liners for incompetence and corruption. The cleric, formerly a devoted Khomeini follower and an early revolutionary during the shah's time, bemoaned the host of social, political, and economic woes afflicting the country—from rising unemployment to growing drug addiction to increasing disregard for the law. No previous internal criticism of the theocratic regime had ever been this scathing. The response by the leadership was mostly dismissive. However, Supreme Leader Khamenei, while complaining that this type of dissent would only embolden the regime's enemies, did acknowledge that he himself had pointed to some of the same shortcomings.

Trust in the power of Islamist ideology has declined even more profoundly as Khomeini's mixture of religion and politics has failed to deliver its promised rewards of prosperity and social justice. Despite a 100 percent rise in average annual oil income since the revolution, most indicators of economic welfare have steadily deteriorated. The so-called misery index (a combination of inflation and unemployment) has reached new highs. Average inflation in the years after the revolution has been at least twice as high as during the 1970s, unemployment has been three times higher, and economic growth is two-thirds lower. As a result, Iran's per capita income has declined by at least 30 percent since 1979. By official admission, more than 15 percent of the population now lives below the absolute poverty line, and private estimates run as high as 40 percent.

A combination of slow growth, double-digit unemployment, high inflation, declining labor productivity, and increasing dependence on oil revenue has thus defied almost all government efforts to put the economy back on track. Although the alarming rate of population growth in the first decade after the revolution has been brought under control, both per capita income and domestic income distribution lag behind official targets. In short, the ailing economy has helped bring the regime's legitimacy further into question. A recent study leaked from Iran's Interior Ministry revealed that nearly 90 percent of the public is dissatisfied with the present government. Of this total 28 percent wants "fundamental" changes in the regime's structure, and 66 percent desires "gradual reforms." Less than 11 percent—most probably those on the government dole—is satisfied with the status quo. Other private polls show an even greater degree of unhappiness with the government.

The combination of these two phenomena—the bankruptcy of Iran's ideology and the failure of its economy—now confronts the Islamic Republic with the worst challenge to its legitimacy yet. The public and the press now openly question the role of Islam—and especially the concept of the *velayat-e faqih*—in a society where people want greater freedom and the rule of law.

A TEHRAN SPRING?

IRAN'S CONSERVATIVE CLERICS are now helplessly witnessing a slow but steady drive toward democratization. Despite the political crackdown, legislative deadlock, and rumors of a coup, two provocative and parallel developments are challenging the mullahs' hegemony and paving the way for the regime's eventual collapse.

The first development relates to the expansion of civil society and the use of civil disobedience to loosen the theocracy's grip on national institutions. Nongovernmental organizations are being formed by the thousands, with and without official permission, to deal with ongoing problems ranging from family planning to drug addiction to pollution. Workers have formed informal (and extralegal) trade unions, and students have organized both Islamic and secular unions of their own. Despite a wave of newspaper closings and press repression, there are now 22 percent more licensed publications than there were in 1998. Furthermore, journalists have found a new haven in cyberspace beyond the authorities' reach. Currently, more than 1.75 million Iranians reportedly have access to the Internet. Even some nonestablishment aya-

tollahs have set up their own Web sites to connect with their flock. Their *fatwas* are now used by dissidents to counter the positions of the ruling clerics.

Street demonstrations, labor strikes, teacher's boycotts, and other forms of civil disobedience (such as taunting the morals police with un-Islamic attire) are increasingly common. For instance, thousands of workers demonstrating against poor working conditions managed to increase this year's official minimum wage. Strikes by teachers resulted in a substantial increase in this year's education budget. Human rights activists have also pushed the authorities to respond to foreign public opinion. According to the latest report by Human Rights Watch, the Islamic Republic may now start cooperating with foreign monitors for the first time. And in a noteworthy victory, the government shelved a bizarre, religiously sanctioned scheme to set up "temporary weddings" after women's groups, politicians, and some clerics denounced it as legalized prostitution. Most recently, several consecutive days of nationwide student protest in mid-November 2002 forced the supreme leader and the head of the judiciary to order an appeals court to expedite review of the death sentence imposed on reformist scholar Hashem Aghajari. The *rahbar* also recommended to judges that they avoid opening themselves up to public criticism in their rulings.

The second important change in Iran is a series of small but significant economic measures that are likely to reduce the oligarchs' economic power and help integrate Iran's oil-dependent economy with the global marketplace. The reduction of the hard-liners' financial support is a critical factor in their declining political clout. Indeed, more than any ideological or religious factor, it is control of the nation's economic resources that has allowed Iran's ruling clerics to hold on to power. Donations by devout Muslims, public and private monopolies in key sectors, special business licenses dispensed through patronage, privileged access to cheap credit and foreign exchange, and even widely reported bank fraud have all helped fund the clerics.

Crucial economic reforms, repeatedly promised by Khatami in the last five years, have partially taken shape in the last few months. This change has occurred largely in response to pressure from foreign institutions such as the International Monetary Fund, the World Bank, and the European Commission, whose approval is necessary for the government's continued access to foreign credit. Although these reforms will not dry up all the hardliner's sources of funding overnight, they can affect them in critical areas. For instance, the legalization of private banking and insurance since early 2000 has opened up new venues for the mobilization and allocation of national savings—and removed them from potential political uses by state banks. The government's efforts to consolidate the country's multiple exchange rates since March 2002 has also bottled up corruption stemming from access to cheaper dollars by privileged institutions or favored cronies. Fiscal reform in late 2001 aimed at lowering corporate income taxes and eliminating tax exemption for so-called religious charitable foundations is expected to increase private investment and level the playing field for potential investors. The government's new law to protect foreign investment and enforce some copyrights may reduce dependence on oil revenue. A successful euro bond issue this past summer has opened up another source of foreign exchange to counter volatility in oil prices.

The government has promised to take a number of further steps in the coming months to privatize state enterprises and further diminish the hard-liners' control of the economy. Replacing the inefficient subsidy system (which takes up some 20 percent of GDP and benefits mostly the urban rich) with a means-tested social safety net would substantially lighten the government's fiscal burden. In addition, further consolidation of the tax code should reduce the more than 50 different fees that various ministries and agencies impose on production and imports, thus cutting collection costs and special sources of finance for pork-barrel projects. The government's plan to enact a value-added tax in lieu of the current uncollected (and uncollectable) income taxes would likely diminish reliance on oil income and also shrink the bureaucracy.

Replacing the current system of quotas and special licenses on imports with tariffs would eliminate the monopolies enjoyed by politically favorite business interests. A comprehensive overhaul of the outdated 1968 commercial code would encourage more transparent and productive ventures, particularly in the small business sector. A revision of the current anti business labor law, enacted when leftist ideologues controlled the Fourth *Majles* in 1990, would encourage new employment. Downsizing the bloated bureaucracy may stop oil income from being invested in politically favored but economically unsound projects. Turning the Tehran Stock Exchange into a self-regulated but politically supervised institution would promote the establishment of mutual funds to attract both domestic and foreign capital.

Finally, Iran's entry into the World Trade Organization (so far blocked by the United States) and the conclusion of a comprehensive trade and cooperation accord with the European Union would shake the entrenched economic mafia to its roots and revolutionize the Iranian economy. The WTO's mantra of free markets is anathema to the Islamic Republic's state-dominated and highly politicized economic system. To qualify for full membership, Iran must make a host of economic changes, ranging from trade liberalization to financial deregulation to copyright protection. These new reforms will undoubtedly meet with severe resistance from vested interests. But the urgent need to find jobs for the millions of unemployed—combined with the paucity of domestic investment, sluggish non-oil exports, and weak foreign-exchange reserves—makes turning to the global economy inevitable.

And this shift will not be possible without fundamental reform.

THE IRAQ CARD

THE INTERNAL CURRENTS shaping Iran's future will undoubtedly be affected by events in the region, particularly the fate of Saddam Hussein's regime in Iraq. At present official Iranian policy, under a principle of "active neutrality", opposes U.S. preemptive action without a United Nations mandate as a "dangerous precedent." In part, the government objects out of a scarcely concealed concern that it could face a similar U.S. challenge one day. Although Tehran would certainly be happy to see Saddam go, Iran's government also worries that the establishment of a pro-American regime in Baghdad would leave Iran encircled by U.S. allies. The ruling clerics are similarly not enthusiastic about the prospect of a free and democratic Iraq, which would surely encourage the Third Force to intensify its reform efforts.

Virtually all Iranians oppose Iraq's partition or potential disintegration for fear that Iraqi Kurds may incite their counterparts in Iran to rise up and agitate for an independent state. Additionally, Iran does not wish to see a long-term decline in the price of oil as Iraq again pumps at full capacity. There is also concern that major international oil companies may invest in Iraqi oil fields at Iran's expense. It is impossible to know at this stage how these various forces will interact.

THE TIPPING POINT

THE ONGOING POLITICAL IMPASSE, economic distress, and social turmoil (not to mention a possible invasion of Iraq) all threaten the survival of the Islamic state. The discontent of the Third Force in particular has created a seemingly unstoppable momentum toward change. In its post revolutionary history, Iran has never been as politically polarized or ideologically divided as it is today.

Khatami's recent belated attempt to reclaim his authority has raised the political temperature several degrees. His open suggestion that he might resign if further stymied by the Council of Guardians or the judiciary, as well as threats of mass resignation from members of the Islamic Participation Front (the largest bloc in the *Majles*), is an ominous warning of a looming constitutional crisis or worse in coming months. Two bills submitted by Khatami to the *Majles* in September 2002, which are now undergoing the long process of ratification, would curb the veto power of the Council of Guardians and give the president legal authority to force hard-line Islamic courts to abide by the constitution. The renewed crackdown by the judiciary on reformist groups is reported to be an attempt to pressure the president to withdraw these bills. Regardless of the ultimate fate of this controversial legislation, its very introduction marks a major turning point in Iran's domestic political dynamics.

Temporary reversals of democratization nevertheless remain likely. But the Iranian people have sown the seeds of change and the country's theocratic rulers cannot postpone their harvest forever. The autocratic and dubiously Islamic concept of the *velayat-e faqih* is clearly in retreat and the oligarchs know that if they do not bend, they will break. A recent open letter signed by more than 125 former *Majles* deputies lambasting the hard-liners made this point abundantly clear.

These developments should not suggest, however, that Washington can determine or even affect the outcome of Iran's political ferment. Events within Iran will dictate the U.S. posture rather than the other way around. Clerical dogmatism cannot be defeated from afar—particularly given Iranians' profound mistrust of outside meddling.

If the United States truly wishes to see a modern, democratic, and peaceful Iran, Washington must follow a cal-culated "wait and see" policy. Neither Bush's anger, nor his empathy, nor even his promise of friendship with democratic forces will be enough to change Iran. Thus as long as U.S. vital national interests are not seriously threatened and Iran is not clearly implicated in anti-American terrorist acts, the United States should refrain from both unsubstantiated accusations and implied threats against the Islamic Republic. Washington would be best served by letting the currently accelerating process of democratization run its course. The theocracy's days are numbered—Iran's own internal currents assure this.

JAHANGIR AMUZEGAR is an international economic consultant. He was Finance Minister and Economic Ambassador in Iran's pre-1979 government.

New Hope for Brazil?

Stanley Gacek

On October 27, 2002, Luiz Inácio Lula da Silva—lathe operator, leader of the independent Brazilian labor movement that emerged in the late 1970s to challenge the military regime, a founder of the Brazilian Workers Party (Partido dos Trabalhadores—PT), and a former congressman from São Paulo—was elected president of Latin America's most powerful state, the world's fifth largest country, with more than 170 million inhabitants, and the ninth largest economy on the globe. His victory was unprecedented in Brazilian history: fifty-two million votes, 61 percent of the total.

> ... the landless movement,... responsible for many of the rural conflicts and occupations of the last decade, expects sweeping agrarian reform to overturn a system in which less than 3 percent of the population control more than 60 percent of the arable land.

The Lula "Era," as it is being called in Brazil, has raised the hopes of leftists throughout the world. The Lula–PT platform calling for sustainable macroeconomic growth and the reduction of poverty and inequality offers a powerful and tangible alternative to the dominant paradigm of globalization. As Lula declared in his inaugural address of January 1, ending unemployment "will be my obsession" and ending hunger must be "a national endeavor." He has pledged to create at least ten million new and decent jobs by the end of his four-year term, as well as guarantee three meals a day for every Brazilian.

Lula's opposition to any version of the Free Trade Area of the Americas (FTAA) that promotes "an economic annexation" of the South to the North augurs a substantial delay if not a fatal blow to any North American Free Trade Agreement expansion in the hemisphere. There can be no FTAA without Brazil. Foreign Affairs Minister Celso Amorim made it clear to U.S. Trade Representative Robert Zoellick that his country would not be rushed into any hemispheric trade agreement—a troubling admonition for Washington, as Brazil and the United States will be co-chairing the FTAA negotiations in 2003. This new relation of forces concerning trade and the world economy led Emir Sader, a well-known University of São Paulo sociologist, to announce "the beginning of the end for the neo-liberal project in Latin America."

Lula's PT is now the largest party in the Chamber of Deputies, moving from 58 to 91 seats (out of a total of 513), as well as the third-largest party in the Senate, with an increase from 8 to 14 (out of 81). Obviously, these numbers are well short of a majority, and the PT has been compelled to maintain and consolidate alliances with other parties of both the left and the center-right.

Nevertheless, Lula's electoral coalition already included other significant parties, and a number of them are represented in his new cabinet, reinforcing the alliance he needs to govern effectively. The PT and the Lula administration have negotiated deftly with the Brazilian Congress. Executive Chief of Staff José Dirceu persuaded the center-right Brazilian Democratic Mobilization Party, which commands one of the largest legislative blocs, to elect leaders amenable to agreements with the new government. This latest negotiation guarantees a pro-Lula majority of 283 seats in the Chamber and 47 in the Senate.

Not surprisingly,... activists see a political opportunity: they want a new labor law that will weaken the posiiton of more conservative trade union leaders.

The Brazilian people decided to give Lula their mandate after two decades of military dictatorship (1964–1985), a return to democracy followed by the sudden death of the civilian president, a grueling presidential impeachment process in the early nineties, and eight years of the administration of Fernando Henrique Cardoso. Cardoso's administration succeeded in reducing inflation, following the prescription of the Washington Consensus, but it also produced growing inequality, higher unemployment, greater violence, and an astronomical increase in debt and interest rates. Much of Brazilian civil society, including business, opted for Lula because of his commitment to greater growth and social stability. Brazilian textile magnate José Alencar of the Liberal Party (PL), Lula's running mate and now vice president, represents the strong belief among national industrialists that macroeconomic stimulus is desperately needed.

Now Lula faces the challenge of managing and satisfying the soaring expectations of his supporters. For example, the landless movement, or Movimento Sem Terra (MST), responsible for many of the rural conflicts and occupations of the last decade, expects sweeping agrarian reform to overturn a system in which less than 3 percent of the population control more than 60 percent of the arable land. Gilmar Mauro, one of the top leaders of the MST, is demanding that the new government immediately expropriate "all lands belonging to owners who owe money to the state, as well as all lands left fallow." He insists that titles must be awarded now to at least a hundred thousand families currently occupying unproductive farms, and promises that the landless movement will continue to pressure the Lula administration until all its demands are satisfied.

A number of union leaders who began as young labor activists in the *novo sindicalismo* movement of the 1970s and 1980s and were founders of both the PT and the Central Única dos Trabalhadores (CUT), Brazil's largest trade union center, representing more than twenty-two million workers, have long advocated a major overhaul of the state-corporatist system of labor relations and union governance. With the presidency of the country now occupied by a metalworker who once challenged a military regime relying on state control of the union movement to repress collective worker action, these unionists say that it is high time to end labor's dependence on the state for support of union structures. Not surprisingly, these activists see a political opportunity: they want a new labor law that will weaken the position of more conservative trade union leaders.

The Brazilian labor movement will also expect Lula to make good on his announced intention to create jobs. By the end of the 1990s, the official unemployment rate (which does not reflect chronic underemployment and the high number of Brazilians permanently discouraged from seeking work) exceeded 7 percent. In São Paulo, Brazil's largest metropolis, 20 percent of the working population is unemployed, making for two million people without jobs. And the number of unregistered, informal workers in the Brazilian economy rose from 42 percent to 58 percent of the total labor force in the last ten years, with more than 4.3 million jobs being eliminated in the formal sector.

Excessive fiscal conservatism and high interest rates will only stanch growth and erode the support from national business on the right and from unions and social movements on the left.

And the Brazilian majority expects Lula to reform the social security regime. The current public system of retirement and disability benefits actually covers only 57 percent of the eligible beneficiaries. It has generated an unsustainable deficit of more than seventy billion *reais* (well over twenty billion dollars), exacerbating Brazil's debt crisis and causing interest rates to climb. The growth of the informal sector erodes the tax base and cuts revenue. Moreover, certain public servants, including top military officers, high-ranking judges, and elite personnel in the executive branch, benefit disproportionately from the system. All told, public sector employees represent 11 percent of social security beneficiaries and receive 45 percent of the benefits.

Social Security Minister Ricardo Berzoini, former president of the São Paulo Bank Employees Union, has called for a more uniform system in which elite public servants will not receive inordinately more than other workers. Although Berzoini is not proposing to make the more egalitarian disbursement retroactive, many military officers and judges argue that any reform will interfere with their "acquired rights." And the leaders of the CUT–affiliated federal employee unions, many of whom supported Lula's electoral campaign and are among the most left-wing activists in the Brazilian labor movement, will resist tampering with the current system.

In addition to the expectations of domestic constituencies, Lula must confront the antagonistic pressures of external interests and institutions. Precisely because the new government will not be a pushover on the FTAA, multinational enterprises and global capital, along with their friends in Washington, will have every motivation to isolate and control Brazil. And with the accumulated public debt rising from 29.2 percent of gross domestic product in 1994 to well over 55 percent at present, a total of approximately 250 billion dollars, the Lula government is particularly vulnerable to the demands of international creditors and financial institutions. The International Monetary Fund's thirty billion dollar loan package negotiated at the end of last year with President Cardoso (once a left-wing sociologist who told big business during his successful 1994 electoral campaign "to disregard everything that I published") imposes high primary budget surplus requirements that may hamstring Lula's program.

In an obvious effort to show both Wall Street and the international financial institutions that the Lula administration will be fiscally responsible, Finance Minister Antônio Palocci has announced a 2003 primary surplus goal of 4.25 percent of GDP, surpassing the IMF's demand of 3.88 percent. In addition, Central Bank Governor Henrique Meirelles, a former chief executive of FleetBoston, has resisted lowering interest rates in an effort to avoid runs on Brazil's currency and an inflationary spiral. (The consumer price index for São Paulo revealed a 2.19 percent increase in January, the highest recorded for any January since 1995.)

Unfortunately, not all the financial analysts are impressed. Although the rating agency Standard and Poor's has praised Palocci and Meirelles for a "prudent and cautious policy stance," it also sent a poisonous message to the markets on January 16, asserting that the Lula administration has "limited room for policy maneuverability in a challenging global and domestic environment."

The new government's understandable attempts to avoid subversion by international finance also threaten to undo Lula's plans for macroeconomic recovery and social justice. Excessive fiscal conservatism and high interest rates will only stanch growth and erode the support from national business on the right and from unions and social movements on the left. Already, leftist militants in the PT have accused the Lula government of selling out to the IMF and the markets.

Academics and journalists have been mixed in their appraisal of Lula's capacity to deal with the formidable domestic and international challenges he faces. *Newsweek* posed the question last October, "Can Lula really lead?" and cited a Wall Street source who claimed that the new president was thoroughly "untested." Peter Hakim, from the Inter-American Dialogue, in a *Washington Post* op-ed, asked whether Lula is the "champion for Brazil's poor majority," challenging "the country's power brokers" and battling "its deep social injustices"—or is he "newly moderated," accepting "market economics" and "pluralist politics"?

They [the PT] succeeded in constructing a political force that is thoroughly open, ecumenical, non-doctrinaire, mass-based, and internally democratic.

In an article appearing in the *New York Review of Books* last December, Kenneth Maxwell of the Council on Foreign Relations claimed that Felipe Gonzalez and the Spanish Socialist Workers Party represent "one of the models for the Brazilian PT," having also managed to shake off a Marxist past and "move to the center ideologically." According to Maxwell, "perhaps 30 percent of the PT call themselves radicals... but most party members have learned to play the democratic game."

Right-wing commentators, such as Constantine Menges of the Hudson Institute, claim to know about a sinister plot being engineered by Lula and the PT that will give new life to the international communist conspiracy. Although Menges's published musings border on the psychotic, they certainly influenced Representative Henry Hyde, chair of the House International Relations Committee. Hyde wrote to President Bush on October 24, warning that Lula was a dangerous "pro-Castro radical who for electoral purposes had posed as a moderate," and likely could form "an axis of evil in the Americas" with Cuba and Hugo Chavez's Venezuela. Such a diabolical alliance would have ready access to Brazil's "30-kiliton nuclear bomb" as well as its "ballistic missiles."

The problem with most of these opinions is that they ignore the remarkable twenty-five year political development of Lula, the PT, and Brazil itself. The trade unionists, human rights advocates, progressive clergy, academics, and other opponents of the military dictatorship who founded the Workers Party more than twenty years ago were dedicated to building a new and unprecedented movement, which would break radically from the Brazilian tradition of top-down populist and *caudilho* politics. They succeeded in constructing a political force that is thoroughly open, ecumenical, non-doctrinaire, mass-based, and internally democratic. Although the PT accepted members who claimed allegiance to revolutionary ideals (many of whom are currently excoriating the government for compromising its socialist principles), it also rejected Leninist vanguardism and "democratic central-ism" as defined by the orthodox communist left. PT militants did not simply "learn to play the democratic game" in Brazil; for all intents and purposes, they *created* it.

In a 1991 press interview in Mexico City, Lula made it clear that the PT never accepted the Soviet model, nor did it have to "renounce" a dogmatic past: "From its birth, the PT criticized Eastern Europe. We criticized the Berlin Wall, state bureaucracy, the absence of union freedom; we defended Walesa from the moment we were founded."

As Johns Hopkins political scientist Margaret Keck has correctly observed, the PT's socialism was never a dog-matic ideology, but rather an "ethical proposal, within which a number of alternative visions of the good society competed... " For Keck, the party's moral aspiration for socialism has been "an aspiration for democracy."

It is also misleading to imply that a "newly moder-ated" Brazilian Workers Party has only recently learned to deal with "pluralist politics" and "market economics" and that Lula is "untested." Over the last twenty years, the Workers Party has built a successful and corruption-free record of governing hundreds of cities, including the world's third-largest, São Paulo, as well as several impor-tant state governments. In addition, hundreds of able PT activists have served in the Brazilian Congress, including Lula himself. None of this could have happened had the PT not respected the principles of democratic pluralism and coped with the realities of a market economy.

As for the "axis of evil" theory, Menges, Hyde, and other fearmongers should be reminded that Lula has spo-ken out forcefully against the spread of nuclear arms and has promised Brazil's total compliance with the Non-Pro-liferation Treaty. Like many other Latin American leaders and statesmen, Lula maintains constructive relations with the current leadership in Cuba and Venezuela. However, he has said that Castro needs to encourage in-ternal democracy, and he has admonished Chavez for "acting like a military officer" rather than "a civilian pol-itician" when dealing with the Venezuelan opposition. Although the anti-Chavez forces have denounced Brazil for sending oil and humanitarian assistance to Venezuela, Lula's special secretary for foreign relations, Marco Aurelio Garcia, recently assured me that the aid is meant to encourage peaceful negotiations and an outcome that will respect Venezuela's democratic institutions and rule of law.

The Lula administration has so far demonstrated good faith in responding to Brazil's domestic constituencies and their demands. The new government has convened a National Social and Economic Council that directly in-volves representatives from all segments of Brazilian civil society-national business, the trade unions, the religious community, cooperatives, and civil and human rights or-ganizations—for the purpose of negotiating an inclusive

social pact. By the end of 2003, the Executive will have presented the Council's proposals to the Congress that address the country's most pressing issues, including so-cial security, tax, and agrarian reform. In addition, the new government is convening a National Forum on La-bor to review all the employment, labor rights and labor relations concerns facing Brazil. After extensive debate and negotiation among government, trade union, and business representatives, the Forum will also have its proposals delivered to the Congress by the end of the year.

This inclusion of Brazilian civil society in serious nego-tiations is something that Lula's predecessors were both unwilling and unable to do. Without question, the task is not an easy one. CUT President João Felicio has already made it clear that social pact discussions do not mean that labor will give in on wages and working conditions. And speaking for the interests of capital, Alencar Burti, presi-dent of the São Paulo Chamber of Commerce, announced, "I am not interested in talking about what I'm supposed to give up." Nevertheless, by involving all representative groups in the decision-making process, the new govern-ment is engaging in constructive dialogue with Brazilian society, as well as with its own current and potential op-position. Lula claims that agrarian reform will go forward during his administration precisely because he and his government will actually sit down with the MST and ne-gotiate.

Several of the new government's ministers have pro-posed specific public policy initiatives. Labor minister Jaques Wagner is considering reducing the standard workweek from forty-four to forty hours in order to in-crease employment. Wagner has spoken of targeted tax incentives and subsidies to employers who hire more workers by reducing hours without cutting wages and benefits. Economist Paul Singer, recently named Secre-tary of the *Economia Solidária* (Solidarity Economy), is pro-posing the massive creation of service, family farm, labor, and microcredit cooperatives as a means of putting thou-sands of Brazilians back to work.

Although the IMF's draconian pressures and Wall Street's insatiable demand for "fiscal responsibility" could spell political and economic disaster, there are several fac-tors working in Brazil's favor. For example, World Bank President James Wolfensohn has openly praised the new government, and has committed funding to Jaques Wag-ner's youth employment programs and to Lula's anti-hun-ger campaign. The plans to multiply domestic credit sources could help reduce the dependence on foreign in-vestment. And the Argentine tragedy should have taught a serious lesson to the world's financial markets and institu-tions: if Lula is not allowed to bring a minimal amount of growth and social stability to Latin America's largest econ-omy, the prospects for security in the hemisphere are bleak.

The international union movement can play an impor-tant role in helping to ameliorate some of the external pressures. For example, workers' pension funds are responsible for literally trillions of dollars in financial markets and can exert substantial influence when it comes to Wall Street's relationship with Brazil. In addition, the labor-friendly social development projects of the Lula administration suggest interesting partnerships with international workers' capital.

More direct contacts between unionists from the United States and Brazil, as well as exchanges involving city and state governments, small agricultural producers and family farmers, educators, musicians, artists, and progressive legislators, can only enhance North American support for Brazil's new democratic hope, as well as help to counter any revanchism on the part of the Bush administration. And the AFL-CIO and its affiliates have a promising opportunity to work with the Lula government to guarantee labor rights to the more than one million Brazilians living and working in the United States.

Lula is well aware of the importance of international union solidarity to his success, as he made clear to hundreds of admirers from the U.S. labor movement who were present at AFL-CIO headquarters to receive him during his visit to Washington last December. He deserves, and needs, many more friends. For the real question is not whether "Lula can lead" or whether "Lula can succeed." The real question is whether the world can afford for Lula to fail.

STANLEY GACEK is a labor attorney and AFL-CIO International Affairs Assistant Director, responsible for the Federation's relations with Latin America and the Caribbean. He has spoken and written extensively on Brazilian labor and politics and has been a friend and adviser to Lula and the PT for more than twenty-two years.

From *Dissent*, Spring 2003. © 2003 by Dissent Magazine.

Latin America's New Political Leaders: Walking on a Wire

"Today's underlying political currents in Latin America are less about ideology and more about a public desire to find leaders who can effectively address everyday problems, and who do so honestly. The formulas of the past—whether 'socialism' in the 1970s or 'neoliberalism' in the 1990s—have been widely questioned, and largely dismissed. With traditional ideas and structures breaking apart, new leaders are being called on to produce results."

MICHAEL SHIFTER

In 1970, Salvador Allende finally ascended to the Chilean presidency on his fourth attempt. The Socialist candidate with a radical agenda won the prize that had long eluded him—only to be overthrown in a generation-defining military coup led by General Augusto Pinochet on September 11, 1973.

Nearly three decades after the Chilean coup, new faces of political leadership are emerging in Latin America, giving form and definition to the next generation. The sharply drawn left–right battles and schisms intrinsic to the cold war no longer drive politics. Rather, today's underlying political currents in Latin America are less about ideology and more about a public desire to find leaders who can effectively address everyday problems, and who do so honestly. The formulas of the past—whether "socialism" in the 1970s or "neoliberalism" in the 1990s—have been widely questioned, and largely dismissed. With traditional ideas and structures breaking apart, new leaders are being called on to produce results.

In 2003, Luiz Inácio da Silva, widely known as "Lula," is the principal focus of regional attention. Lula was re-soundingly elected Brazil's president after he had failed, like Allende, in three previous attempts. Predictably, as the leader of the Workers' Party took the reins of Latin America's largest and most significant country, all eyes have turned to the region's new, charismatic leader—and the most promising experiment for social renewal.

Will Lula balance Brazil's mounting demands for greater social justice and equality with the formidable constraints and pressures imposed by international financial institutions? Will the United States not only recognize the need for but also help support Lula's alternative social policies? In short, will Lula succeed in charting a new path?

Brazil's passing of the guard from President Fernando Henrique Cardoso to Lula epitomizes the shift in Latin America in the political winds, and in political leadership, over the course of the past decade. Cardoso, not long ago considered the quintessential leftist intellectual, served two terms as president of a center-right government whose main achievements included democratic continuity and economic stability. At the same time, Cardoso's government accumulated a huge public debt, failed to achieve sustained growth, and despite improvements in many social indicators, could not make a significant dent in Brazil's vast disparities in wealth. Brazilians—from the poor to a sizeable share of the business community—in the end clamored for political change.

A SOUR MOOD

For Latin America, Lula's election to the Brazilian presidency comes at a critical moment. Stories abound in publications such as *The Economist, Financial Times,* and *Newsweek* about a backlash against neoliberalism, a resurgent populism, and a decided turn to the left in Latin America. The high hopes and expectations that accompanied the early 1990s, when free trade and spreading democracy were widely touted, are a fading memory.

The regional economic picture is especially gloomy and dramatic. According to the World Bank, Latin America in 2002 suffered its most disastrous economic performance in nearly two decades, with a negative growth rate of 1.1 percent. In Argentina and Venezuela the outlook is notably grim, marked by stunning economic declines and social disintegration. Experts argue that even under the most sanguine scenarios, any substantial economic rebound in these countries will take years, if not decades. Moreover, since 1998, for the region as a whole, per capita income has dropped some 0.3 percent per year. Although projections for 2003 are slightly more upbeat, with the regional growth rate possibly reaching 2 percent, it is difficult to imagine that such a modest performance will bring much relief to Latin America's acute social conditions.

Public opinion survey data also offer little to cheer about. According to the respected Latino-barómetro, which has conducted a comparative survey annually since 1995, public dissatisfaction with government performance has steadily risen because of stubbornly poor and disappointing economic results. The inability of governments in many countries to reduce mounting unemployment and crime has also fueled enormous citizen discontent. Although the survey findings reveal that most Latin Americans continue to see democracy as the preferred political model and that they have not given up on market-oriented prescriptions for economic difficulties, these views are largely arrived at by default and are far from a ringing endorsement of the core ideas that were expected to deliver improvements in citizens' lives. Frustration with economic policies that have yielded few tangible benefits runs deep throughout the region.

Despite so many profound problems, elections remain the accepted way to select leaders. Viewed in wider historical perspective, this adherence to the democratic norm is remarkable and cause for optimism. But the signs suggest voters are groping for fresh answers and alternatives to the status quo. The profile of a Latin American leader—perhaps exemplified by Cardoso—who is an internationalist and blends technical or academic prowess and sophistication with a knack for politics—appears to be receding. "Populist" or "leftist" may be the most con-venient and common terms employed to characterize new leaders who deviate from this model, and who represent the promise of a more thoroughgoing and committed social agenda. But there are scant resources available to fund meaningful social programs. And at a time that has seen the questioning and dissolution of traditional political structures and ideas, these labels obscure more than they illuminate.

BRAZIL'S "AMERICAN DREAM"

Lula differs fundamentally from many of the other political leaders who have recently emerged on Latin America's political stage. What sets him apart is his longstanding dedication to building the Workers' Party (PT), widely regarded not only as the most solid and coherent political party in Brazil, but as among the most effective parties in Latin America. Lula's two-decade project—born under a Brazilian dictatorship that began in the 1960s and only came to an end in 1985—is especially noteworthy in a region where political parties are in deep crisis, broadly discredited, and perceived as ossified and corrupt.

Although the PT has yet to be tested at the national level, it has had ample—and successful—governing experience at the state and especially municipal levels in Brazil. The PT is far from the amorphous movements and inchoate groupings that often pose as political parties in other Latin American countries. Yet it is not monolithic but composed of factions that range from decidedly hardline to pragmatic.

Lula himself has undergone a remarkable political evolution. Over the years his positions on a variety of issues have moderated. In the 2002 election campaign he was referred to as "Lula lite" because of the alliances he made with more conservative factions, a move that would have been difficult to imagine just a few years ago. Moreover, his personal story is extraordinary, embodying the "American dream" with his rise from poverty through the ranks of the powerful metalworkers' union and eventually to party leader. It also testifies to the vitality of Brazil's democracy and the gradual loosening of its rigid social structures.

Lula's call for a "social pact"—agreements involving government, business and labor unions aimed at assuring economic and political stability—signals a shift to an alternative political arrangement, perhaps more in line with European concepts. Still, it will not be easy for Lula to contend with pressures from members of his own base, who see the PT in power as "their turn." Demands made by the international financial community will similarly test Lula's political talents. His administration's success may hinge less on the ideological direction Lula decides to pursue than on the competence of his team and the ability to devise sound and coherent policies that deliver growth and some social progress while maintaining economic stability.

CHÁVEZ'S VENEZUELA: REVOLUTION OR CHAOS?

Nearly four years before Brazilians elected Lula as president, the majority of Venezuelans put their confidence in Hugo Chávez. Chávez, a former paratrooper who led a failed military coup in February 1992, was the product and beneficiary of a traditional political party system that had collapsed from decades of corruption and mismanagement. Since the early 1980s, no Latin American country has suffered the precipitous economic decline and social decay that has struck Venezuela, made all the more egregious when viewed against the country's substantial oil wealth. A succession of failed administrations and a protracted period of party unresponsiveness led to a sharp public repudiation. Chávez, a superb orator who issued a scathing indictment of the old order, took advantage of the political vacuum.

While there is still some question about which of the "two" Lulas will ultimately leave his mark on Brazil's presidency, there is little doubt that Chávez's authoritarian instincts are stronger than his democratic tendencies. He has often defied and violated the constitution and has constantly lashed out at key elements of Venezuelan civil society, such as the media, the Roman Catholic Church, and unions. Gabriel Garciá Marquez, Colombia's Nobel Prize–winning author, wrote an especially prescient profile of Chávez just before he took office in February 1999: "I was overwhelmed by the feeling that I had just been traveling and chatting pleasantly with two opposing men. One to whom the caprices of fate had given an opportunity to save his country. The other, an illusionist, who could pass into the history books as just another despot."

It is clear that honesty and effectiveness—

not rigid formulas or sweeping blueprints—

are what the current generation is looking for.

Although Chávez has been variously described as a "populist," "revolutionary," and "leftist," he has been so in rhetoric only. He seems, rather, a throwback, a figure who more closely resembles Argentina's legendary Juan Domingo Perón of the 1950s and 1960s. His actions and politics have been erratic and inept, with detrimental results for Venezuela. (The poorest Venezuelans, Chávez's core constituency, have suffered most from the country's prolonged drift and chaos.) Moreover, his confrontational style, with the charged rhetoric about Venezuela's "rancid oligarchy," has only exacerbated social tensions.

In 2002, Venezuela endured unprecedented political and social polarization, along with extraordinary bitterness and mistrust. The government and its supporters jockeyed for position with opposition forces—made up of an array of nongovernmental and professional groups and diverse political figures organized under the umbrella Democratic Coordinator—that mounted a strike to press for Chávez's ouster, either through his resignation or through new elections. With each side digging in, the prospects for any reconciliation seemed negligible. In the unfolding crisis, critical factors included Chávez's ability to figure out how to keep the oil flowing. For the world's fifth-largest supplier of oil, the fate of the petroleum sector and the state-run oil company Petróleos de Venezuela, SA, is decisive.

Another key factor is the Venezuelan military, which has also been badly divided between forces that support and those that oppose Chávez. As the violence has escalated and threatens to erupt into something that could conceivably resemble a civil war, many observers have tried to anticipate how, and at what point, the armed forces might react.

When this crisis is resolved. Venezuela will need considerable time to heal the deep wounds that have for many years run through society and have been aggravated under Chávez. But an opposition that lacks effective political leadership, a clear strategic focus, and any vision for a post-Chávez Venezuela—especially ideas about how to unite the country, including Chávez's mostly poor constituency—does not augur well for a swift reconciliation. Chávez's ability to hold on to a core base of support in the context of such a disastrous performance underscores the depth of rage and mistrust many Venezuelans feel toward an old order in need of deep-seated reform.

ECUADOR'S CONSUMMATE OUTSIDER

Ecuador also has seen a former coup plotter elected to the presidency. Like Chávez, Lucio Gutiérrez was a former army lieutenant colonel who tried to topple a democratically elected government. Joining with Ecuador's powerful indigenous movement and other military officials, Gutiérrez, unlike Chávez, succeeded in his January 2000 attempt. Although short lived, the coup catapulted Gutiérrez onto Ecuador's political stage and, less than three years later he was, to the surprise of many, overwhelmingly elected president.

Although Gutiérrez expressed admiration for Chávez at the time of the coup, he has since sought to distance himself from the controversial Venezuelan president. The initial signs suggest that, unlike Chávez, Gutiérrez will attempt to reach out and engage with his country's broad-based civil society and managerial and entrepreneurial sectors. The new Ecuadoran president's announcement of his intention to seek an agreement with the International Monetary Fund, his acceptance of the country's "dollarization" scheme (under which the United States dollar became the official currency in January 2000), and his emphasis on fiscal discipline do not signal a shift to the left or the beginning of a resurgent populism.

Yet Gutiérrez is, in one key sense, similar to Chávez. He too is the product of the wholesale rejection of his country's traditional political class, which has generally failed to deliver tangible benefits to vast sectors of the population. That Gutiérrez—as well as his rival in the second round of voting, banana magnate Álvaro Noboa—are the consummate political "outsiders" shows that most Ecuadorans have little appetite for "more of the same" (Ecuador's politics has been notably unsettled; Gutiérrez is the country's fifth president in six years). To be politically successful, he will have to produce results for the country's most marginal group, especially the poorest sectors and a sizable indigenous population.

At the same time, Gutiérrez faces formidable challenges. He inherits an enormously difficult economic situation. Although he can count on the important support of the Confederation of Indigenous Peoples of Ecuador, Latin America's strongest and most coherent indigenous confederation, to pursue his reform agenda, Gutiérrez will nonetheless have to deal with a broad array of parties represented in a remarkably fragmented Congress. In a country distinguished by notoriously sharp geographic, ethnic, and social divisions, Gutiérrez, a political neophyte, will find it difficult to forge alliances and build coalitions to accomplish his goals of reducing corruption, making the Ecuadoran economy more competitive, and working toward greater social equality.

THE EXPECTATIONS GAME

In taking on such a mammoth political task, Gutiérrez may find the experiences of other current Latin American presidents instructive. With their parties failing to enjoy a majority in their respective legislatures, Mexican President Vicente Fox (in office since December 2000) and Peruvian President Alejandro Toledo (elected to the post in June 2001) have had some difficulty constructing coalition governments to implement their agendas.

After more than seven decades of single-party authoritarian rule, Mexico expressed its profound desire for change by voting for a former Coca-Cola executive and National Action Party governor from Guadalajara. But Fox has failed to satisfy the high—perhaps unrealistic—expectations that were generated by his administration. To be sure, Mexico's political system has opened up considerably—indeed, irreversibly. And the Fox government has escaped the many serious charges of corruption that had been leveled against previous Institutional Revolutionary Party (PRI) administrations. After more than two years in office, his approval rating is still roughly 50 percent.

Criticism has revolved around Fox's inability to "get things done," most notably in fiscal reform as well as other areas. The Mexican president has struggled to strike deals and work effectively with a Congress still dominated by the PRI. Fox's cabinet and advisers are capable, but have not come together as an effective team. Still open to question is Fox's ability to articulate a clear program for the country, devise a political strategy, and instill the necessary discipline to achieve his main policy aims.

Peru's Toledo came to office following an eight-month transition government in Peru that had been preceded by the highly corrupt and decade-long authoritarian regime of President Alberto Fujimori and his national security chief, Vladimiro Montesinos. Unlike Fox, Toledo—born in poverty and proudly indigenous—had virtually no political background, apart from leading the opposition against the previous government, which he did with great courage. In addition, his political party, Péru Posible, is very heterogeneous and does not have programmatic coherence or discipline.

Toledo's lack of political experience and struggle to gain some traction have exacted a heavy toll. After his first year in office, Toledo's approval rating had dropped from a high of nearly 70 percent to below 20 percent. Toledo promised considerably more—"thousands of jobs"—than he has been able to deliver, frustrating many Peruvians. He has had difficulty making decisions and outlining a coherent policy course. In June 2002, Toledo's failure to consult with local officials about the privatization of electrical companies in Peru's second-largest city of Arequipa helped spark widespread protests, with huge political costs. Toledo's self-inflicted difficulties have been compounded by the ferocity of opposition politics and signs that remnants of the Fujimori regime may be trying to sabotage the current administration.

Frustration with economic policies

that have yielded few tangible benefits

runs deep throughout the region.

Toledo's slide—and the mounting protests and unrest in the streets—are all the more remarkable in light of the country's 4 percent growth rate in 2002, the region's highest. In addition, the Toledo government has vigorously pursued human rights abuses and corruption that occurred under previous governments. After resolving a paternity suit and recognizing his daughter in October 2002, Toledo, who had mishandled the controversy, recovered some public support. The change did not, however, translate into an electoral victory for his party in Peru's regional elections in November.

The ability to master coalition politics and shape a governing consensus will also be the crucial question in Bolivia, under the administration of Gonzalo Sánchez de Lozada, who assumed office in August 2002. Sánchez de Lozada, who previously served as president from 1993 to 1997 and presided over wide-ranging reforms, including innovative capitalization and privatization schemes, faces an even more monumental task than he did in his first term. His relationship with his coalition partner, Jaime Paz Zamora of the Movement of the Revolutionary Left, is strained. Most significantly, Sánchez de Lozada will have to learn how to work with Evo Morales, the popular indigenous figure and leader of the numerous and increasingly mobilized coca growers. Morales, who is pressing for major social changes and could upset Bolivia's political system, registered stunningly strong support in the last election. As in Ecuador, the indigenous population in Bolivia is an important actor in shaping the country's politics.

Also in August 2002, Colombia, the only Latin American country in the midst of a civil conflict, elected a new leadership when Álvaro Uribe succeeded Andrés Pastrana as president. Uribe won a resounding victory and secured an impressive mandate in the May 2002 elections. He promised to implement his vision of "democratic security," applying a firm hand to the country's armed actors. With the collapse of peace talks in February 2002 and the hardening of domestic and international public opinion, Colombians have rallied around their new president's plan to make the country's security forces more effective. Five months into his term, Uribe enjoyed a 75 percent approval rating, the highest of any elected leader in Latin America.

Although Uribe successfully obtained public backing for his agenda, it is uncertain whether that backing will be fleeting or enduring. Much, of course, will depend on his administration's ability to reassert government authority and control over the country and protect Colombians against widespread violence. Colombia's fiscal imbalance is also serious, and it will be hard to adopt the familiar formula for greater spending restraint within a "war economy." Sharp cutbacks could well increase the risk of greater social polarization. Still, Uribe's leadership style—he is indefatigable and has built a highly disciplined cabinet and team of advisers—has broad appeal in Columbia, and he appears to have taken welcome initiative and generated some momentum. To the delight of most Colombians, Uribe is genuinely in charge.

Fundamental strategic concerns loom, however. Even under the most optimistic scenario, it will take Colombia years to substantially reverse its long-term deterioration in public order, much less to pursue meaningful social reform and reconciliation. Uribe will have to manage high public expectations and will need to think through—beyond adopting a series of new, albeit risky, security measures—how to end the conflict and successfully incorporate the country's varied armed actors into Colombian society. Whether he will be able to exhibit the necessary flexibility, vision, and commitment to human rights standards under such extraordinarily difficult conditions remains a major question.

WHAT HAPPENED TO POLITICAL RENEWAL?

Rarely has Latin America witnessed the sort of implosion of a major country like that which occurred in Argentina in December 2001. As a result of a series of converging factors both internal and external, the government led by President Fernando de la Rúa collapsed,

followed by a foreign debt default of some $140 billion. The country lived through a rapid succession of presidents and tremendous uncertainty until Congress finally settled on Eduardo Duhalde, a Peronist, as Argentina's leader.

Duhalde's political skills have received mixed reviews as he has struggled to gain the necessary authority to govern the country effectively. According to the current timetable, Argentines will select Duhalde's successor in April 2003. Yet few analysts believe the new president will be able to restore the severely damaged legitimacy and credibility of the country's traditional political class; that is a longer-term challenge. Today, political parties and leaders are held responsible for Argentina's meltdown, an attitude reflected in the plummeting levels of public confidence in survey after survey. It is instructive that Carlos Menem, Argentina's wily two-term president who brought the country economic stability in the 1990s but has been accused of massive corruption and disregard for democratic institutions, is a serious contender for his old job. His support suggests that, for at least some Argentines, a proven ability to get the job done may trump deep ethical qualms.

Perhaps even more striking is the prospect that Alan Garcia, who presided over Peru's unremitting economic chaos and political violence in the late 1980s, has a good chance of returning to his old job in 2006. In 2001 Garcia, a crafty politician and superb communicator, received more than 48 percent of the vote in the second round run-off election against Toledo. And in November, his American Popular Revolutionary Alliance was the big winner in Peru's regional elections. Further, the headline of a *New York Times* article in December 22, 2002—"Peru's Former President Plots His Return to Power"—suggests a possible comeback as well by Fujimori, who resigned in disgrace and is in exile in Japan. For those who have stood for reform and renovation, few developments would be more dispiriting. While this prospect is hardly reassuring, it highlights the continuing search for political leadership in Latin America.

IS ANYONE PAYING ATTENTION?

Latin America's new cast of political leaders clearly reflects profound changes within the region's complex societies. The landscape is evolving in fundamental ways, conditioned by globalization—its benefits and downsides—and substantially influenced as well by the United States. And the emerging strategic concept, developed in Washington in response to the terrorist attacks on September 11, 2001 and manifested in shifting priorities and resources, has a direct bearing on Latin America's political dynamics. The region's prevailing political uncertainties pose a crucial test for United States policy in the hemisphere.

The key question is whether the United States can be positively and wisely engaged in Latin American affairs so that it can help support leaders committed to democratic politics and economic policies that are market oriented and emphasize a strong social agenda. Such leaders understandably resist the failed formulas for economic reform championed by Washington but are nonetheless generally interested in a cooperative relationship with the United States.

The initial signs from the United States in the post–September 11 period are not encouraging. The principal effect appears to have been distraction, as senior government officials have been consumed by urgent actions and decisions regarding the Middle East and the global effort to combat terrorism. Latin America has seldom been a high priority for the United States, but the lack of high-level sustained attention and engagement has coincided with unprecedented turbulence and deepening crises in many key countries. In the context of globalization, more is at stake for the United States in the region than in the past.

Illustrations of this indifference, or neglect, abound. Perhaps the most dramatic case is Argentina, where the United States treated the country's crisis largely as a fiscal matter, overlooking critical political and foreign policy implications. Former Treasury Secretary Paul O'Neill's remarks were seen as particularly callous. (In an August 2001 interview with CNN, for example, O'Neill said: "We're working to find a way to create a sustainable Argentina, not just one that continues to consume the money of the plumbers and carpenters in the United States who make $50,000 a year and wonder what in the world we're doing with their money.")

It is not surprising that, according to the Latinobarómetro, anti-American sentiment—that is, criticism of United States policies—is appreciably higher in Argentina than anywhere else in the region. Although it would be a stretch to attribute Argentina's political precariousness to the passivity of the United States government at a critical moment, greater political engagement might have reinforced American backing of democratic leaders.

In Venezuela, too, the absence of any strategic engagement is striking, especially in view of a possible war with Iraq and the fact that Venezuela accounts for roughly 15 percent of crude oil imports to the United States. American ineptitude was evident in its handling of both the botched coup in April 2002 and the public call for "early elections" in December 2002. The tacit approval of an unconstitutional act in the former case—the State Department initially failed to express any concern about Chávez's forced ouster—and the public association with opposition forces in the latter have done little to enhance United States credibility on the democracy question in Latin America. (The image of Venezuelan generals standing behind a business leader they had tried to install as president had especially chilling echoes of previous episodes of military takeovers in Latin America, including the American-backed Chilean coup of 1973.) In the current crisis, Washington has also failed to take full advantage of regional institutions to press for a constitutional resolution.

In Mexico it is difficult to separate Fox's political fortunes from his administration's relationship with the United States. This is especially so in light of the expectations raised by the personal relationship between Fox and Bush and the excitement that accompanied the state dinner for Fox held in Washington just days before the terrorist attacks. Without assigning any responsibility, it is undeniable that Mexico's lower place on the American foreign policy agenda in the post–September 11 period—and growing friction between the two countries—have created some political fallout for Fox. Although bilateral cooperation on myriad issues remains significant, Mexico's domestic politics are unusually sensitive to American actions and decisions.

In country after country, it is clear that what the United States does—or fails to do—is germane to the national political scene, and often affects the prospects of particular leaders. In his first year in office, for example, Toledo went up in opinion polls only on two occasions: first, when President George W. Bush visited Lima and second, when the United Sates Congress approved Andean trade legislation extending preferences and giving products from Peru, Colombia, Bolivia, and Ecuador greater access to United States markets. The long-anticipated free trade agreement between the United States and Chile, signed in December 2002, gave a political boost to Chile's president, Ricardo Lagos. The Bush administration's bailouts of Uruguay ($1.5 billion) and Brazil ($30 billion), which sought to contain contagion from Argentina's collapse and buttress Brazil at a moment of political uncertainty, also yielded political dividends in those countries. And in a telling twist, a critical statement made by the United States ambassador in Bolivia regarding the coca growers' candidate, Evo Morales, nearly succeeded in getting Morales elected! ("I want to remind the Bolivian electorate that if they vote for those who want Bolivia to return to exporting cocaine, that will seriously jeopardize any future aid to Bolivia from the U.S.")

Some in Washington, taking advantage of the transformed climate following September 11, have referred to a new "axis of evil" in Latin America, with Lula and Gutiérrez joining Chávez and, of course, the stalwart Fidel Castro. Such hysteria even found its way into a disturbing October 24 letter to President Bush from Republican Representative Henry Hyde, the chairman of the House International Relations Committee, who warned that Lula was a dangerous "pro-Castro radical who for electoral purposes had posed as a moderate." If the United States government were to fashion policies based on such simplistic, Manichaean terms, the Latin American "axis of evil" would risk becoming a self-fulfilling prophecy.

A more apt phrase that captures the political dynamic in much of Latin America is "axis of upheaval." The region's turmoil offers an opportunity for the United States to avoid the sloppiness that often results from neglect—or the hard-line unilateralism that tends to flow from a "good guy versus bad guy" mentality. Positive, sustained, high-level engagement could translate into greater sensitivity to the problems facing Latin American countries. What is required from Washington is increased flexibility that gives struggling leaders in the midst of enormously difficult circumstances more room to maneuver and undertake new social and economic policies. How the Bush administration deals with the Lula presidency is likely to be a major test in this regard.

But the chief responsibility for shaping Latin America's political future rests squarely with a fresh set of political leaders. Their task will be far from easy. Latin America's old problems, such as endemic poverty and fragile institutions, persist and are only aggravated by heightened expectations held by ever-expanding segments of the region's population. Globalization's liberating forces have made this possible. But globalization's dark side is also keenly evidence in the region, with a widening schism between intensifying social demands and political institutions in disarray.

In seeking to navigate their way through these complex challenges, Latin America's new political leaders will be put to a severe test. It is clear that honesty and effectiveness—not rigid formulas or sweeping blueprints—are what the current generation is looking for.

MICHAEL SHIFTER, *a* Current History *contributing editor, is vice president for policy at the Inter-American Dialogue and an adjunct professor at Georgetown University.*

The Many Faces of Africa
Democracy Across a Varied Continent

JOEL D. BARKAN

A decade ago, seasoned observers of African politics including Larry Diamond and Richard Joseph argued that the continent was on the cusp of its "second liberation." Rising popular demand for political reform across Africa, multiparty elections, transitions of power in several countries, and negotiations toward a new political framework in South Africa led these experts to conclude that the prospects for democratization were good. Today, these same observers are not so sure. They describe Africa's current experience with democratization in terms of "electoral democracy," "virtual democracy," or "illiberal democracy," and are far more cautious about predicting what is to come. What is the true state of African democracy? And what is its future?

Governance Before the 1990s

Africa's first liberation was precipitated by the transition from colonial to independent rule that swept much of the continent, except the south, between 1957 and 1964. The West hoped that the transition would be to democratic rule, and more than 40 new states with democratic constitutions emerged following multiparty elections that brought new African-led governments to power. The regimes established by this process, however, soon collapsed or reverted to authoritarian rule—what Samuel Huntington has termed a "reverse wave" of democratization. By the mid-1960s, roughly half of all African countries had seen their elected governments toppled by military coups.

In the other half, elected regimes degenerated into one-party rule. In what was to become a familiar scenario, nationalist political parties formed the first governments. The leaders of these parties then destroyed or marginalized the opposition through a combination of carrot-and-stick policies. The result was a series of clientelist regimes that served as instruments for neo-patrimonial or personal rule by the likes of Mobutu Sese Seko in Zaire or Daniel Arap Moi in Kenya—regimes built around a political boss, rather than founded in a strong party apparatus and the realization of a coherent program or ideology.

This pattern, and its military variant (as with Sani Abacha in Nigeria), became the modal type of African governance from the mid-1960s until the early 1990s. These regimes depended on a continuous and increasing flow of patronage and slush money for survival; there was little else binding them together. Inflationary patronage led to unprecedented levels of corruption, unsustainable macroeconomic policies that caused persistent budget and current account deficits, and state decay, including the decline of the civil service. Most African governments still struggle with this structural and normative legacy, which has obstructed the process of building democracy.

Decade of Democratization?

Africa's second liberation began with the historic 1991 multiparty election in Benin that resulted in the defeat of the incumbent president, an outcome that was replicated in Malawi and Zambia in the same year. The results of these elections raised expectations and created hopes for the restoration of democracy and improved governance across the continent. By the end of 2000, multiparty elections had been held in all but five of Africa's 47 states—Comoros, the Democratic Republic of Congo, Equatorial Guinea, Rwanda, and Somalia.

Along with the new states of the former Soviet Union, Africa was the last region to be swept by the so-called "third wave" of democratization, and as with many of the successor states of the former Soviet Union, the record since has been mixed. In stark contrast to the democratic transitions that occurred in the 1970s and 1980s in Southern and Eastern Europe and Latin America (excluding Mexico), most African transitions have not been marked by a breakthrough election that definitively ended an authoritarian regime by bringing a group of political reformers to power. While this type of transition has occurred in a small number of states, most notably Benin and South Africa, the more typical pattern has been a process of protracted transition: a mix of electoral democracy and political liberalization combined with elements of authoritarian rule and, more fundamentally, the perpetuation of clientelist rule. In this context, politics is a three-cornered struggle between authoritarians, patronage-seekers, and reformers. Authoritarians attempt to retain power by permitting greater liberalization and elections while selectively allocating patronage to those who remain loyal. Meanwhile, patronage-seekers attempt to obtain the spoils of office via electoral means, as reformers pursue the establishment of democratic rule. The boundaries between the first and second of these groups, and sometimes between the second and third, can be blurred because political alignments are very fluid. Liberal democracy is unlikely to be consolidated until reformers ascend to power.

Politics is a three-cornered struggle between authoritarians, patronage-seekers, and reformers.

The result is what Thomas Carothers has termed a "gray zone" of polities, describing countries where continued progress toward democracy beyond elections is limited and where the consolidation of democracy, if it does occur, will unfold over a long period, perhaps decades. This characterization does not necessarily mean that the third wave of democratization is over in Africa. Rather, we should expect Africa's democratic transitions to be similar to those of India or Mexico. In the former, the party that led the country to independence did not lose an election for three decades, and periodic alternation of power between parties did not occur until after 40 Years. In the latter, the end of one-party rule and its replacement by an opposition committed to democratic principles played out over five elections spanning 13 years rather than a single founding election. Such appears to be the pattern in Africa, where two-thirds of founding and second elections have returned incumbent authoritarians to power, but where each iteration of the electoral process has usually resulted in a significant incremental advance in the development of civil society, electoral fairness, and the overall political process.

That many African polities fall into the gray zone is confirmed by the most recent annual *Freedom in the World* survey conducted by Freedom House. Of the 47 states that comprise sub-Saharan Africa, 23 were classified by the survey as "partly free" based on the extent of their political freedoms and civil liberties. Only eight (Benin, Botswana, Cape Verde, Ghana, Mali, Mauritius, Namibia, and South Africa)

were classified as "free" while 16, including eight war-torn societies (Angola, Burundi, the Democratic Republic of Congo, Ethiopia, Eritrea, Liberia, Rwanda, and Sudan) were deemed "not free."

The overall picture revealed by these numbers is sobering. Less than one-fifth of all African countries were classified as free, and of these, only two or three (Botswana, Mauritius, and perhaps South Africa) can be termed consolidated democracies. On the other hand, if one excludes states in the midst of civil war, one-fifth of Africa's countries are free, one-fifth not free, and three-fifths fall in-between. That is to say, four-fifths of those not enmeshed in civil war are partly free or free, a significant advance over the continent's condition a decade ago. Only a handful are consolidated democracies, but few are harsh dictatorships of the type that dominated Africa from the mid-1960s to the beginning of the 1990s. As noted by Ghana's E. Gyimah-Boadi, "Illiberalism has persisted, but is not on the rise. Authoritarianism is alive in Africa today, but is not well."

Optimists and Realists

The current status of democracy in Africa varies greatly from one country to the next, and one should resist generalizations that apply to all 47 of the continent's states; one size does not fit all. Notwithstanding this reality, those who track events in Africa have divided themselves into two distinct camps: optimists and realists. Those in the United States who take an optimistic view—mainly government officials involved in efforts to promote democratization abroad, former members of President Bill Clinton's administration responsible for Africa, members of the Congressional Black Caucus, and the staff of some Africa-oriented nongovernmental organizations—trumpet the fact that multiparty elections have been held in nearly 90 percent of all African states. They note that most African countries have now held competitive elections twice and that some, including Benin, Ghana, and Senegal, have held genuine elections three times, at least one of which has resulted in a change of government. The optimists further note that the quality of these elections has improved in some countries, both in terms of efficiency and of fairness. Electoral commissions seem to have been more independent, even-handed, and professional in recent elections than in the early 1990s. Opposition candidates and parties have

greater freedom to campaign and have faced less harassment from incumbent governments. The presence of election observers, both foreign and domestic, is now widely accepted as part of the process. Perhaps most significant, citizen participation in elections has been fairly high, averaging just under two-thirds of all registered voters.

Recognizing that elections are a necessary but insufficient condition for the consolidation of democracy, the optimists also point to advances in several areas, listed below in their approximate order of accomplishment. First, there has been a re-emergence and proliferation of civil society organizations after their systematic suppression during the era of single-party and military rule. Second, an independent and free press has also re-emerged, spurred on by the privatization of broadcast media in several countries. Third, members of the legislature have increasingly asserted themselves in policymaking and overseeing the executive branch. Fourth, the judiciary and the rule of law have been strengthened in countries such as Tanzania, and human rights abuses have also declined. Fifth, there have been new experiments with federalism—the delegation or devolution of authority from the central government to local authorities—to enhance governmental accountability to the public and defuse the potential for ethnic conflict, most notably in Nigeria but also in Ethiopia, Ghana, Tanzania, Uganda, and South Africa. One or more of these trends, especially the first two and perhaps the third, can be found in most African countries that are not trapped in civil war.

Optimists also point to less exclusive membership in the governing elite, which has expanded into the upper-middle sector of society far more than during the era of authoritarian rule. In country after country, repeated multiparty elections have resulted in significant turnover in the national legislatures and local government bodies, sometimes as high as 40 percent per cycle. While the quality of elected officials at the local level remains poor, members of national legislatures are younger, better educated, and more independent in their political approach than the older generation they have displaced. Although further research is needed to confirm any major change in the composition of these bodies, new politicians and legislators also appear more likely than their predecessors to be democrats and to focus on issues of public policy and less likely to be patronage seekers.

Finally, public opinion across Africa appears to prefer democracy over any au-

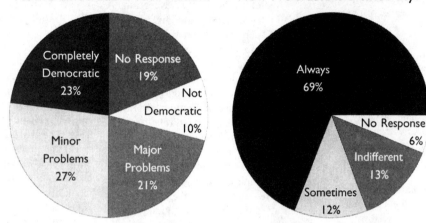

THE PEOPLE'S PERSPECTIVE

How Democratic is Your Nation?

Completely Democratic 23%
No Response 19%
Not Democratic 10%
Minor Problems 27%
Major Problems 21%

How Preferable is Democracy?

Always 69%
No Response 6%
Indifferent 13%
Sometimes 12%

These two figures, which represent the attitudes of citizens of African nations toward democracy, suggests a strong preference for democratic rule but uncertainty as to its prevalence.

Afrobarometer at www.afrobarometer.org

thoritarian alternatives. Surveys undertaken for the Afrobarometer project in 12 African countries between 1999 and 2001 found that a mean of 69 percent of all respondents regarded democracy as "preferable to any other kind of government," while only 12 percent agreed with the proposition that "in certain situations, a non-democratic government can be preferable." Moreover, 58 percent of all respondents stated that they were "fairly satisfied" or "very satisfied" with the "way democracy works" in their country.

Realists—who criticize what they contend was a moralistic approach to US foreign policy by the Clinton administration and disparage the use of democratization as a foreign policy goal—take a far more cautious view of what is occurring on the continent. Considering the same six developments that optimists cite as examples of democratic advances, realists note that all six are present in fewer than six countries. They also see much less progress than the optimists when nations are considered one by one. First, regular multiparty elections across the continent have resulted in an alternation of government in only one-third of the countries that have held votes. Moreover, only about one-half of these elections have been regarded as free and fair, with results accepted by those who have lost. It is also debatable in most of these countries whether recent elections have been of higher quality than those held in the early 1990s.

Second, although the re-emergence of civil society and the free press is a significant advance from the era of authoritarian rule when both were barely tolerated or systematically suppressed, civil society remains very weak in Africa compared to other regions and is concentrated in urban areas. Political parties are especially weak and rarely differentiate themselves from one another on the basis of policy. Apart from the church, farmers' organizations, or community self-help groups in a smattering of countries (such as Kenya, Cote d'Ivoire, and Nigeria), civil society barely exists in rural areas where most of the population resides. The press, especially the print media, is similarly concentrated in urban areas and thus reaches a relatively small proportion of the entire population. Only the broadcast media penetrates the countryside, but it is largely state-owned. Although private broadcasting has grown in recent years, especially in television and FM radio, stations cater almost exclusively to urban audiences. With a few exceptions, AM and short- and medium-wave radio—the chief sources of information for the rural population—remain state monopolies.

Third, while the legislature holds out the promise of becoming an institution of countervailing power in some countries, it remains weak and has rarely managed to effectively check executive power. Fourth, the judicial system in most countries is ineffectual, either because its members are corrupt or because it has too few magistrates and too poor an infrastructure to keep pace with the number of cases. Human rights abuses also continue, though less frequently and with less intensity than a decade ago. Fifth, Africa's experiments with federalism, though apparently suc-

cessful, are confined to six states. Finally, the extent to which Africans have internalized democratic values is hard to judge. Although the Afrobarometer surveys indicate broad support for democracy, the results also suggest that such support is "a mile wide and an inch deep." An average of only 23 percent of respondents in each country described their country as "completely democratic."

Both optimists and realists are correct in their assessments of what is occurring in Africa. But how can both views be valid? The answer is that each presents only one side of the story. On a continent where the record of democratization is one of partial advance in over one-half of the cases, those assessing progress toward democratization, or lack of it, tend to dwell either on what has been accomplished or on what has yet to be achieved. These divergent assessments are proverbial examples of those who view the glass as either half-full or half-empty. Optimists and realists also draw their conclusions from slightly different samples. Whereas optimists focus mainly on states that are partly free or free, realists concentrate on states that are partly free or not free.

Optimists and realists are also both right because there are several Africas rather than one. In fact, at least five Africas cut across the three broad categories of the Freedom House survey. First are the consolidated and semi-consolidated democracies—a much smaller group of counties than those classified as free. This category presently consists of only two or three cases, such as Botswana, Mauritius, and perhaps South Africa. The second group consists of approximately 15 aspiring democracies, including the remaining five classified by Freedom House as free but not yet consolidated democracies plus roughly 10 classified as partly free where the transition to democracy has not stalled. All these states have exhibited slow but continuous progress toward a more liberal and institutionalized form of democratic politics. In this group are Benin, Ghana, Madagascar, Mali, Senegal, and possibly Kenya, Malawi, Tanzania, and Zambia. Third are semi-authoritarian states, countries classified as partly free where the transition to democracy has stalled. This category consists of approximately 13 countries including Uganda, the Central African Republic, and possibly Zimbabwe. Fourth are countries that are not free, with little or no prospect for a democratic transition in the near future. About 10 countries make up this group, including Cameroon, Chad, Eritrea, Ethiopia, Rwanda, and Togo. Finally, there are the states mired in civil war, such as Angola, Congo, Liberia, and Sudan. Each of these five Africas presents a different context for the pursuit of democracy.

Inhibiting Democracy

Several conditions peculiar to the continent make Africa a difficult place to sustain democratic practice. They explain why Africa lags behind other regions in its extent of democratic advance, why political party organizations are weak, and why the ties between leaders and followers are usually based on clientelist relationships. These conditions in turn create pressures for more and more patronage, a situation that undermines electoral accountability and leads to corruption.

Africa is the poorest of the world's principal regions: per capita income averages US$490 per year. This condition does not affect the emergence of democracy but does impact its sustainability. On average, democracies with per capita incomes of less than US$1,000 last 8.5 years while those with per capita incomes of over US$6,000 endure for 99. The reasons for this are straightforward. Relatively wealthy countries are better able to allocate their resources to most or all groups making claims on the state, while poor countries are not. The result is that politics in a poor country is likely to be a zero-sum game, a reality that does not foster bargaining and compromise between competing

interests or a willingness to play by democratic rules.

Almost all African countries remain agrarian societies. With few exceptions like South Africa, Gabon, and Nigeria, 65 percent to 90 percent of the national populations reside in rural areas where most people are peasant farmers. Consequently, most Africans maintain strong attachments to their places of residence and to fellow citizens within their communities. Norms of reciprocity also shape social relations to a much greater degree than in urban industrial societies. In this context, Africans usually define their political interests—that is to say, their interests as citizens vis-a-vis the state—in terms of where they live and their affective ties to neighbors, rather than on the basis of occupation or socio-economic class.

Several conditions peculiar to the continent make Africa a difficult place to sustain democratic practice.

With the exceptions of Botswana and Somalia, all African countries are multi-ethnic societies where each group inhabits a distinct territorial homeland. Africans' tendency to define their political interests in terms of where they live is thus accentuated by the fact that residents of different areas are often members of different ethnic groups or sub-groups.

Finally, African states provide much larger proportions of wage employment, particularly middle-class employment, than states in other regions do. African states have also historically been large mobilizers of capital, though to a lesser extent recently. Few countries have given rise to a middle class that does not depend on the state for its own employment and reproduction. In this context, people seek political office for the resources it confers, for their clients' benefit, and for the chance to enhance their own status. In the words of a well known Nigerian party slogan, "I chop, you chop," literally, "I eat, you eat."

Likely Scenarios

Given these realities, what is the future of democratization on the continent? The answer varies greatly by type of polity. Over the next decade, the small handful of states that are currently classified as consolidated or semi-consolidated democracies is likely to gain perhaps six additional members. This category, however, will remain the smallest of the five Africas because of the limited number of polities that can be realistically considered as candidates. These include countries classified as free in the Freedom House surveys but that have yet to experience a turnover, much less a double turnover, of elected government. Even Botswana and South Africa have yet to pass this test.

Within the current category of aspiring democracies, we can reasonably expect a process of further political liberalization and growth of civil society, including the emergence of interest group politics that will challenge continuing clientelist arrangements. Civil society will also become better organized in the countryside. The legislature and judiciary are likely to become stronger and more independent. Greater electoral turnover of government will occur simply because some countries, such as Kenya, Tanzania, and Zambia, have adopted term limits at the presidential level. Some members of this category will be promoted to the ranks of consolidated democracies. All or nearly all should see their performance improve.

Among semi-authoritarian states where democratization has stalled, the prospects for progress are much worse. This Africa consists of approximately 13 polities, all of which are in the lower range of the partly free category. As a result, there is a limited measure of political space open for the growth of civil society and the press on the one hand, and competitive electoral politics on the other. One should not write off this category, however. Rather, one should acknowledge that democratization in these states will be an especially protracted process.

The fourth and fifth Africas—those countries classified as not free and those mired in civil war—are obviously the least promising. The immediate challenges in countries classified as not free is halting human rights abuses and beginning a process of political liberalization, the first steps toward democratization. The first challenge for countries mired in civil war is to stop the fighting and reconstitute the state. State decay is probably the greatest problem for these two groups and must be addressed before any meaningful process of democratization can begin.

In conclusion, one can be cautiously optimistic about the prospects for further democratization in between one-third and one-half of the states on the African continent, but one must be realistic about the prospects in the remainder, especially the bottom one-third of countries. Future progress is likely to be incremental, played out over a long period of at least a decade or more. Moreover, progress will not occur on a linear basis through a series of well-defined stages but unevenly and haltingly. Further progress, especially in the third Africa of semi-authoritarian states and to a limited extent in the fourth category of states ranked as not free, will depend on the

outcome of a continuous three-cornered struggle between authoritarians, patronage-seekers, and reformers. One must also expect some erosion or reversals among states currently regarded as semi-consolidated democracies and aspiring ones.

Finally, it is important that US policy-makers and those of other established democracies seeking to support the process appreciate these constraints and are honest about the varied nature of the challenge. The most difficult task will be to nurture the process in states falling into the third and fourth categories. Indeed, there is already a tendency to over-celebrate progress in the first and second Africas while retreating from the challenges in the third and fourth groups. Such an approach will undermine

both the accomplishments to date and the prospects for further democratic advance. Critics of a foreign policy that stresses democratization should also reflect on the fact that the development of democratic systems in the West evolved over a period of more than a century. While one can appreciate the sense of impatience among many realists, this is no time to walk away from the Kenyas, Ugandas, and Zimbabwes, no matter how halting the democratic process seems.

JOEL D. BARKAN is Professor of Political Science at the University of Iowa and Fellow at the Woodrow Wilson International Center for Scholars.

NGOs and the New Democracy

The False Saviors Of International Development

SANGEETA KAMAT

Conservatives and liberals agree that globalization is hastening civil society's coming of age. Liberals consider civil society the only countervailing force against an unresponsive, corrupt state and exploitative corporations that disregard both environmental issues and human rights. Meanwhile, conservatives celebrate the awakening of civil society as proof of the beneficial effects of globalization for the development of democracy. Thus, in the debate on development and the state, left and right appear to converge on the side of civil society. In advancing this proposition, the dynamic rise of non-governmental organizations (NGOs) is offered as proof of the self-organizing capacity of civil society and the consequent redundancy of the state.

The global phenomenon of NGOs reflects the new policy consensus that these groups are de facto agents of democracy rather than end-products of a thriving democratic culture. This is evident in the astonishing speed with which NGOs have emerged in countries on the verge of establishing democracies. The leading role ascribed to NGOs foretells a reworking of democracy in ways that coalesce with global capitalist interests. Global policy institutions are actively enlisting NGOs in the economic reform process, but in doing so, they undercut their popular role as forces of democratization.

Current debate on the role of NGOs points to the dangers of replacing the state as the representative of democracy.

Given expanding market economies and shrinking states, NGOs fill a growing void by responding to the needs and demands of the poor and marginalized sections of society. Pointing to this emergent trend, development analysts caution that, unlike governments and state bureaucracies, there are no mechanisms by which NGOs can be made accountable to the people they serve. Instead, analysts suggest that a balanced partnership between states and NGOs can best serve the interests of society.

[NGOs'] dependence on external funding and compliance with funding agency targets raise doubts about whether their accountability lies with the people or with funding agencies.

Much of the current discussion on NGOs focuses on issues of improving NGO accountability, autonomy, and organizational effectiveness. However, as Robert Hayden's recent essay in the *Harvard International Review* ("Dictatorships of Virtue?" Summer 2002) illustrated, NGO autonomy is a mirage that obscures the interests of powerful states, national elites, and private capital. If NGO autonomy is indeed a myth, the more relevant task at hand is to understand the nature of developing states' de-

pendency on NGOs as well as the effects on development and democracy.

The evolution of community based organizations (CBOs) is illustrative of the changed environment in which NGOs operate and the grave implications of the new scenario for development, democracy, and political stability. CBOs are locally based organizations seen as the champions of "bottom up" or "pro-people" development. They have been particularly vulnerable to the unexpected patronage of donor agencies. CBOs emerged in the post–World War II period between the 1960s and 1980s in response to the failure of developmental states to ensure the basic needs of the poor. For the most part, the leaders of CBOs were socially conscious middle class citizens, many of whom had been active in women's or radical left movements of the post-independence period but later became disenchanted with leftist political parties and movements. The CBOs promoted a "development with social justice" approach, and established political rights and awareness campaigns alongside health and livelihood projects.

Donor NGOs such as the United Kingdom's Oxfam were eager to fund CBOs directly because these organizations were more committed and effective in reaching the poor than were the governments of developing countries. The nature of their work requires CBOs to interact with local communities on a daily basis, building relationships of cooperation and trust to understand local needs and tailor projects that respond to those needs. Consequently, CBOs tend to have intimate working relations with the people of the community, some of whom are paid staff of the CBO. The work of numerous such activists and organizations in India—which political scientist Rajni Kothari identified as "non-party political formations"—was looked upon suspiciously by the state.

This early history of CBOs signified the birth of pluralist democratic cultures in many developing countries but has been ignored in the current policy environment characterized by free market reform and the dismantling of the social democratic state apparatus. With the imposition of structural adjustment programs and neoliberal economic policies in Africa, Latin America, and South Asia, CBOs have become useful and even essential to the functioning of international donor institutions. The lack of state infrastructure, combined with the decline in state entitlements to the poor, has led donor agencies to channel greater amounts of aid to CBOs and NGOs rather than to state governments. In fact, the *Financial Times* reported in July 2001 that the United Kingdom is increasingly inclined to fund locally based NGOs directly, bypassing its own NGOs such as Oxfam.

The parallel between a "minimalist" state and the exponential increase in community development NGOs has led development theorist Geoff Wood to conclude that the phenomenon is analogous to "franchising the state." Financial institutions both recommend the withdrawal of state support from the social sector and allocate aid to community-based NGOs for those very same social services. This phenomenon indicates that the expansion of the NGO sector has been externally induced by foreign policy decisions. This dual policy of aid insitutions undermines the credibility of NGOs, formed during their early community-based operations, as homegrown constituents of a thriving political culture, independent of patronage from state and international institutions. Their dependence on external funding and compliance with funding agency targets raise doubts about whether their accountability lies with the people or with funding agencies.

The Evolution of CBOs

This influx of money, combined with pressure to lead when the state is absent, has forced NGOs and CBOs in particular to restructure their operations to suit the new partnership with First World donor agencies. In this process, the organizational ethic that distinguishes CBOs as democratic and more representative of the popular will than other types of NGOs is being dismantled. CBOs have an active membership base among the particular community in which they work, be it urban slum-dwellers or poor farmers. These "target" or "client" groups at the local level are themselves involved in decision-making processes and provide organizational direction often through a complex tiered system that involves members from the smallest unit (such as village or hamlet) to the larger district level. This form of direct democracy enthralls donor agencies but also inconveniences them. On the one hand, it locates the unique strengths of NGOs, which, as outlined by the World Bank in 1998, include "their ability to reach poor communities and remote areas, promote local participation, operate at low cost, identify local needs, build on local resources." On the other hand, direct democracy is inconvenient to international donor agencies because of its "limited replicability, self-sustainability, managerial and technical capacity, narrow context for programming, and politicization."

In order to better serve the needs of donor agencies, funding is directed toward non-membership CBOs or what the World Bank has designated as "operational NGOs"-groups that operate within poor constituencies but are not organizations of the poor. Operational NGOs are thus organizations "engaged primarily in design, facilitation, and implementation of development sub-projects," and they have been explicitly designated as the preferred recipients of World Bank funding. As a consequence, the nature of NGO activity at the local level has shifted significantly. The implementation of projects calls for training in specific skills rather than a more general education that involves analysis of social and economic policies and processes. In other words, these developments have compelled CBOs to adopt a narrowly economic and apolitical approach to working with the poor. The logical consequence of funding flows is that CBOs that have no local support or participation have sprung up overnight. Stephen Commins, a World Bank social policy analyst, admits that the Bank now faces the problem of assessing whether a local organization really does have broad based support or whether it is a "bringo"—"bring your own NGO." But the donor community continues to ignore its own warnings about both the growing disconnect between the people and NGOs and the resulting crisis of credibility.

The change in focus of NGO activity impacts the organizational character of NGOs as well. The shift toward a managerial and functional approach to development has led to a more professional orientation to the extent that professionally trained staffs constitute a significant component of the leadership in CBOs today. The change in leadership has an enduring impact on the political capabilities of NGOs because a technical staff tends to regard its work as apolitical and disconnected from larger social and economic processes, such as structural adjustment or international debt policies, even when they directly impact the poor. More often than not, technical personnel adopt a functionalist problem-solving approach to social issues of inequality and poverty that translates into paternalism toward the poor. In other words, the professionalization of community-based NGOs and their subsequent depoliticization represent two sides of the same coin and produce a common set of effects.

Neoliberal "Empowerment"

This new emphasis on project implementation at the local level results in a focus on individual capacities to minimize the social and political causes of poverty. The apolitical and managerial approach to community development draws upon the liberal notion of empowerment in which the poor are encouraged to find entrepreneurial solutions to their basic needs. This entrepreneurial notion of empowerment (not unlike the US

NGO EXPLOSION

Growth of International NGOs in the 1990s

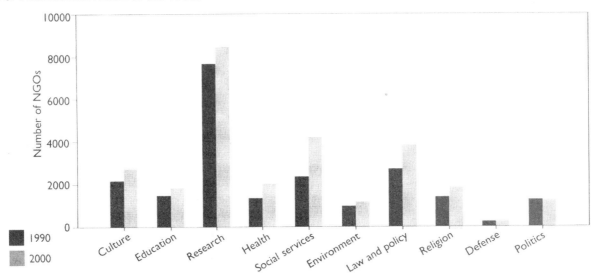

As the figure above indicates, the 1990s witnessed a large increase in the number of international NGOs, particularly amoung social service NGOs. Though the last century saw astonishing growth in international NGOs, from 1,083 in 1914 to more than 37,000 in 2000, nearly 20 percent of the international NGOs in existence today were formed after 1990.

United Nations Human Development Report 2002

motto of "pulling oneself up by one's bootstraps") is altogether different from the understanding of empowerment for social justice that characterized the work of CBOs in the post–World War II development period. In the current use of the term empowerment, the individual is posited as both the problem and the solution to poverty, diverting attention from the issue of the state's redistribution or global trade policies. On the other hand, the "development with social justice" approach involves educating the poor in terms of both social and economic policies and their own political rights. This strategy is similar to the one used in the women's rights and environmental justice movements, which aim at empowering individuals to change their societies.

The partnership between NGOs and new economic institutions has thus enhanced NGO activity by separating NGOs from their original mandate to organize the poor against state and elite interests. This is a clear case in which market demand determines supply. Operational NGOs emerge and flourish to meet the demand of international aid agencies, thereby restructuring political engagement at the local level in com-

pletely new ways. CBOs are increasingly engaged in empowering the poor to become active in their own development, which is much in the spirit of the World Bank's own conception of empowerment. The World Bank's Participation Sourcebook explains: "As the capacity of poor people is strengthened and their voices begin to be heard, they become 'clients' who are capable of demanding and paying for goods and services from government and private sector agencies. …We reach the far end of the continuum when these clients ultimately become the owners and managers of their assets and activities."

The popularity of micro-credit programs among NGO projects can be understood within this context where the state is no longer responsible for creating employment, and the poor are expected to strengthen their own capacities toward livelihood security. Micro-credit programs are well suited to the neoliberal economic context in which risks are shifted to the individual entrepreneur—usually poor women who are forced to compete among themselves in a restricted, uneven, and fluctuating market environment. The promise of livelihood

security thus translates into optimal utilization of one's own capacities and resources.

The neoliberal notion of empowerment leads unmistakably to the marketization of social identities and relations. Individualizing the process of empowerment where each individual has to build his or her capacities to access the marketplace reduces the concept of public welfare to one of private interest. The identity of the "citizen" is reduced to that of a "client," such that the solution to social inequality requires individuals to build their capacities to access the marketplace. Public welfare is reduced to an aggregate of individual gains, and the social democratic notion that public welfare is something that must prevail over and above private gain ceases to exist. Questions of public goods and services or of distributional issues are ignored in this version of empowerment and participation. The democratization that NGOs represent is thus more symbolic than substantive. For the most part, they are engaged in producing a particular kind of democracy that coincides with and can function within a neoliberal economic context.

Studies conducted independently by scholars in different countries have confirmed the phenomena of both NGO professionalization and depoliticization at the grassroots level and agree that there has been a remarkably rapid shift both in the organizational character of NGOs and in the nature of their work. For instance, Miraftab traces the evolution of Mexican NGOs from organizations geared toward "deep social change through raising consciousness, making demands, and opposing the government," to organizations aiming at the "incremental improvement of the poor's living conditions through community self-reliance." In my research in Western India, I found a similar transition from consciousness-raising and political-organizing work to an emphasis on skills-training for economic livelihood projects. In each case, community-based NGOs moved away from empowerment programs that involved political organization of the poor and education about unfair state policies or unequal distribution of resources. Instead, NGOs have adopted a "skills training" approach to mitigate poverty and inequality by providing social and economic inputs based on a technical assessment of the capacities and needs of the community.

Operational NGOs that establish instrumental relations with their constituencies allow development experts to proceed as if the demands of the people are already known and pre-defined—demands such as roads, electricity, literacy, mid-day meals, birth control for women, micro-credit, and poultry farming. Empowerment and participation are simulated by NGOs and their donor agencies even as their practices are increasingly removed from the meaning of these terms. As a result, grassroots organizations that do nor function within the "operational NGO" formula of simply managing development projects in a technical and professional manner and instead politicize social and economic issues of livelihood security, health, water, and education are delegitimized as anti-national and anti-development. Dubbed "anti-globalization" movements by the popular media, these organizations are actually invested in making globalization work for the poor. Economist and Nobel Laureate Amartya Sen eloquently argues that the work of political NGOs and other organizations forces us to reckon with questions of redistribution and equity within and between states that are largely ignored by current development policies. The one outstanding exception to this general negligence is the debt relief granted to highly indebted poor countries, a program that was itself made possible by an NGO campaign.

Another issue that has been neglected in the discussion of NGOs is the rise of religious conservatism in many developing countries. While the NGO sector in these countries represents a significant counter to the religious right, corporatized NGOs disconnected from the popular base are significantly constrained in their capacity to intervene in this emergent political crisis. To recover the value and ethics that underlie social and economic development, it is necessary to examine donor patronage of the NGO sector, the depoliticization of CBOs, and the ascendancy of religious and cultural nationalisms as interconnected processes.

SANGEETA KAMAT is Professor at the University of Massachusetts Center for International Education.

UNIT 5

Population, Development, Environment, & Health

Unit Selections

Key Points to Consider

- How do current population trends represent a departure from the past?

- How does environmental degradation affect developing countries?

- What are the consequences of the hazardous waste trade?

- Was the World Summit on Sustainable Development a success? Explain your answer.

- What are the constraints on dispensing new drug therapies for diseases in poor countries?

 **Links: www.dushkin.com/online/**
These sites are annotated in the World Wide Web pages.

Earth Pledge Foundation
  http://www.earthpledge.org
EnviroLink
  http://envirolink.org
Greenpeace
  http://www.greenpeace.org
Linkages on Environmental Issues and Development
  http://www.iisd.ca/linkages/
Population Action International
  http://www.populationaction.org
The Worldwatch Institute
  http://www.worldwatch.org

The developing world's population continues to increase at an annual rate that exceeds the world average. The average fertility rate (the number of children a woman will have during her life) for developing countries is 3.1, while in sub-Saharan Africa it is as high as 5.8. Although growth has slowed somewhat, world population is currently growing at the rate of approximately 80 million per year, with most of this increase taking place in the developing world. Increasing population complicates development efforts, puts added stress on ecosystems, threatens food security, and affects population migration patterns.

World population surpassed 6 billion toward the end of 1999 and, if current trends continue, could reach 9.8 billion by 2050. Even if, by some miracle, population growth were immediately reduced to the level found in industrialized countries, the developing world's population would continue to grow for decades. Approximately one-third of the population in the developing world is under the age of 15, with that proportion jumping to 43 percent in the least developed countries. The population momentum created by this age distribution means that it will be some time before the developing world's population growth slows substantially. Some developing countries have achieved progress in reducing fertility rates through family planning programs, but much remains to be done. At the same time, reduced life expectancy, especially related to the HIV/AIDS epidemic, is having a significant demographic impact in parts of the developing world.

More than a billion people live in absolute poverty, as measured by a combination of economic and social indicators. As population increases, it becomes more difficult to meet the basic human needs of the developing world's citizens. Indeed, food scarcity looms as a major problem among the poor as population increases, production fails to keep pace, demand for water increases, per capita cropland shrinks, and prices rise. Larger populations of poor people also place greater strains on scarce resources and fragile ecosystems. Deforestation for agriculture and fuel has reduced forested areas and contributed to erosion, desertification, and global warming. Intensified agriculture, particularly for cash crops, has depleted soils. This necessitates increased fertilization, which is costly and also produces runoff that contributes to water pollution.

Economic development, regarded by many as a panacea, has not only failed to eliminate poverty but has exacerbated it in some ways. Ill-conceived economic development plans have diverted resources from more productive uses. There has also been a tendency to favor large-scale industrial plants that may be unsuitable to local conditions. Where economic growth has occurred, the benefits are often distributed inequitably, widening the gap between rich and poor. If developing countries try to follow Western consumption patterns, sustainable development will be impossible. Furthermore, economic growth without effective environmental policies can lead to the need for more expensive clean-up efforts in the future.

Divisions between North and South on environmental issues became more pronounced at the 1992 Rio Conference on Environment and Development. The conference highlighted the fundamental differences between the industrialized world and

developing countries over causes of and solutions to global environmental problems. Developing countries pointed to consumption levels in the North as the main cause of environmental problems and called on the industrialized countries to pay most of the costs of environmental programs. Industrialized countries sought to convince developing countries to conserve their resources in their drive to modernize and develop. Divisions have also emerged on the issues of climate and greenhouse gas emissions. The Johannesburg summit on sustainable development, a follow-up to the Rio conference, grappled with many of these issues, achieving some modest success in focusing attention on water and sanitation needs.

Rural-to-urban migration has caused an enormous influx of people to the cities, lured there by the illusion of opportunity and the attraction of urban life. In reality, opportunity is limited. Nevertheless, most choose marginal lives in the cities rather than a return to the countryside. As a result, urban areas in the developing world increasingly lack infrastructure to support this increased population and also have rising rates of pollution, crime, and disease. Additional resources are diverted to the urban areas in an attempt to meet increased demands, often further impoverishing rural areas. Meanwhile, food production may be affected, with those remaining in rural areas having to choose either to farm for subsistence because of low prices or to raise cash crops for export.

Poverty and urbanization also contribute to the spread of disease. Environmental factors account for about one-fifth of all diseases in developing countries and also make citizens more vulnerable to natural disasters. Hazardous waste also represents a serious health and environmental risk. The HIV/AIDS epidemic has focused attention on public health issues, especially in Africa. Africans account for 70 percent of the more than 40 million AIDS cases worldwide. Aside from the human tragedy that this epidemic creates, the development implications are enormous. The loss of skilled and educated workers, the increase in the number of orphans, and the economic disruption that the disease causes will have a profound impact in the future. Meanwhile, the development and availability of drugs to treat AIDS and other common diseases in the developing world is constrained by patent and profitability concerns.

The Population IMPLOSION

Be careful what you wish for. After decades of struggling to contain the global population explosion that emerged from the healthcare revolution of the 20th century, the world confronts an unfamiliar crisis: rapidly decreasing birthrates and declining life spans that might set back the progress of human development.

By Nicholas Eberstadt

It may not be the first way we think of ourselves, but almost all of us alive today happen to be children of the "world population explosion"—the momentous demographic surge that overtook the planet during the course of the 20th century. Thanks to sweeping mortality declines, human numbers nearly quadrupled in just 100 years, leaping from about 1.6 or 1.7 billion in 1900 to about 6 billion in 2000.

This unprecedented demographic expansion came to be regarded as a "population problem," and in our modern era problems demand solutions. By century's end, a worldwide administrative apparatus—comprised of Western foundations and aid agencies, multilateral institutions, and Third World "population" ministries—had been erected for the express purpose of "stabilizing" world population and was vigorously pursuing an international anti-natal policy; focusing on low-income areas where fertility levels remained relatively high.

To some of us, the wisdom of this crusade to depress birthrates around the world (and especially among the world's poorest) has always been elusive. But entirely apart from its arguable merit, the continuing preoccupation with high fertility and rapid population growth has left the international population policy community poorly prepared to comprehend (much less respond to) the demographic trends emerging around the world today—trends that are likely to transform the global population profile significantly over the coming generation. Simply put, the era of the worldwide "population explosion," the only demographic era within living memory, is coming to a close.

Continued global population growth, to be sure, is in the offing as far as the demographer's eye can see. It would take a cataclysm of biblical proportions to prevent an increase in human numbers between now and the year 2025. Yet global population growth can no longer be accurately described as "unprecedented." Despite the imprecision of up-to-the-minute estimates, both the pace and absolute magnitude of increases in human numbers are markedly lower today than they were just a few years ago. Even more substantial decelerations of global population growth all but surely await us in the decades immediately ahead.

In place of the population explosion, a new set of demographic trends—each historically unprecedented in its own right—is poised to reshape, and recast, the world's population profile over the coming quarter century. Three of these emerging tendencies deserve special mention. The first is the spread of "subreplacement" fertility regimens, that is, patterns of childbearing that would eventually result, all else being equal, in indefinite population decline. The second is the aging of the world's population, a process that will be both rapid and extreme for many societies over the coming quarter century. The final tendency, perhaps the least appreciated of the three, is the eruption of intense and prolonged mortality crises, including brutal peacetime reversals in health conditions for countries that have already achieved relatively high levels of life expectancy.

For all the anxiety that the population explosion has engendered, it is hardly clear that humanity will be better served by the dominant demographic forces of the post-population-explosion era. Nobody in the world will be untouched by these trends, which will have a profound impact on employment rates, social safety nets, migration patterns, language, and education policies. In particular, the impact of acute and extended mortality setbacks is ominous. Universal and progressive peacetime improvements in health conditions were all but taken for granted in the demographic era that is now concluding; they no longer can be today, or in the era that lies ahead.

Where Have All the Children Gone?

2000 Fertility Rates

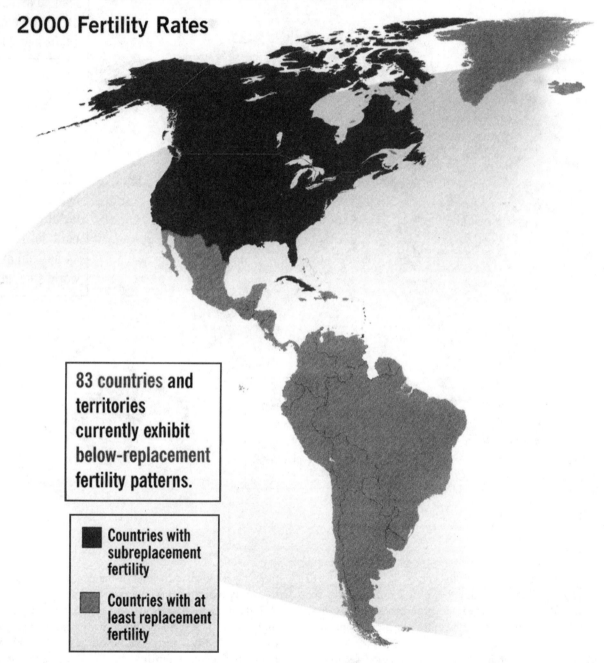

83 countries and territories currently exhibit below-replacement fertility patterns.

■ Countries with subreplacement fertility

▨ Countries with at least replacement fertility

(continued on next page)

Source: U.S. Bureau of the Census, International Data Base

THE GLOBAL BABY BUST

In arithmetic terms, the 20th-century population explosion was the result of improvements in health and the expansion of life expectancy. Human life expectancy at birth is estimated to have doubled or more between 1900 and 2000, shooting up from approximately 30 years to nearly 65 years. Population growth rates accelerated radically thanks to the concomitant plunge in death rates. Despite tremendous population growth, rough calculations suggest that the world's population would be over 50 percent larger today in the absence of any other demographic changes.

The world's population currently totals about 6 billion, rather than 9 billion or more, because fertility patterns also

Where Have All the Children Gone? 2000 Fertility Rates *Continued*

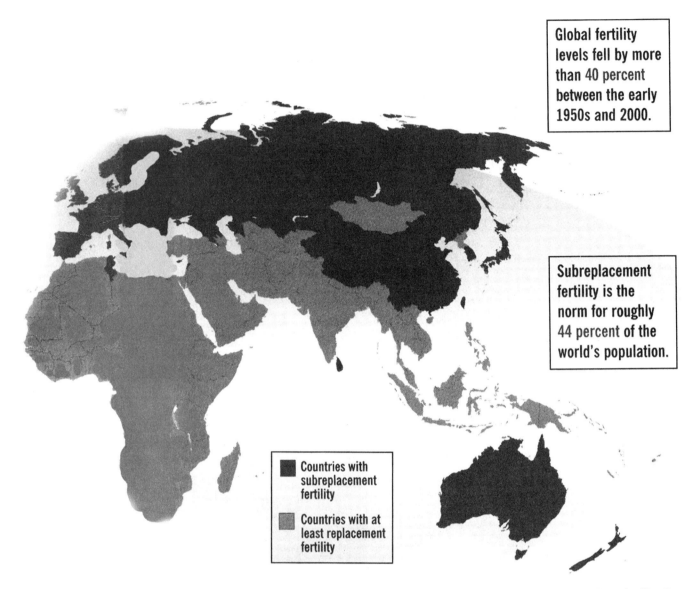

Global fertility levels fell by more than 40 percent between the early 1950s and 2000.

Subreplacement fertility is the norm for roughly 44 percent of the world's population.

- ■ Countries with subreplacement fertility
- ■ Countries with at least replacement fertility

Source: U.S. Bureau of the Census, International Data Base

changed over the course of the 20th century. And of all those diverse changes, without question the most significant was secular fertility decline: sustained and progressive reductions in family size due to deliberate birth control practices by prospective parents.

Within the full sweep of the human experience, secular fertility decline is very; very new. It apparently had not occurred in any human society until about two centuries ago in France. Since that beginning, secular fertility decline has spread steadily, if unevenly, embracing an ever rising fraction of the global population. In the final decades of the 20th century, subreplacement fertility made especially commanding advances: According to estimates and projections by the U.S. Census Bureau and the United Nations Population Division, fertility levels for the world as a whole fell by more than 40 percent between the early 1950s and the end of the century—a drop equivalent to over two births per woman per lifetime.

Indeed, subreplacement fertility has suddenly come amazingly close to describing the norm for childbearing the world over. In all, 83 countries and territories are thought to exhibit below-replacement fertility patterns today [see map on this page]. The total number of persons inhabiting those countries is estimated at nearly 2.7 billion, roughly 44 percent of the world's total population.

Secular fertility decline originated in Europe, and virtually every population in the world that can be described as of European origin today reports fertility rates below the replacement level. But these countries and territories today currently account for only about a billion of the over 2.5 billion people living in "subreplacement regions." Below-replacement fertility is thus

The Population Explosion Fizzles
Fertility Rates For Populous Low-Income Countries: 1975, 2000, 2025

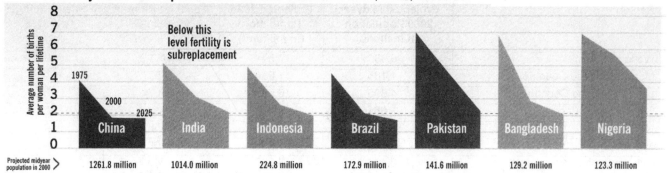

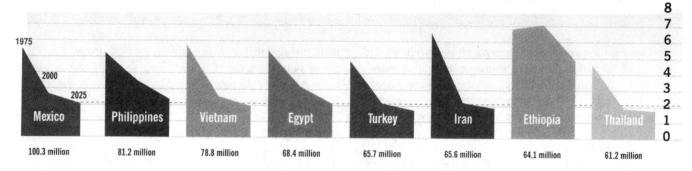

CHARTS BY TRAVIS C. DAUB

Source: U.S. Bureau of the Census, International Data Base, United Nations Population Division, World Population Prospects (New York: United Nations, 1998)
Note: 1975 rates interpolated from estimated 1970/75 levels. 2000 and 2025 rates are projected.

no longer an exclusively—nor even a predominantly—European phenomenon. In the Western Hemisphere, Barbados, Cuba, and Guadeloupe are among the Caribbean locales with fertility rates thought to be lower than that of the United States. Tunisia, Lebanon, and Sri Lanka have likewise joined the ranks of subreplacement fertility societies.

The largest concentration of subreplacement populations, however, is in East Asia. The first non-European society to report subreplacement fertility during times of peace and order was Japan, whose fertility rate fell below replacement in the late 1950s and has remained there almost continuously for the last four decades. In addition to Japan, all four East Asian tigers—Hong Kong, the Republic of Korea, Singapore, and Taiwan—have reported subreplacement fertility levels since at least the early 1980s. By far the largest subreplacement population is in China, where the government's stringent antinatal population control campaign is entering its third decade.

The singularity of the Chinese experience, however, should not divert attention from the breadth and scale of fertility declines that have been taking place in other low-income settings. A large portion of humanity today lives in countries where fertility rates are still above the net replacement level, but where secular fertility decline is proceeding at a remarkably rapid pace.

A glance at the 15 most populous developing countries illustrates the magnitude of fertility change over the last quarter century [see graph "The Population Explosion Fizzles"]. These countries account for about three quarters of the current population of the "less developed regions," and three fifths of

the total world population. In addition to China, Thailand is believed to be below the replacement level. Three other countries (Brazil, Iran, and Turkey) are thought to be just barely above the replacement level. Another four (Bangladesh, Indonesia, Mexico, and Vietnam) are slightly higher. Today, in other words, nine of the 15 largest developing countries are believed to register fertility levels lower than those that characterized the United States as recently as 1965. And over the last quarter century, fertility decline in this set of countries has been pronounced: In eight of those 15, fertility dropped by over half.

The regions where fertility levels remain highest, and where fertility declines to date have been most modest, are sub-Saharan Africa and the Islamic expanse to its north and east—more specifically, the Middle East. Those areas encompassed a total population of about 900 million in 2000, less than a fifth of the estimated total for less developed regions, and a bit under a seventh of the world total. Even for this grouping, however, the image of uniformly high "traditional" fertility patterns is already badly outdated. A revolution in family formation patterns has begun to pass through these regions. In 2000, in fact, the overall fertility level for North Africa—the territory stretching from Western Sahara to Egypt—was lower than the U.S. level of the early 1960s. Perhaps even more surprisingly, secular fertility decline appears to be unambiguously in progress in a number of countries in sub-Saharan Africa. For instance, Kenya's total fertility rate is believed to have dropped by almost four births per woman over the past 20 years.

Go Forth and Multiply

Population Projections from 2000

Region	Projected Midyear 2000 Population (in millions)	Projected 2025 Population (in millions)	Absolute Change 2000-2025 (in millions)	Percent Change 2000-2025
World	6,080	7,841	1,761	29
More Developed Countries	1,186	1,239	53	4
Less Developed Countries	4,895	6,602	1,707	35
Sub-Saharan Africa	661	1,071	410	62
North Africa	145	203	58	40
Middle East	171	280	109	64
Asia (excluding Middle East)	3,444	4,387	943	27
Latin America and the Caribbean	520	671	151	29

Source: U.S. Bureau of the Census, International Data Base

The remarkable particulars of today's global march toward smaller family size fly in the face of many prevailing assumptions about when rapid fertility decline can, and cannot, occur. Poverty and illiteracy (especially female illiteracy) are widely regarded as impediments to fertility decline. Yet, very low income levels and very high incidences of female illiteracy have not prevented Bangladesh from more than halving its total fertility rate during the last quarter century. By the same token, strict and traditional religious attitudes are commonly regarded as a barrier against the transition from high to low fertility. Yet over the past two decades, Iran, under the tight rule of a militantly Islamic clerisy, has slashed its fertility level by fully two-thirds and now apparently stands on the verge of subreplacement.

For many population policymakers, it has been practically an article of faith that a national population program is instrumental, if not utterly indispensable, to fertility decline in a low-income setting. Iran, for instance, achieved its radical reductions under the auspices of a national family planning program. (In 1989, after vigorous doctrinal gymnastics, the mullahs in Tehran determined that a state birth control policy would indeed be consistent with the Prophet's teachings.) But other countries have proven notable exceptions. Brazil has never adopted a national family planning program, yet its fertility levels have declined by well over 50 percent in just the last 25 years.

What accounts for the worldwide plunge in fertility now underway? The honest and entirely unsatisfying answer is that nobody really knows—at least, with any degree of confidence and precision. The roster of contemporary countries caught up in rapid fertility decline is striking for the absence of broad, obvious, and identifiable socioeconomic thresholds or common preconditions. (Reviewing the evidence from the last half century, the strongest single predictor for any given low-income country's fertility level is the calendar year: The later the year, the lower that level is likely to be.) If you can find the shared, underlying determinants of fertility decline in such disparate countries as the United States, Brazil, Sri Lanka, Thailand, and Tunisia, then your Nobel Prize is in the mail.

Two points, however; can be made with certainty. First, the worldwide drop in childbearing reflects, and is driven by, dramatic changes in desired family size. (Although even this observation only raises the question of why personal attitudes about these major life decisions should be changing so commonly in so many disparate and diverse locales around he world today.) Second, it is time to discard the common assumption, long championed by demographers, that no country has been modernized without first making the transition to low levels of mortality and fertility. The definition of "modernization" must now be sufficiently elastic to stretch around cases like Bangladesh and Iran, where very low levels of income, high incidences of extreme poverty, mass illiteracy, and other ostensibly "non-modern" social or cultural features are the local norm, and where massive voluntary reductions in fertility have nevertheless taken place.

SEND YOUR HUDDLED MASSES ASAP

Barring catastrophe, the world's total population can be expected to grow substantially over the coming quarter century: U.S. Census Bureau projections for 2025 would place global population at over 7.8 billion, almost 30 percent larger than today. Yet, due to declining fertility, population growth is poised to decelerate markedly over the coming generation. The projected annual rate of world population growth in 2025 is just under 0.8 percent, considerably slower than the current projected rate of 1.3 percent, and far below the estimated 2.0 percent annual growth rate of the late 1960s. The great global birth wave will have crested and begun ebbing by 2025. In fact, by those projections, slightly fewer babies will be born worldwide in the year 2025 than in any year over the previous four decades.

The prospective pace of population growth for the different regions of the world is highly uneven over the coming generation [see table above]. The most dramatic increases will occur in sub-Saharan Africa, followed by countries in North Africa and the Middle East. By 2025 more people may be living in Africa than in all of today's "more developed countries" taken together.

172

This Old World

Percent of Population 65 and Above (Projected)

Country	2000	2025
Algeria	4.0	7.1
Brazil	5.3	11.3
China	7.0	13.5
Ethiopia	2.8	2.7
India	4.6	7.8
Iraq	3.1	4.3
Saudi Arabia	2.6	5.6
South Africa	4.8	9.6
Germany	16.2	23.1
Japan	17.0	27.6
Russia	12.6	18.5
Spain	16.9	23.5
United States	12.6	18.5

Source: U.S. Bureau of the Census, International Data Base

The natural growth of population in the more developed countries has essentially ceased. The overall increase in population for 2000 in these nations is estimated at 3.3 million people, or less than 0.3 percent. Two thirds of that increase, however, is due to immigration; the total "natural increase" amounts to just over 1 million. Over the coming quarter century, in the U.S. Census Bureau's projections, natural increase adds only about 7 million people to the total population of the more developed countries. And after the year 2017, deaths exceed births more or less indefinitely. Once that happens, only immigration on a scale larger than any in the recent past can forestall population decline. (The specter of population decline in more developed countries looms even larger if the United States, with its relatively high fertility level and relatively robust inflows of immigrants, is taken out of the picture. Excluding the United States, total deaths already exceed total births by almost half a million a year.)

For Europe as a whole (including Russia), the calculated long-term volume of immigration required to avert overall population decline is nearly double the recent annual level—an average of 1.8 million net newcomers a year, versus the roughly one million net entrants a year in the late 1990s. To prevent an eventual decline in the size of the 15 to 64 grouping (often termed the "working-age" population), Europe's net migration will have to nearly quadruple to a long-term average of about 3.6 million a year. Migration of this magnitude would change the face of Europe: By 2050, under these two scenarios, the descendants of present-day non-Europeans will account for approximately 20 to 25 percent of Europe's inhabitants.

Even more dramatic are the prospects for Japan, where current net migration levels are close to zero. To maintain total population size, Japan would have to accept a long-term average of almost 350,000 newcomers a year for the next 50 years, and it would need nearly twice that number to keep its working-age population from shrinking. Under the first contingency, over a sixth of Japan's 2050 population would be descendants of present-day *gaijin* (foreigners); under the second contingency, that group would account for nearly a third of Japan's total population.

Europe and Japan will not lack immigration candidates in the years ahead. If Europe's needed immigration flows continue to come largely from North Africa, the Middle East, sub-Saharan Africa, and South Asia, those migrants will account for only about 3 to 7 percent of the population growth in their home countries. By the same token if Japan, for reasons of history and affinity, relies upon China and Southeast Asia for all its new national recruits, it will require just 2 to 4 percent of those countries' total envisioned population increase over the next 25 years. And as long as a huge income gap separates these more developed and less developed locales, there will be a compelling motive for such migration.

The issue clearly will not be supply, but rather demand. Will Western countries facing population decline opt to let in enough outsiders to stabilize their domestic population levels? Major and sustained immigration flows will entail correspondingly consequential long-term changes in a country's ethnic composition, with accompanying social alterations and adjustments. Such inflows will also require a capability to assimilate newcomers, so that erstwhile foreigners (and their descendants) can become true members of their new and chosen society.

The current outlook for "replacement migration" varies dramatically within the more developed regions. Throughout Europe, vocal (but still marginal) antiforeign political movements have taken the stage in recent years, while more tolerant sectors of the public have worried about the impact of immigration on their welfare states. Yet the continent, populated as it has been by successive historical flows of peoples, possesses traditions and capacities of assimilation that are not always fully appreciated.

The situation looks very different for Japan, where no major influxes of newcomers have been recorded over the past thousand years, and where the delicate distinctness of the Japanese *minzoku* (race) is a matter of intense, if not always enunciated, public consciousness. Despite reforms in Japanese immigration laws, a community of ethnic Koreans in Japan—many of them fourth-generation residents of the country—still does not enjoy Japanese citizenship. Indeed, Japan naturalizes fewer foreigners each year than tiny Switzerland.

It is extraordinarily difficult to imagine any circumstances under which the Japanese public might acquiesce in "replacement migration." Socially and politically, long-term demographic decline seems likely to be a much more acceptable alternative. But these are the only two choices, and over the coming decades all the more developed countries must decide between them. For all societies with long-term fertility rates significantly below the replacement level, the only alternative to an eventual decline of the total population—or of key age groups within that total population—is steady and massively enhanced immigration.

A GRAY WORLD

The world's population is set to age markedly over the coming generation: The longevity revolution of the 20th century has foreordained as much. The tempo of social aging, however, has been accelerated in many countries by extremely low levels of fertility. In 2025, there will likely remain a few pockets of the world in which populations will remain as youthful as those from earlier historical epochs. For example, the median age in sub-Saharan Africa in 2025 will be just 20 years, that is, as many people would be under 20 as over 20. (Such a profile probably characterized humanity from the Neolithic era up until the Industrial Revolution.) Throughout the rest of the world, however, the phenomenon of aging will transform the structure of national populations, often acutely.

Population aging will be most pronounced in today's more developed countries. By the U.S. Census Bureau's estimates, the median age for this group of countries today is about 37 years. In 2025, the projected median age will be 43.

Due to its relatively high levels of fertility and immigration (immigrants tend to be young), the population of the United States is slated to age more slowly than the rest of the developed world. By 2025, median age in the United States will remain under 39 years. For the rest of the developed world, minus the United States, median age will be approximately 45 years. And for a number of countries, the aging process will be even further advanced [see table "This Old World"].

Current developed countries grew rich before they grew old; many of today's low-income countries, by contrast, look likely to become old first.

In Germany, for example, the projected median age in the year 2025 is 46. Greece and Bulgaria are both ascribed median ages in excess of 47. Japan would have a median age of over 49. In this future Japan, more than a fifth of the citizenry would be over 70 years of age, and nearly one person in six would be 75 or older. In fact, persons 75 and older would outnumber children under 15 years of age.

Population aging, of course, will also occur in today's less developed regions. Current developed countries grew rich before they grew old; many of today's low-income countries, by contrast, look likely to become old first. One of the most arresting cases of population aging in the developing world is set to unfold in China, where relatively high levels of life expectancy, together with fertility levels suppressed by the government's resolute and radical population control policies, are transforming the country's population structure. Between 2000 and 2025, China's median age is projected to jump by almost 9 years. This future China would have one-sixth fewer children than contemporary China, and the 65-plus population would surge by over 120 percent, to almost 200 million. These senior citizens would account for nearly a seventh of China's total population.

Caring for the elderly will inexorably become a more pressing issue for China under such circumstances, but nothing remotely resembling a national pension system is yet in place in that country. Even with rapid growth over the next quarter century, China will still be a poor country in 2025. Coping with its impending aging problem promises to be an immense social and economic issue for this rising power.

DEATH MAKES A COMEBACK

Given the extraordinary impact of the 20th century's global health revolution, well-informed citizens around the world have come to expect steady and progressive improvement in life expectancies and health conditions during times of peace.

Unfortunately, troubling new trends challenge these happy presumptions. A growing fraction of the world's population is coming under the grip of peacetime retrogressions in health conditions and mortality levels. Long-term stagnation or even decline in life expectancy is now a real possibility for urbanized, educated countries not at war. Severe and prolonged collapses of local health conditions during peacetime, furthermore, is no longer a purely theoretical eventuality. As we look toward 2025, we must consider the unpleasant likelihood that a large and growing fraction of humanity may be separated from the planetary march toward better health and subjected instead to brutal mortality crises of indeterminate duration.

In the early post-World War II era, the upsurge in life expectancy was a worldwide phenomenon. By the reckoning of the U.N. Population Division, in fact, not a single spot on the globe had a lower life expectancy in the early 1970s than in the early 1950s. And in the late 1970s only two places on earth—Khmer Rouge-ravaged Cambodia and brutally occupied East Timor—had lower levels of life expectancy than 20 years earlier. In subsequent years, however, a number of countries unaffected by domestic disturbance and upheaval began to report lower levels of life expectancy than they had known two decades earlier. Today that list is long and growing. U.S. Census Bureau projections list 39 countries in which life expectancy at birth is anticipated to be at least slightly lower in 2010 than it was in 1990. With populations today totaling three quarters of a billion people and accounting for one eighth of the world's population, these countries are strikingly diverse in terms of location, history, and material attainment.

This grouping includes the South American countries of Brazil and Guyana; the Caribbean islands of Grenada and the Bahamas; the Micronesian state of Nauru; 10 of the 15 republics of the former Soviet Union; and 23 sub-Saharan African nations. As might be surmised from the heterogeneity of these societies, health decline and mortality shocks in the contemporary world are not explained by a single set of factors, but instead by several syndromes working simultaneously in different parts of the world to subvert health progress.

Russia has experienced a prolonged stagnation and even decline in life expectancy, and its condition illuminates the problems facing some of the other former Soviet republics [see graph "The Days of Our Lives"]. After recording rapid postwar reductions in mortality in the 1950s, Russian mortality levels stopped falling in the 1960s and began rising for broad groups of the population. By 1990, overall life expectancy at birth in Russia was barely as high as it had been 25 years earlier. With the end of communist rule in 1991, Russia suffered sudden and severe declines in mortality, from which it has not yet fully recovered. By 1999, overall life expectancy at birth in Russia had regressed to the point where it had been four decades earlier.

Although many aspects of Russia's continuing health crisis remain puzzling, it appears that lifestyle and behavioral risks—including heavy smoking and extremely heavy drinking—figure centrally in the shortening of Russian lives. A weak and rudderless public health system, combined with apparent indifference in Moscow to the nation's ongoing mortality crisis, also compromises health progress. Although Russia is an industrialized society with an educated population and a large indigenous scientific-technical cadre, such characteristics do not automatically protect a country from the sorts of health woes that have befallen the Russian Federation.

In sub-Saharan Africa, a different dynamic drives mortality crises: the explosive spread of the HIV/AIDS epidemic. In its most recent report, the Joint United Nations Programme on HIV/AIDS (UNAIDS) estimated that 2.8 million died of AIDS in 1999, 2.2 million in sub-Saharan Africa alone. UNAIDS also reported that almost 9 percent of the region's adult population is already infected with the disease. By all indications, the epidemic is still spreading in sub-Saharan Africa. As of 2000, UNAIDS projected that in several sub-Saharan countries, a 15-year-old boy today faces a greater than 50 percent chance of ultimately dying from AIDS—even if the risk of becoming infected were reduced to half of current levels.

HIV/AIDS may not be the only plague capable of wrenching down national levels of life expectancy over the coming quarter century.

Given sub-Saharan Africa's disappointing developmental performance and conspicuously poor record of governance over the post-independence period, the pervasive failure in this low-income area to contain a deadly but preventable contagion may seem tragic but unsurprising. Yet it is worth noting that the AIDS epidemic appears to have been especially devastating in one of Africa's most highly developed and best-governed countries: Botswana.

Unlike most of the region, Botswana is predominantly urbanized; its rate of adult illiteracy is among the subcontinent's very lowest; and over a generation in which sub-Saharan economic growth rates were typically negative, Botswana's was consistently positive. Yet despite such promising statistics, Botswana's population has been decimated by HIV/AIDS over

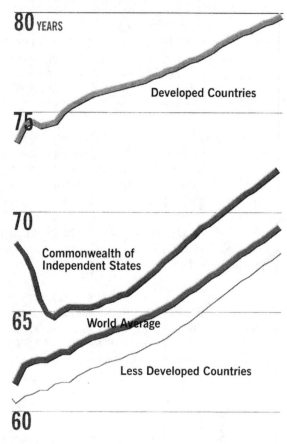

The Days of Our Lives
Life Expectancy at Birth, 1990-2025

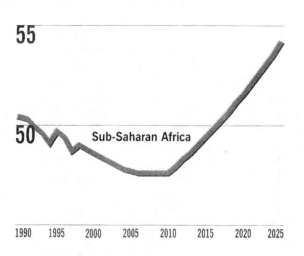

Source: Latest projections, U.S. Bureau of the Census, International Data Base

the last decade. Between 1990 and 2000, life expectancy in Botswana plummeted from about 64 years to about 39 years,

Want to Know More?

Over the last two centuries, world population trends have been dominated by what is known as "the demographic transition": the progressive and seemingly inexorable shift in society after society from a regimen of high birthrates and death rates to one of low death and birthrates. One of the best introductions to this powerful, but still mysterious, phenomenon is Jean-Claude Chesnais's *The Demographic Transition: Stages, Patterns, and Economic Implications* (New York: Oxford University Press, 1992).

Detailed global population projections are available from a number of sources, but perhaps the two most authoritative of these are the U.S. Bureau of the Census and the United Nations Population Division (UNPD). The Census Bureau's latest outlook is available online through its International Data Base. The UNPD's projections are published biennially. The latest update, **World Population Prospects: The 2000 Revisions** (New York: UNPD, forthcoming) is due out shortly.

For a summary of what is currently know about the global contours and determinants of sub-replacement fertility, read **"Below Replacement Fertility,"** a double issue of the UNPD's *Population Bulletin of the United Nations* (Nos. 40/41, dated 1999 but published in 2000) that offers a collection of expert assessments. The economic and

policy implications of social aging are examined in a series of publications and papers by the Orginisation for Economic Cooperation and Development (OECD), including **"Maintaining Prosperity in an Ageing Society"** (Paris: OECD, 1998) and **"Reforms for an Ageing Society"** (Paris: OECD, 2000). The UNPD report *Replacement Migration: Is It a Solution to Declining and Ageing Populations?* (New York: **UNPD**, March 2000), available online, examines the potential of migration to forestall depopulation and social aging. Demetrios G. Papademetriou and Kimberly A. Hamilton argue that Japan must end its restrictive immigration policies if it is to ensure its place as a global leader in *Reinventing Japan: Immigration's Role in Shaping Japan's Future* (Washington: Carnegie Endowment for International Peace, 2000).

On health setbacks in Russia and other post-communist countries, see Nicholas Eberstadt's *Prosperous Paupers and Other Population Problems* (New Brunswick: Transaction Publishers, 2000) and **"Russia: Too Sick To Matter?"** *(Policy Review*, June/July 1999).

•For links to relevant Web sites, as well as a comprehensive index of related FOREIGN POLICY articles, access **www.foreignpolicy.com.**

---

that is to say, by almost a quarter century. Recent projections for 2025 envision a life expectancy of a mere 33 years. If this projection proves accurate, Botswana will have a much lower life expectancy 25 years from now than it had nearly half a century ago.

One of the disturbing facets of the Botswanan case is the speed and severity with which life expectancy projections have been revised downward. Assuming most recent figures are accurate, as recently as 1994 expert demographers were overestimating Botswana's life expectancy for 2000 by about 30 years. Such abrupt and radical revisions raise the question of whether similar brutal adjustments await other sub-Saharan countries— or, for that matter, countries in other regions of the world. This question cannot be answered with any degree of certainty today, but we would be unwise to dismiss it from consideration. HIV/ AIDS may not be the only plague capable of wrenching down national levels of life expectancy over the coming quarter century. Twenty-five years ago, HIV/AIDS had not even been identified and diagnosed.

Surprisingly, sub-Saharan Africa's AIDS catastrophe is not projected to alter the region's population totals dramatically. That speaks to the extraordinary power of high fertility levels. Given the region's current and prospective patterns of childbearing, the subcontinent's population totals in 2025 may prove

to be unexpectedly insensitive to the scope or scale of the disasters looming ahead. Yet it is the mortality patterns that will do much to define the quality of life for those human numbers— and to circumscribe their economic and social potential.

THE SHAPE OF THINGS TO COME

Looking toward 2025, we must remember that many 20th-century population forecasts and demographic assessments proved famously wrong. Depression-era demographers, for example, incorrectly predicted depopulation for Europe by the 1960s and completely missed the "baby boom." The 1960s and 1970s saw dire warnings that the "population explosion" would result in worldwide famine and immiseration, whereas today we live in the most prosperous era humanity has ever known. In any assessment of future world population trends and consequences, a measure of humility is clearly in order.

Given today's historically low death rates and birthrates, however, the arithmetic fact is that the great majority of people who will inhabit the world in 2025 are already alive. Only an apocalyptic disaster can change that. Consequently, this reality provides considerable insight into the shape of things to come.

By these indications, indeed, we must now adapt our collective mind-set to face new demographic challenges.

A host of contradictory demographic trends and pressures will likely reshape the world during the next quarter century. Lower fertility levels, for example, will simultaneously alter the logic of international migration flows and accelerate the aging of the global population. Social aging sets in motion an array of profound changes and challenges and demands far-reaching adjustments if those challenges are to be met successfully. But social aging is primarily a consequence of the longer lives that modern populations enjoy. And the longevity revolution, with its attendant enhancements of health conditions and individual capabilities, constitutes an unambiguous improvement in the human condition. Pronounced and prolonged mortality setbacks portend just the opposite: a diminution of human well-being, capabilities, and choices.

It is unlikely that our understanding of the determinants of fertility, or of the long-range prospects for fertility, will advance palpably in the decades immediately ahead. But if we wish to inhabit a world 25 years from now that is distinctly more humane than the one we know today, we would be well advised to marshal our attention to understanding, arresting, and overcoming the forces that are all too successfully pressing for higher levels of human mortality today.

Nicholas Eberstadt holds the Henry Wendth Chair in Political Economy at the American Enterprise Institute in Washington, D.C.

Local difficulties

Greenery is for the poor too, particularly on their own doorstep

WHY should we care about the environment? Ask a European, and he will probably point to global warming. Ask the two little boys playing outside a newsstand in Da Shilan, a shabby neighbourhood in the heart of Beijing, and they will tell you about the city's notoriously foul air: "It's bad—like a virus!"

Given all the media coverage in the rich world, people there might believe that global scares are the chief environmental problems facing humanity today. They would be wrong. Partha Dasgupta, an economics professor at Cambridge University, thinks the current interest in global, future-oriented problems has "drawn attention away from the economic misery and ecological degradation endemic in large parts of the world today. Disaster is not something for which the poorest have to wait; it is a frequent occurrence."

Every year in developing countries, a million people die from urban air pollution and twice that number from exposure to stove smoke inside their homes. Another 3m unfortunates die prematurely every year from water-related diseases. All told, premature deaths and illnesses arising from environmental factors account for about a fifth of all diseases in poor countries, bigger than any other preventable factor, including malnutrition. The problem is so serious that Ian Johnson, the World Bank's vice-president for the environment, tells his colleagues, with a touch of irony, that he is really the bank's vice-president for health: "I say tackling the underlying environmental causes of health problems will do a lot more good than just more hospitals and drugs."

The link between environment and poverty is central to that great race for sustainability. It is a pity, then, that several powerful fallacies keep getting in the way of sensible debate. One popular myth is that trade and economic growth make poor countries' environmental problems worse. Growth, it is said, brings with it urbanisation, higher energy consumption and industrialisation—all factors that contribute to pollution and pose health risks.

In a static world, that would be true, because every new factory causes extra pollution. But in the real world, economic growth unleashes many dynamic forces that, in the longer run, more than offset that extra pollution. As chart 5 " Dangers old and new" makes clear, traditional environmental risks (such as water-borne diseases) cause far more health problems in poor countries than modern environmental risks (such as industrial pollution).

Rigged rules

However, this is not to say that trade and economic growth will solve all environmental problems. Among the reasons for doubt are the "perverse" conditions under which world trade is carried on, argues Oxfam. The British charity thinks the rules of trade are "unfairly rigged against the poor", and cites in evidence the enormous subsidies lavished by rich countries on industries such as agriculture, as well as trade protection offered to manufacturing industries such as textiles. These measures hurt the environment because they force the world's poorest countries to rely heavily on commodities—a particularly energy-intensive and ungreen sector.

Mr Dasgupta argues that this distortion of trade amounts to a massive subsidy of rich-world consumption paid by the world's poorest people. The most persuasive critique of all goes as follows: "Economic growth is not sufficient for turning environmental degradation around. If economic incentives facing producers and consumers do not change with higher incomes, pollution will continue to grow unabated with the growing scale of economic activity." Those words come not from some anti-globalist green group, but from the World Trade Organisation.

Another common view is that poor countries, being unable to afford greenery, should pollute now and clean up later. Certainly poor countries should not be made to adopt American or European environmental standards. But there is evidence to suggest that poor countries can and should try to tackle some environmental problems now, rather than wait till they have become richer.

This so-called "smart growth" strategy contradicts conventional wisdom. For many years, economists have observed that as agrarian societies industrialised, pollution increased at first, but as the societies grew wealthier it declined again. The trouble

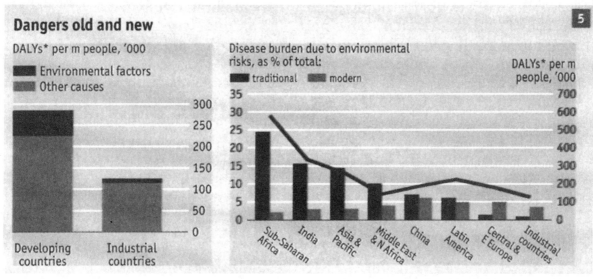

Dangers old and new

DALYs* per m people, '000
- ■ Environmental factors
- ▬ Other causes

Developing countries / Industrial countries

Disease burden due to environmental risks, as % of total:
- ■ traditional ▬ modern

DALYs* per m people, '000

Sub-Saharan Africa, India, Asia & Pacific, Middle East & N Africa, China, Latin America, Central & E Europe, Industrial countries

Source: World Bank

*Disability Adjusted Life Years - number of years lived with disability and years lost to premature death.

is that this applies only to some pollutants, such as sulphur dioxide, but not to others, such as carbon dioxide. Even more troublesome, those smooth curves going up, then down, turn out to be misleading. They are what you get when you plot data for poor and rich countries together at a given moment in time, but actual levels of various pollutants in any individual country plotted over time wiggle around a lot more. This suggests that the familiar bell-shaped curve reflects no immutable law, and that intelligent government policies might well help to reduce pollution levels even while countries are still relatively poor.

Developing countries are getting the message. From Mexico to the Philippines, they are now trying to curb the worst of the air and water pollution that typically accompanies industrialisation. China, for example, was persuaded by outside experts that it was losing so much potential economic output through health troubles caused by pollution (according to one World Bank study, somewhere between 3.5% and 7.7% of GDP) that tackling it was cheaper than ignoring it.

One powerful—and until recently ignored—weapon in the fight for a better environment is local people. Old-fashioned paternalists in the capitals of developing countries used to argue that poor villagers could not be relied on to look after natural resources. In fact, much academic research has shown that the poor are more often victims than perpetrators of resource depletion: it tends to be rich locals or outsiders who are responsible for the worst exploitation.

Local people usually have a better knowledge of local ecological conditions than experts in faraway capitals, as well as a direct interest in improving the quality of life in their village. A good example of this comes from the bone-dry state of Rajasthan in India, where local activism and indigenous know-how about rainwater "harvesting" provided the people with reliable water supplies—something the government had failed to do. In Bangladesh, villages with active community groups or concerned mullahs proved greener than less active neighbouring villages.

Community-based forestry initiatives from Bolivia to Nepal have shown that local people can be good custodians of nature. Several hundred million of the world's poorest people live in and around forests. Giving those villagers an incentive to preserve forests by allowing sustainable levels of harvesting, it turns out, is a far better way to save those forests than erecting tall fences around them.

To harness local energies effectively, it is particularly important to give local people secure property rights, argues Mr Dasgupta. In most parts of the developing world, control over resources at the village level is ill-defined. This often means that local elites usurp a disproportionate share of those resources, and that individuals have little incentive to maintain and upgrade forests or agricultural land. Authorities in Thailand tried to remedy this problem by distributing 5.5m land titles over a 20-year period. Agricultural output increased, access to credit improved and the value of the land shot up.

Name and shame

Another powerful tool for improving the local environment is the free flow of information. As local democracy flourishes, ordinary people are pressing for greater environmental disclosure by companies. In some countries, such as Indonesia, governments have adopted a "sunshine" policy that involves naming and shaming companies that do not meet environmental regulations. It seems to achieve results.

Bringing greenery to the grass roots is good, but on its own it will not avert perceived threats to global "public goods" such as the climate or biodiversity. Paul Portney of Resources for the Future explains: "Brazilian villagers may think very carefully and unselfishly about their future descendants, but there's no reason for them to care about and protect species or habitats that no future generation of Brazilians will care about."

That is why rich countries must do more than make pious noises about global threats to the environment. If they believe that scientific evidence suggests a credible threat, they must be willing to pay poor countries to protect such things as their tropical forests. Rather than thinking of this as charity, they should see it as payment for environmental services (say, for carbon storage) or as a form of insurance.

In the case of biodiversity, such payments could even be seen as a trade in luxury goods: rich countries would pay poor countries to look after creatures that only the rich care about. Indeed, private green groups are already buying up biodiversity "hot spots" to protect them. One such initiative, led by Conservation International and the International Union for the Conservation of Nature (IUCN), put the cost of buying and preserving 25 hot spots exceptionally rich in species diversity at less than $30 billion. Sceptics say it will cost more, as hot spots will need buffer zones of "sustainable harvesting" around them. Whatever the right figure, such creative approaches are more likely to achieve results than bullying the poor into conservation.

It is not that the poor do not have green concerns, but that those concerns are very different from those of the rich. In Beijing's Da Shilan, for instance, the air is full of soot from the many tiny coal boilers. Unlike most of the neighbouring districts, which have recently converted from coal to natural gas, this area has been considered too poor to make the transition. Yet ask Liu Shihua, a shopkeeper who has lived in the same spot for over 20 years, and he insists he would readily pay a bit more for the cleaner air that would come from using natural gas. So would his neighbours.

To discover the best reason why poor countries should not ignore pollution, ask those two little boys outside Mr Liu's shop what colour the sky is. "Grey!" says one tyke, as if it were the most obvious thing in the world. "No, stupid, it's blue!" retorts the other. The children deserve blue skies and clean air. And now there is reason to think they will see them in their lifetime.

A Dirty Dilemma: The Hazardous Waste Trade

Zada Lipman

Since the 1980s, exporters of hazardous waste have targeted developing countries. Some of this waste is destined for dumping or disposal, while other waste is directed to resource recovery, recycling, or reuse. To protect developing countries from the dangers associated with hazardous waste, the international community adopted the Basel Convention on the Transboundary Movements of Hazardous Wastes and their Disposal, which first regulated and then banned the trade of hazardous waste. Although lauded as a landmark for global democracy and environmental justice, the ban has created a dilemma for developing countries with large recycling industries that rely on hazardous waste imports for their continued operation.

Environmental problems arising from the disposal of hazardous waste in developing countries did not gain international attention until the late 1980s, when several incidents of dumping were reported in African nations. One of the most serious cases occurred in 1987. Several thousand tons of highly toxic and radioactive waste, labeled "substances relating to the building trade," were exported from Italy to Koko, Nigeria, and stored in drums in a backyard. Many of these drums were damaged and leaking; workers packing the drums into containers for retransport to Italy suffered severe chemical burns and partial paralysis, and land within a 500-meter radius of the dump site was declared unsafe. The Italian government eventually accepted the return of the waste, and the Nigerian government has since imposed the death penalty on the waste importers. In 1988, Guinea-Bissau was offered a US$600 million contract—four times its gross national product—to dispose of 15 million tons of toxic waste over five years. The contract was never concluded because of public concern within Guinea-Bissau, but many similar arrangements were reported in the 1980s in countries such as Namibia, Guinea, Sierra Leone, and Haiti. In some cases, dumping took place with the consent of the government in question, while in other cases it was part of an illegal operation. Since then, numerous incidents of dumping in developing countries have been reported throughout the world.

Logic of the Market

Although precise estimates of the worldwide generation of hazardous waste are difficult to obtain, the United Na-

tions Environment Programme (UNEP) estimated in 1992 that approximately 400 million metric tons of hazardous waste were generated annually, with 80 percent of this waste coming from countries in the Organisation for Economic Cooperation and Development (OECD). This figure is likely to be significantly higher today.

The disposal of hazardous waste has become a major issue for countries that are large waste-generators. Before the dangers associated with disposal were understood, most of this waste was deposited in landfills, causing serious problems for surrounding areas. A well-documented example is the "Valley of the Drums" in Kentucky, a seven-acre site with 17,000 drums of hazardous waste that has contaminated nearby soil and water. As a result of incidents like this, most developed countries introduced stringent environmental and safety measures for the disposal of hazardous waste. This trend led to increasingly limited and costly disposal options in developed countries.

Developing countries became targets for waste generators—mostly developed countries—since they provided disposal options for a mere fraction of the equivalent cost in the state of origin. According to a study by Katharina Kummer in *International Management of Hazardous Wastes*, disposal costs for hazardous waste in developing countries in 1988 ranged from US$2.50 to US$50 per ton, compared with costs of US$100 to US$2,000 per ton in OECD countries. The cost of incineration was even higher, at US$10,000 for one ton of hazardous waste in the United Kingdom. The lower disposal costs in developing countries generally stem from low or nonexistent environmental standards, less stringent laws, and an absence of public opposition due to a lack of information concerning the dangers involved. Given these considerations, the economic logic for exporting hazardous waste to developing countries is indisputable.

The Basel Convention

International concerns about the export of hazardous waste to developing countries led to the negotiation of the 1989 Basel Convention, which became binding in 1992. As of August 2001, 148 countries had ratified the Convention. Unfortunately, the United States, which generates approximately 60 percet of the world's hazardous waste, has yet to ratify.

The Basel Convention itself does not ban the transboundary movements of hazardous waste, except to Antarctica. Rather, it seeks to control and limit the movement of waste based on a process of prior informed consent. Hazardous waste exports cannot proceed unless the pertinent authorities in the recipient and transit countries are notified in advance and provide written consent. Any movement of hazardous waste without a movement document or prior notification is illegal under the Basel Convention. This Convention requirement applies to both hazardous waste exported for final disposal and waste exported for recycling. The Convention also requires parties to prohibit the import of hazardous wastes when it is likely that the waste will not be managed in an environmentally sound manner.

The Basel Convention initially focused on protecting developing countries from hazardous waste dumping by developed countries. But by 1992, at the First Meeting of the Conference of the Parties (COP-1), concerns had already shifted to hazardous waste traded for recycling or recovery. Jim Puckett of the Basel Action Network estimates that from 1980 to 1988, only 36 percent of hazardous waste exports to developing countries were destined for recycling. In 1992 these exports had risen to 88 percent, and in 2001 they are likely to constitute over 95 percent of all waste exports.

The lack of a distinction between "waste" and "products" in the Convention and its vague criteria for "hazardous" allowed the continued export of hazardous waste to developing countries for recycling on the basis that the toxic substances exported were commodities rather than wastes. Some exports of waste are for "sham" recycling, but even when recycling takes place, this waste presents environmental and health risks to developing countries that lack the technology to handle the waste safely. For instance, Greenpeace and the Basel Action Network have not encountered a single hazardous waste recycling facility in a non-OECD country that does not cause serious pollution. In addition, Greenpeace estimates that more than 2.5 million tons of hazardous waste were exported to developing countries between 1989 and 1994. Dissatisfaction with the Basel Convention resulted in over 100 developing countries unilaterally imposing regional or national waste-import bans.

The control regime imposed by the Basel Convention was a compromise between developing countries, which favored a total ban on the transboundary movements of hazardous waste, and developed countries, which wanted a more flexible control regime. In 1994, developing countries finally gained sufficient international support to achieve a total ban on hazardous waste exports, when the parties to the Basel Convention decided to ban all exports of hazardous wastes from OECD to non-OECD (largely developing) countries. To ensure that the ban was legally binding, it was adopted as an amendment to the Convention in 1995. At the same time, a newly added paragraph recognized that exports of hazardous waste to developing countries are highly unlikely to be managed in an environmentally sound manner consistent with the Convention. Clearly, protection of developing countries was the rationale for this amendment. However, the ban has not yet received the requisite number of ratifications to come into force. As of August 2001, there were 26 ratifications—out of the minimum of 62—and only half of these were from developing countries, which had been the chief initiators and presumed benefactors of the ban. However, the European Union has ratified the amendment and introduced regulations to ban exports to developing countries, a decision that binds its 15 member states.

Destiny's Landfill?

Developing countries with large recycling and reclamation industries are concerned that they will be deprived of resources if the ban comes into force. This concern is exacerbated by the uncertainty as to which types of waste are subject to the ban. A major weakness of the Basel Convention is its failure to provide clear definitions of hazardous waste. Waste characterized as hazardous and subject to the ban, as well as waste not covered by the Convention (including a large percentage of internationally traded metals and secondary raw materials), were classified into two lists and adopted as Annexes to the Convention in 1998, which made them legally binding. Waste is primarily classified according to its degree of hazard, with its value as a secondary raw material also taken into account. Wastes that have yet to be classified include items of economic importance such as zinc, lead and copper compounds, lithium, and spent catalysts.

The ban will likely have a considerable impact on recycling industries in developing countries. In particular, a future decision to classify lead-acid batteries as hazardous will affect countries such as India and the Philippines, which rely on imported lead-acid batteries for a significant proportion of their lead requirements. The demand for lead in developing countries in Southeast Asia is also increasing due to rising demand for batteries in motor vehicles, telecommunications, and computer equipment. As Jonathan Kreuger points out in his book, *International Trade and the Basel Convention*, if the ban proceeds, lead ingots will have to be bought to supplement the output of the domestic recycling industry which will itself become reliant on domestic supplies or on imports from other non-OECD countries. A 1999 United Nations Conference on Trade and Development (UNCTAD) study into lead-acid batteries found that a Philippine secondary lead smelter that provides 80 percent of the country's refined lead output may need to close if feedstock requirements were to become unavailable.

Of course, any financial benefits that the recycling of hazardous waste may provide should be offset against the costs to human health and to the environment. Most developing countries lack the capacity to handle hazardous waste safely. The UNCTAD study identified thousands of small battery-reconditioning shops in major cities throughout the Philippines that were located in busy streets, often adjacent to fast-food vendors. Workers did not wear protective clothing and often dismantled batteries with their bare hands. Reconditioners routinely dumped diluted sulfuric acid down street drains or behind their premises. Lead plates were then sold to licensed smelters, and the residues from the smelters, which can have a lead content of over 90 percent, were dumped in a river, in the countryside, or behind the smelters. Greenpeace research in the Philippines into imports of lead-acid batteries has revealed that even legitimate hazardous-waste recycling operations promoted by the Philippine government are in many cases creating residual waste more toxic than the original product.

Mercury waste is also subject to the ban. The danger associated with the reprocessing of mercury wastes is well illustrated by a notorious incident in South Africa. Thor Chemicals, a British company, established one of the world's largest mercury-reprocessing plants in South Africa and began importing mercury wastes in 1986. Over the next eight years, the plant imported thousands of tons of waste from the United States and Europe. In 1988, mercury contamination 1,000 times higher than the World Health Organization's standards was discovered in a river about 50 kilometers from the Thor plant. Subsequent samples of soil at the reprocessing site revealed high levels of mercury contamination. Mercury, which can be absorbed into the body through food, air, or skin contact, is linked to many neurological problems causing symptoms such as trembling, loss of muscle control, headaches, mental confusion, nausea, and hair loss. Long-term exposure can lead to a coma and eventually to death. Workers at the Thor plant claim that they were not warned about the dangers of working with mercury. They continued to work at the plant because unemployment was high and Thor "paid the highest wages." This comment encapsulates the painful dilemma for workers in developing countries—poverty or pollution.

By 1992, two Thor workers had died of mercury poisoning, while many others were permanently disabled. Tests conducted on workers in 1992 revealed that almost 30 percent were at risk of severe mercury poisoning. The families of the deceased workers sued Thor in a British court and were awarded almost US$2 million. In comparison, compensation for injured workers has been paltry. In 1997, 20 former Thor workers were paid approximately US$1.3 million in an out-of-court settlement by Thor's head office in Britain. A second class-action lawsuit in 1999 involving 20 workers resulted in an award of approximately US$400,000 split amongst the plaintiffs—an amount barely sufficient to cover all the medical expenses. In the interim, the plant site remains an ecological time bomb. Although the plant was shut down in 1994, 1,000 tons of stockpiled waste remains on the site in leaking barrels. A South African Commission of Inquiry has recommended the incineration of the stockpiled wastes at standards far below those in developed countries. Environmental groups have opposed the recommendations of the Commission and have called on companies that originally exported the waste to South Africa to reclaim it.

Out of the Waste Land

A ban on hazardous-waste exports to developing countries is the simplest control measure to implement, as well as being morally justifiable. Countries that benefit from industrialization should also bear the full weight of its burdens. This principle is exemplified by the "polluter pays" principle, by which countries that are the primary generators of hazardous waste have the responsibility to deal with the waste at its source rather than exporting it to developing countries. As the case studies illustrate, most developing countries do not have efficient hazardous-waste disposal facilities. They also lack the requisite skill to evaluate the risks posed by hazardous waste or, where unsafe disposals have taken place, to institute monitoring systems and remedial strategies to mitigate widespread contamination and loss of workers' lives and health. Greenpeace believes that the ban has succeeded in preventing a global environmental disaster and claims that hazardous waste exports to developing countries have diminished dramatically since the ban was adopted, notwithstanding the fact that it has not yet entered into force.

However, many problems with the ban have yet to be resolved. For example, not all developing countries are in favor of a total ban. Attempts by developed countries to impose a ban on countries that oppose it not only suggest paternalism, but also infringe on the sovereign right of these countries to consent to the import of hazardous waste. Cash-starved economies and corruption at the government level also leave open the possibility of illegal imports of hazardous wastes. Regardless of the legality of such actions, developing countries may continue to import hazardous waste as long as it remains profitable. A ban on imports of hazardous waste may put thousands of workers in developing countries out of work; it also denies developing countries the opportunity to import cheap materials. There is an undeniable tension between the justice of banning hazardous waste exports that cause harm to people in a developing country and the realization that doing so may endanger the very livelihood of these people.

Although it is impossible to ignore the ethical problems that arise in imposing a ban on exports of hazardous waste to developing countries, the alternative is an escalating waste colonization process that would further degrade the health of impoverished people and stressed environments. This process is well-illustrated by developments in the global shipbreaking industry, which is not regulated by the Basel Convention and its ban amendment. This industry has relocated from the developed world to countries such as India, Bangladesh, and China. Thousands of workers toil under the most arduous conditions without any safety precautions. Apart from the dangers posed by exposure to toxic chemicals, it is estimated that one in four of these workers will contract cancer from the asbestos on board these waste vessels.

A Way Forward

International investment is necessary to assist developing countries in establishing their own industries and to relieve them from their dependency on hazardous waste imports. As developing countries continue to industrialize, they will require additional funding and technology to build environmentally safe recycling and disposal plants for any waste that they generate. Developed countries have a responsibility under the Basel Convention to provide training for developing countries in managing hazardous waste and to facilitate technology transfers. The underlying rationale is that if developing countries are trained to adopt clean production technology from the outset, they can avoid the mistakes made by developed countries.

Technology transfers, however, may not be a panacea for problems associated with the handling of hazardous waste. Greenpeace has observed that the challenges faced by developing nations are not just a matter of know how and technology; the successful export of the developed world's environmental knowledge would, instead, also require the export of an entire social structure. This claim may be an overstatement. It is more likely that as developing countries improve economically, social and environmental reforms will follow. The requirement that exporters adhere to global environmental standards for admission to international markets may provide the necessary incentive for developing countries to upgrade their environmental performance. Yet to impose these requirements without providing sufficient funding and technology will result in greater poverty and hardship for these countries.

At the same time, it is essential for all responsible developed countries to ratify the ban amendment to ensure its lasting success. It is also in the long-term interest of developing countries to ratify the ban. The developed world must provide more assistance in order to persuade developing countries to give up a hazardous livelihood chosen only out of the fear of having no livelihood at all.

ZADA LIPMAN is Associate Director of the Centre for Environmental Law, Macquarie University, Australia and a Barrister of the Supreme Court of New South Wales.

Ensuring Environmental Sustainability

'Undoing the Damage We Have Caused'

By Horst Rutsch, for the *Chronicle*

The World Summit on Sustainable Development, held in Johannesburg, South Africa, from 26 August to 4 September, concluded with world leaders declaring that the "deep fault line" between rich and poor posed a major threat to global prosperity and stability. In response to these challenges, the Summit set specific global targets in poverty reduction, clean water and sanitation, and infant mortality, and also addressed related problems in agriculture, biodiversity, climate change, renewable energy and trade. The Johannesburg Summit, the biggest-ever United Nations conference, with 191 countries participating and over 21,340 accreditations, brought together 104 heads of State and Government.

Adopting the Johannesburg Declaration on Sustainable Development and reaffirming their commitment to Agenda 21, which was adopted ten years earlier in Rio de Janeiro, world leaders stated that although globalization had created new opportunities, enabling the rapid integration of markets and increasing the mobility of capital and investment flows, benefits and costs were unevenly distributed. "We risk the entrenchment of these global disparities", the Summit acknowledged, "and unless we act in a manner that fundamentally changes their lives, the poor of the world may lose confidence in their representatives and the democratic systems to which we remain committed, seeing their representatives as nothing more than sounding brass or tinkling cymbals". The Declaration

noted that the global environment continued to suffer from the loss of biodiversity, depletion of fish stocks, advancing desertification, worsening climate change, more frequent and devastating natural disasters, and increasingly vulnerable developing countries.

The Summit also adopted a wide-ranging Implementation Plan, which aims to tackle many of these challenges by 2015 and calls for:

- halving the proportion of the world's population who live on less than $1 a day;

- halving the number of people living without safe drinking water or basic sanitation;

- reducing mortality rates for infants and children under five by two thirds; and

- reducing maternal mortality by three quarters.

The Implementation Plan also calls for: "with a sense of urgency" a substantial increase in the use of renewable sources of energy, although it sets no specific targets; implementation of a new global system for classification and labelling of chemicals; and restoration of depleted fish stocks. It urges States that have not yet done so to ratify "in a timely manner" the Kyoto Protocol to the United Nations Framework Convention on Climate Change.

Other provisions address a comprehensive range of environmental and development issues, such as agriculture, biodiversity, climate change, energy and trade. The Plan also supports the development of Africa and small island States.

The President of the Summit, South African President Thabo Mbeki, in his closing statement urged that in response to all the voices heard at the conference, heads of State and Government should return to the world with the conviction "to undo the damage we have caused". The Summit's Secretary-General, Nitin Desai, said he hoped that the "Johannesburg plus 15" Conference would be able to say that measures promised during the Summit had led to a new dynamic, and that countries had lived up to their goals. All of that was possible, he said, if the decisions already made were taken seriously, adding that that was why the Summit had been called the "Summit for Action".

Partnerships and Commitments

- The United States will invest $970 million over the next three years on water and sanitation projects. It pledged $90 million for sustainable agriculture programmes and up to $43 million in energy projects next year, and will spend $53 million on forests in 2002-2005. It further pledged to spend $2.3 billion through 2003 on health programmes, including the Global Fund against HIV/AIDS.
- The European Union announced a $700-million energy initiative; and its "Water for Life" initiative seeks to engage partners to meet water and sanitation goals, primarily in Africa and Central Asia.
- The Asian Development Bank announced a $5-million grant to the United Nations Centre for Human Settlements (UN-Habitat) and $500 million in fast-track credit for the Water for Asian Cities Programme.
- The world's nine major energy companies signed a range of agreements with the United Nations to facilitate technical cooperation for sustainable energy projects in developing countries. The South African energy utility, Eskom, announced a partnership to extend modern energy services to neighbouring countries.

The Summit's high-level segment, held from 2 to 4 September, heard more than 100 world leaders address a wide range of issues, among them: the principle of common but differentiated responsibilities; the need to address the inequities of globalization; combating HIV/AIDS; changing unsustainable patterns of consumption and production; the importance of regional cooperation in achieving the goals of sustainable development; and the correlation between poverty and environmental degradation. The removal of agricultural subsidies, the transfer of environmentally sound technologies and the need for open markets for developing-world products featured prominently in most statements, which emphasized that subsidies to agricultural producers in the developed countries were detrimental to many developing-country markets.

How Foreboding Is the Future?

Five young citizens addressed the Johannesburg Summit during the opening of the high-level segment, and presented to world leaders a list of challenges—inspired, written and voted on by some 400 children from eighty countries—representing their hopes and fears for the future of the planet.

Scary Statistics

By Lauren Kansley
Global Youth Reporter from South Africa

Almost 5 million children die each year from preventable causes. Environmental hazards kill the equivalent of a jumbo jet full of children every 45 minutes. These scary statistics have spurred the World Health Organization (WHO) to launch a new movement to try and tackle the crisis and reduce by two thirds the number of deaths of under-five-year-olds by 2015.

Under WHO Director-General Dr. Gro Harlem Brundtland, the movement is busy mobilizing partners, such as key organizations and Governments, to achieve results in six areas: household water quality and availability; hygiene and sanitation; indoor and outdoor air pollution; disease vectors such as mosquitoes; chemicals; and accidents.

According to Dr. Brundtland, the provision of healthy environments for children would be one of the highest social and political priorities of the decade. "Our top priority must be in investing in the future of children, a group that is particularly vulnerable to environmental hazards." She identified "hazards" as being dangers present in the environment in which children live, learn and play. She added that increased industrialization, explosive urban population growth and lack of pollution control were just a few added factors that lack affect children's lives.

Poor children were most at risk because poverty further aggravated the environmental hazards.

Another issue underlined by delegations was the need to set time-bound targets for the use of renewable energy. Energy had to be provided to the 2 billion people who lacked access, speakers said, without increasing pollution and changing the climate. Some suggested a global target of 15 per cent renewable energy by 2010, with industrial countries taking the lead. Sustainable development could not be achieved if sources of energy were not renewable or efficient, they stressed.

A number of speakers addressed global climate change, with representatives of small island developing States, in particular, stressing the dire impact of sea-level rise on their very survival. Small island nations, noted one speaker, should not disappear due to the "greed" of the industrialized world. For many of them, another speaker said, time—"the most precious non-renewable resource"—was running out.

The high-level segment also included four round-table events during which heads of State and Government held discussions with heads of United Nations specialized agencies, as well as with representatives of intergovemmental agencies, non-governmental organizations (NGOs) and other major groups. Canada and the Russian Federation announced their intention to ratify the Kyoto Protocol, which raised the prospect that it could come into force without the participation of the United States, which has long opposed it. The Protocol would set the first binding restrictions on emissions of carbon dioxide and other heat-trapping greenhouse gases by the industrialized nations.

In plenary sessions, government delegations, major groups, specialized agencies, United Nations funds and programmes, intergovernmental organizations, NGOs and business representatives discussed partnerships in the five priority areas outlined by Secretary-General Kofi Annan prior to the Summit: water, energy, health, agricultural productivity and biodiversity. Numerous partnerships were launched to undertake initiatives aimed at achieving various goals within the priority areas, with the clearest achievements being made in water and sanitation.

From *UN Chronicle,* Number 4, 2002, pp. 50-51. © 2002 by UN Chronicle. Reprinted by permission.

Withholding the Cure

by N. A. Siegal

Bill Haddad helped create the U.S. Peace Corps and now works as a volunteer for a generic drug manufacturer, Cipla Ltd., based in India. Haddad's mission is to bring inexpensive medication to dying populations in the Third World. Wherever he goes, he carries an eyedropper filled with a drug called Neverapine, which is known to reduce the risk of transmitting HIV/AIDS from an HIV-positive pregnant woman to her unborn child.

The bottle of Neverapine costs Cipla less than forty cents to produce, Haddad says, and each one contains about 200 doses. The mother is given one drop when she goes into labor, the baby is delivered by Caesarian section, and the newborn is given a single drop within forty-eight hours of birth. Using this method, the risk of infecting the newborn is reduced by 90 percent, says Haddad.

Haddad gives away his eyedroppers to pediatric hospitals at no cost and donates them to physicians in small villages in sub-Saharan Africa where AIDS has so devastated the population that only the very old and the very young survive. Yet many countries, such as Kenya, Ghana, and South Africa, won't let him.

"It's Dickensian: children who are under five dying without one dose of medications that we could give them for free."

"Because of alleged patent laws or other political barriers, I can't do it in most of the sub-Saharan countries and in half the Latin American countries," he says. "We produce the drugs legally, and the international law says we can do it. The companies have public statements that say, 'We won't prevent you.' But then, some piece of paper arrives and stops you.

The piece of paper is usually a legal notice informing him of an injunction by one of the major brand-name pharmaceutical companies in the United States or other countries that develop and patent new medications, Haddad says. The notice usually states that his life-saving efforts violate intellectual property laws.

To Haddad, each one is a death certificate for scores of infants. "Every trip I'm on, I run into 100 people who die," he says. "I don't have the words to describe what's really happening. It's Dickensian: children who are under five dying without one dose of medications that we could give them for free. You send the drugs for free and they don't get distributed. Frustrating would be the mild word for it."

It wasn't supposed to be this way. Less than two years ago, the nations of the world assembled at the World Trade Organization's 2001 meeting in Doha, Qatar, seemed to agree that treating poor people in the world's indigent nations was a higher priority than maintaining the intellectual property claims of multinational corporations. At Doha that November, 142 nations signed a historic agreement that allowed governments to override international patent protections and produce generic versions of medicines for a small fraction of the price charged by brand-name drug makers.

The pact, known as the Doha Declaration, was good news for the more than thirty million Africans living with HIV and AIDS, and millions more people in developing countries dying of illnesses that are treatable or curable in the West (such as measles, influenza, asthma, various types of cancer, and digestive diseases). But since it was signed, the United States, the European Union, Canada, Japan, and other wealthy countries have attempted to undermine that agreement, say nongovernmental organizations such as Médècins Sans Frontières (Doctors Without Borders), Oxfam, and Africa Action.

"They're seeking about four types of restrictions," says Asia Russell, who directs international policy at Health GAP, an activist organization founded in 1999 to campaign for increased access to affordable medication in developing countries. These restrictions "would significantly compromise" the ability of poor countries to provide urgent health care to their people, she says. U.S. trade negotiators have suggested limiting the list of illnesses covered under the Doha pact to AIDS, tuberculosis, malaria, and future equivalent pandemics. The United States and other countries, such as Japan and Switzerland, have also suggested restricting the number of countries that could benefit from the pact by strictly defining what constitutes a poor country. Western nations have also proposed a series of economic "tests" that would determine which countries could legally import or export generics. And finally, pharmaceutical companies and Western countries want to ensure that generic versions of brand-name drugs are not diverted to rich countries, where they could undercut potential profits.

"If you want new medicines, you have to have intellectual property protections," says Mark Grayson, a spokesman for the Pharmaceutical Research and Manufacturers of America, the trade group for the leading U.S. drug and biotechnology companies. "Right now, many people believe that Doha has to be applicable to all diseases. If it's all diseases, then you have to limit the countries."

The Pharmaceutical Research and Manufacturers of America, the twenty-fourth most powerful lobbying group in the country, according to *Fortune* magazine, agrees that the process of getting cheap medications to the world's dying populations has stalled. But Grayson argues that Haddad and nongovernmental organizations (NGOs) are the ones to blame.

Grayson's members don't want companies like Cipla in India and other generic drug makers in Brazil and China to be able to profit by producing cheap versions of drugs that cost U.S. pharmaceutical companies as much as $800 million to develop, he says. Instead of focusing on getting the drugs to the poorest of the poor, he argues, Haddad and the NGOs are trying to give manufacturers of generics the power to expand their reach into the global market.

"People want to use this for industrial policy," Grayson said. "They want to put generic versions of drugs on the market everywhere, including in the United States, and make money that way, rather than providing drugs to the poorest countries."

The pharmaceutical patent debate is expected to be a central issue during the next round of international trade negotiations at the World Trade Organization in Cancún, Mexico, from September 10 to September 14.

"They [the pharmaceuticals] are trying to figure out how to do the absolute minimum you can do to keep to the basic premise of this agreement but not provide the

needed medicines," says James Love, director of the Consumer Project on Technology, a nonprofit group based at Ralph Nader's Center for the Study of Responsive Law in Washington, D.C. "It's all about making it hard to do, to make it burdensome—whatever they can do to slow the process."

Access to medicines and other life-saving technologies has become, in the last few years, one of the most explosive issues in the debate over globalization. Before 1994, there were no uniform international trade barriers to manufacturing or distributing generic versions of drugs invented in the West. The World Trade Organization that year created a set of laws governing intellectual property protections, known as the Trade-Related Aspects of Intellectual Property Rights, or TRIPS. It gave twenty-year patent protection for "all inventions, whether of products or processes, in almost all fields of technology"—including medications.

That means that if a U.S. company were to develop a cure for cancer today, for example, a generic version wouldn't be available for twenty years. This provision would allow the company that develops the cure the opportunity to recoup its investment. Anyone who couldn't afford the brand-name medication, priced for maximum profit, would be out of luck.

Under TRIPS, there was relief for developing nations, however: A country could break a patent monopoly if it followed strict guidelines allowing them to issue what's known as a "compulsory license." Article 31 of the agreement listed seven ways a country could grant such a license, one of which was to declare a national public health emergency.

On the African continent, where 2.2 million people died of AIDS in 2001, according to the World Health Organization, it made sense to override patent protections. With such a compulsory license, African nations could get a triple combination "cocktail" from India's Cipla for $300 a year, or from other generic companies in Brazil and Thailand for as little as $250 annually per patient instead of buying the medicine from U.S. drug makers that charge about $12,000 to $15,000 each year per patient.

And yet, in 1997, when South African President Nelson Mandela signed into law a measure to issue a compulsory license for his country, thirty-nine of the world's largest pharmaceutical companies, including Bristol-Myers Squibb, GlaxoSmithKline, and Merck, sought and received an injunction to block it. The same companies later filed lawsuits against the South African government, charging it with patent violations. Their stance created enormous negative publicity, so the companies dropped their suit.

But no country was willing to take the political risk of attempting to issue a compulsory license after South Africa, says Rachel Cohen, U.S. director of the Campaign for Access to Essential Medicines, a project of Doctors Without Borders. "They wanted assurance that they wouldn't face political consequences if they tried to issue a compulsory license," she says. That explains why Haddad has had such a hard time distributing his AIDS medicine.

The South Africa case helped put the conflict over international property law and public health concerns on the front burner for the next round of international trade negotiations that November in Doha.

"It was because of the fallout from that South Africa case that so much momentum was gained to support the victory at Doha," says Health GAP's Russell.

At Doha, trade officials crafted language to clarify TRIPS and to reassert the rights of poor nations to issue compulsory licenses. The text said that TRIPS "can and should be interpreted in a matter of WTO members' rights to protect public health and in particular to ensure access to medicines for all."

"This was a landmark moment," says Cohen. "For the first time in WTO history, all of the participants said public health takes priority over these other commercial interests. This was a huge, huge victory for developing countries."

"8,500 people are dying every day in Africa of AIDS, and many more infected children are being born who, without Neverapine, will die the same way."

Yet many issues were still unresolved as the trade negotiators departed Qatar. For one thing, it was unclear how countries would be able to make effective use of compulsory licenses if they didn't have the infrastructure or resources to manufacture generics at home. Nor was it clear whether they could import them.

Paragraph six of the Doha Declaration instructed participating countries to "find an expeditious solution to the problem and report to the General Council before the end of 2002." For the last twenty months, trade negotiators have been trying to do just that, even as they missed the 2002 year-end deadline.

In the meantime, the United States announced an interim moratorium plan, pledging not to challenge any WTO members that broke trade rules to export drugs to a country in need.

Robert Zoellick, the U.S. trade representative, and the U.S. pharmaceutical giants said they wanted to make it possible for India, Brazil, and other countries with some manufacturing capacity to export, as long as there were limits set. Zoellick proposed a "disease list" to define the types of medications that could be included in compulsory licenses.

To activist groups, the list was laughable. At first, it included only AIDS, tuberculosis, and malaria. Then, the U.S. added nineteen other illnesses—all of them possible "pandemics." A report prepared by Dr. Mary Moran of Doctors Without Borders found that the listed illnesses didn't even "correlate with the major causes of morbidity and mortality in Africa." Of the twenty-two on the list, nine were diseases for which there was no existing treatment—thus, there was no patent the countries could override.

Also, the "approved list" was restricted to infectious diseases such as yellow fever and dengue fever, cholera, hepatitis, and the plague. Cardiovascular diseases were excluded although, according to the World Health Organization, they are the second biggest cause of death on the African continent, taking about 985,000 lives a year. Some 544,000 Africans die of various forms of cancer, and another 336,000 fall to respiratory diseases, including pneumonia—neither of which was covered. Also not included were digestive diseases and diabetes, which are treatable with prescription medications but which cause some 434,000 deaths in Africa annually.

Mouanodji Mbaissouroum, a cardiologist in the central hospital at N'Djamena, Chad, told *The Wall Street Journal* that he writes many prescriptions for medicines to relieve hypertension that he knows will never be filled, either because the medicine isn't sold in Chad or because it is priced way beyond his patients' means.

"People in the developed world think AIDS and malaria and communicable diseases are the biggest problems in Africa," Dr. Mbaissouroum told the *Journal*. "But we also suffer the same illnesses as rich people do."

James Love, of the Consumer Project on Technology, says U.S. negotiators may be backing off from attempts to limit the list of diseases and are now looking to keep a lid on the number of countries that could import drugs to only the most indigent. Those excluded could be developing nations such as the Philippines and Malaysia, because they aren't considered so poor that they would need such broad protections.

On the pharmaceutical side, the argument is about property rights: Limiting the scope of the international agreement to safeguard inventions is essential to keeping pharmaceutical companies in business and to ensuring future research and development for new life-saving drugs and medical technologies, the drug companies say.

"The profit is what has ended up giving us these medicines we have now," says Grayson. "Without profit, you won't have investment."

Grayson estimates that U.S. drug companies put about 20 percent of their revenues into research and development. Currently, he says, U.S. drug companies invest about $800 million developing a single medication. In 2002, he says, they spent an estimated $32 billion to discover and produce new medicines.

The pharmaceuticals industry was the most profitable sector of the economy last year, according to the 2003 Fortune 500 list of the world's top companies. Pfizer alone made $9.13 billion in profits, and Merck made $7.15 billion.

"The pharmaceutical industry is, fairly stated, a profit-making group—and that doesn't mean price-gouging or 'I don't care about the Third World,' " says Neil Sweig, a financial analyst who tracks pharmaceutical companies for Fulcrum Global Partners in New York. "If prices plunge or if prices are too low or if drugs are given away at no charge, it represents a very serious problem in the ability of global brand-name companies to go forward with research and development that runs into the billions for the companies.

In 2003 alone, the World Health Organization predicts that there will be some ten million deaths in developing countries from malaria, tuberculosis, acute respiratory infections, and diarrhoeal diseases. There are safe, effective drugs to treat these diseases—for people who can afford them. (While the United States says it has agreed that malaria and tuberculosis should be covered by the Doha accord, nothing formal has been agreed to yet.) Millions more poor people are dying of treatable cancers, diabetes, digestive diseases, and sleeping sicknesses. These people could be served by the science and technologies that are now prolonging lives in the West.

Bill Haddad has been involved in this, the other drug war, for many years. A former investigative journalist for the *New York Herald Tribune*, he exposed a tetracycline cartel that prevented the drug from being distributed in Latin America and became so interested in the subject that he quit journalism to work as a generics' advocate. He became chairman and CEO of the Generic Pharmaceutical Industry Association, and in 1994 created his own company, the United States Research and Development Corporation. In 1999, he founded Biogenerics, Inc., which produced generics, and a year later he went to work for Cipla as a volunteer to reduce the price of AIDS medications in Africa.

In April, he attended the European Commission conference in Brussels and gave a presentation in which he listed the casualties: 8,500 people are dying every day in Africa of AIDS, and many more infected children are being born who, without Neverapine, will die the same way.

"I can't go to some of these places anymore, the places people want you to see with your own eyes. It's horrendous," says Haddad. "Patents need to be protected, but they're so exaggerated in what prices they need. They continue to raise the prices because they have no conscience. It looks like a snowball's chance in hell that these powerful forces will back down and let anything happen."

N. A. Siegal is a New York-based reporter who has contributed to The Progressive *since 1997.*

From *The Progressive*, September 2003, pp. 29-33. © 2003 by The Progressive, 409 E. Main Street, Madison, WI 53703, www.progressive.org. Reprinted by permission

UNIT 6
Women and Development

Unit Selections

Key Points to Consider

- What factors can lead to enhancing the status of women? What prevents wider educational opportunities for women in the developing world?

- Why might a larger role for women help to resolve conflicts?

- What effect have foreign aid programs had on gender?

- How did the genocide in Rwanda change women's roles in society?

- What is the "true clash of civilizations"?

 **Links: www.dushkin.com/online/**
These sites are annotated in the World Wide Web pages.

African Women Global Network
http://www.ohio-state.edu/org/awognet/

WIDNET: Women in Development NETwork
http://www.focusintl.com/widnet.htm

WomenWatch/Regional and Country Information
http://www.un.org/womenwatch/

There is widespread recognition of the crucial role that women play in the development process. Women are critical to the success of family planning programs, bear much of the responsibility for food production, account for an increasing share of wage labor in developing countries, and are acutely aware of the consequences of environmental degradation. Despite their important contributions, however, women lag behind men in access to health care, nutrition, and education while continuing to face formidable social, economic, and political barriers.

Women's lives in the developing world are invariably difficult. Often female children are valued less than male offspring, resulting in higher female infant and child mortality rates. In extreme cases, this undervaluing leads to female infanticide. Those females who do survive face lives characterized by poor nutrition and health, multiple pregnancies, hard physical labor, discrimination, and perhaps violence.

Clearly, women are central to any successful population policy. Evidence shows that educated women have fewer and healthier children. This connection between education and population indicates that greater emphasis should be placed on educating women. In reality, female school enrollments are lower than those of males for reasons having to do with state priorities, family resources that are insufficient to educate both boys and girls, female socialization, and cultural factors. Although education is probably the largest single contributor to enhancing the status of women and thereby promoting development, access to education is still limited for many women. Sixty percent of the children worldwide who are not enrolled in school are girls. Higher status for women also has benefits in terms of improved health, better wages, and greater influence in decision making.

Women make up a significant portion of the agricultural workforce. They are heavily involved in food production from planting to cultivation, harvesting, and marketing. Despite their agricultural contribution, women frequently do not have adequate access to advances in agricultural technology or the benefits of extension and training programs. They are also discriminated against in land ownership. As a result, important opportunities to improve food production are lost when women are not given access to technology, training, credit, and land ownership commensurate with their agricultural role.

The industrialization that has accompanied the globalization of production has meant more employment opportunities for women, but often these are low-tech, low-wage jobs. The lower labor costs in the developing world that attract manufacturing facilities are a mixed blessing for women. Increasingly, women are recruited to fill these production jobs because wage differentials allow employers to pay women less. On the other hand, expanding opportunities for women in these positions contributes to family income. The informal sector, where jobs are smaller-scale, more traditional, and labor-intensive, has also attracted more women. These jobs are often their only source of employment due to family responsibilities or discrimination. Clearly, women also play a critical role in economic expansion in developing countries. Nevertheless, women are often the first to feel the effects of an economic slowdown.

The consequences of the structural adjustment programs that many developing countries have had to adopt have also fallen disproportionately on women. As employment opportunities have declined because of austerity measures, women have lost jobs in the formal sector and faced increased competition from males in the informal sector. Cuts in spending on health care and education also affect women, who already receive fewer of these benefits. Currency devaluations further erode the purchasing power of women.

Because of the gender division of labor, women are often more aware of the consequences of environmental degradation. Depletion of resources such as forests, soil, and water are much more likely to be felt by women who are responsible for collecting firewood and water and who raise most of the crops. As a result, women are an essential component of successful environmental protection policies but are often overlooked in planning environmental projects.

Enhancing the status of women has been the primary focus of recent international conferences. The 1994 International Conference on Population and Development focused attention on women's health and reproductive rights and the crucial role that these issues play in controlling population growth. The 1995 Fourth World Conference on Women held in Beijing, China, proclaimed women's rights to be synonymous with human rights. These developments represent a turning point in women's struggle for equal rights. International conferences not only focus attention on gender issues, but also provide additional opportunities for developing leadership and encouraging grassroots efforts to realize the goal of enhancing the status of women. Greater political involvement may also increase women's participation in efforts to resolve conflict. Some argue that women have certain qualities that are more likely to facilitate compromise and the peaceful resolution of conflicts. Moreover, women may find that changes in societies that are recovering from conflict may alter traditional gender roles giving women greater influence in society.

There are some indications that women have made progress in certain regions of the developing world. In some parts of the Middle East, several factors have contributed to the erosion of some of the restrictions on women, but the region's women still struggle to escape the status of second class citizens. The countries of the Southern African Development Community have made substantial but uneven progress in meeting the targets for women's political participation set at the 1995 Beijing Conference on Women. In India, a constitutional amendment has set aside one-third of village council and chief positions for women, and a percentage of those positions must go to lower-caste women. In Latin America, women have made significant gains in attaining equal pay, access to reproductive health, and protection from violence, but there are differences between urban and rural women as well as differences in women's status among countries in the region.

Empowering Women

Lori S. Ashford

Thanks to the growing activism and influence of women's rights advocates around the world, the situation of women has moved to the forefront of both national and international population policy debates. Since the Cairo and Beijing conferences, there has been greater discussion of gender issues, and of the differences in men's and women's socially defined roles. Governments and donor agencies increasingly acknowledge that women's inferior status hinders development and support policies and programs to reduce gender inequalities. Those concerned about the negative effects of population growth also see a connection between enhancing a woman's status within the family and society and increasing her control over childbearing.

No government could deny that women deserve higher status and better opportunities.

The agreements adopted at the population and development conference (ICPD) in Cairo and the women's conference in Beijing called for equal participation and partnerships between men and women in nearly all areas of public and private life. As innovative—even revolutionary—as these notions seem, they met with little dissent during the conferences themselves. In spite of the diversity of cultures represented, no government could deny that women deserve higher status and better opportunities. Beyond the conference halls, however, putting these ideas into practice requires overcoming deeply rooted cultural values and ways of life.

Understanding Gender

Gender refers to the different roles that men and women play in society, and the relative power they wield. Gender roles vary from one country to another, but almost everywhere, women face disadvantages relative to men in the social, economic, and political spheres of life. Where men are viewed as the principal decisionmakers, women often hold a subordinate position in negotiations about limiting family size, contraceptive use, managing family resources, protecting family health, or seeking jobs.

Inequalities between men and women are closely linked to women's health–making the issue of gender pertinent to discussions on how to improve reproductive health. Gender differences affect women's health and well-being throughout the life cycle:[1]

- Before or at birth, parents who prefer boys may put girls at risk of sex-selective abortions (where technology is available to identify the sex) or infanticide.
- Where food is scarce, girls often eat last, and usually less than boys.
- Girls may be less likely than boys to receive health care when they are ill.
- In some countries, mainly in Africa, girls are subjected to female genital cutting.
- Adolescent girls may be pressured into having sex at an early age—within an arranged marriage, by adolescent boys proving their virility, or by older men looking for partners not infected with STIs.

Table 1

Fertility and Access to Reproductive Health Care Among the Poorest and Richest Women, Selected Countries, 1990s

Country	Births per woman (TFR)*		Prenatal care (Percent of pregnant women)		Births attended by skilled staff (Percent of deliveries)	
	Poorest fifth	Richest fifth	Poorest fifth	Richest fifth	Poorest fifth	Richest fifth
Bolivia	7.4	2.1	39	95	20	98
Cameroon	6.2	4.8	53	99	32	95
Guatemala	8.0	2.4	35	90	9	92
India	4.1	2.1	25	89	12	79
Indonesia	3.3	2.0	74	99	21	89
Morocco	6.7	2.3	8	74	5	78
Vietnam	3.1	1.6	50	92	49	99

*TFR (total fertility rate) is the average total number of births per woman given prevailing birth rates.

Note: Women were ranked according to their household assets.

Source: A. Tinker, K. Finn, and J. Epp, *Improving Women's Health: Issues and Interventions* (World Bank, June 2000): 10.

- Married women may be pressured by husbands or families to have more children than they prefer, and women may be unable to seek or use contraception.
- Married and unmarried women may be unable to deny sexual advances or persuade partners to use a condom, thereby exposing themselves to the risk of STIs.
- In all societies, women are more likely than men to experience domestic violence. Women may sustain injuries from physical abuse by male partners or family members, and the fear of abuse can make women less willing to resist the demands of their husbands or families.

A mixture of cultural and social factors explain women's lack of power in protecting themselves or their daughters from these health threats. These factors include women's limited exposure to information and new ideas, ignorance of good health practices, limited physical mobility, and lack of control over money and resources.[2] In some South Asian and Middle Eastern countries, for example, women's use of health care services is inhibited by cultural restrictions on women traveling alone or being treated by male health care providers.

The Poverty Connection

Gender disadvantages intertwine with poverty. Poverty is strongly linked to poor health, and women represent a disproportionate share of the poor. In all regions of the world, including wealthier regions, reproductive health is worse among the poor (see Table 1). Women in the poorest households have the highest fertility, poorest nutrition, and most limited access to skilled pregnancy and delivery care, which contribute to higher maternal and infant death rates. Women's disadvantaged position also perpetuates poor health, an inadequate diet, early entry into motherhood, frequent pregnancies, and a continued cycle of poverty.[3]

Women's low socioeconomic status also makes them more vulnerable to physical and sexual abuse. Unequal power in sexual relationships exposes women to coerced sex (see Box 1), unwanted pregnancies, and sexually transmitted infections. Impoverishment can also lead some women into commercial sex.[4] Thus, women's access to and control over resources can give women greater control of their sexuality, which is fundamental to controlling their fertility and improving their health.

Reducing Gender Inequalities

Recent reports from the World Bank show that reducing gender inequalities can bring about greater economic prosperity and help reduce poverty. One study found that a 1 percent increase in women's secondary schooling results in a 0.3 percent increase in economic growth.[5] In addition, the strong links between women's status, health, and fertility rates make gender equality a critical strategy for policies to improve health and stabilize population growth.

Women's control over their sexuality is central to population and health concerns.

The empowerment of women is seen as a key avenue for reducing the differences between the sexes that exist in nearly all societies. Empowerment refers to "the process by which the powerless gain greater control over the circumstances of their lives."[6] It means not only greater control over resources but also greater self-confidence and the ability to make decisions on an equal basis with men. Empowerment of women also requires that men are aware of gender inequalities and are willing to question traditional definitions of masculinity.

In many societies, these concepts may be threatening to men, who are accustomed to having authority in the household, the community, the economy, and national politics. The concepts may also be frightening to women, who may fear the implications of these changes for their personal lives.[7] For these reasons, and because concepts of women's rights, empowerment, and gender equality are still relatively new in many places, progress in advancing women's rights has been modest.

Box 1

Violence Against Women

Violence against women is the issue that perhaps best illustrates the connection between women's rights and women's health, and the tragic consequences of women's inferior position. Once thought to be only a private matter, violence against women has gained visibility as a serious public policy and public health concern.

Violence against women (also referred to as gender-based violence) occurs in nearly all societies, within the home or in the wider community, and it is largely unpoliced. It may include female infanticide, incest, child prostitution, rape, wife-beating, sexual harassment, wartime violence, or harmful traditional practices such as forced early marriage, female genital cutting, and widow or bride burning. A recent study published by the Center for Health and Gender Equity and Johns Hopkins University estimated that one in three women worldwide suffers from some form of gender-based violence.

Domestic violence is the most common form of gender-based violence, and it is most often perpetrated by a boyfriend or husband against a woman (see figure). Psychological abuse almost always accompanies physical abuse, and the majority of women who are abused by their partners are abused many times. Many women tolerate the abuse because they fear retaliation by their spouse or extended family, or both, if they protest. Women's vulnerability to violence is reinforced by their economic dependence on men, widespread cultural acceptance of domestic violence, and a lack of laws and enforcement mechanisms to combat it.

Although women's control over their sexuality is central to population and health concerns, the extent to which sexual activity

is forced or coerced has only recently been addressed. Most coerced sex takes place between people who know each other—spouses, family members, or acquaintances. One-quarter to one-half of domestic violence cases involve forced sex. Coercion also takes place against children and adolescents in more developed and less developed countries. Statistics on rape suggest that between one-third and two-thirds of rape victims around the world are younger than 16.

What is the connection between gender-based violence and reproductive health? Violence against women is rooted in unequal power between men and women. It affects women's physical, mental, economic, and social well-being. It can lead to a range of health problems. Since girls are more often subjected to coerced sex than boys, they are at risk of becoming infected with sexually transmitted infections (STIs) at a younger age than are boys. Some STIs can lead to pelvic inflammatory disease, infertility, and AIDS. Forced and unprotected sex also leads to unintended pregnancies, abortions, and unwanted children. The experience of abuse puts women at greater risk of mental health problems, including depression, suicide, and alcohol and drug abuse. Ultimately, these outcomes have negative consequences for the whole society, not just the women who are victims of such violence.

After a series of international conferences and conventions in the 1990s called for eliminating all forms of violence against women, many countries strengthened laws and enforcement mechanisms related to domestic violence. Much of the pressure to change laws and community standards has come from nongovernmental organizations, and particularly women's groups.

These groups are at the forefront of efforts to combat violence against women through grassroots activism, lobbying, and working with women survivors of violence. Ending the violence requires community-level action and, ultimately, changes in the values that lead to the subjugation of women.

Women Reporting Physical Assault by Male Partner, Selected Studies, 1990s

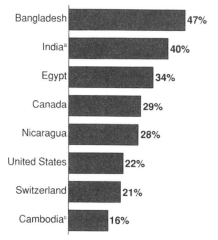

Country	%
Bangladesh	47%
India[a]	40%
Egypt	34%
Canada	29%
Nicaragua	28%
United States	22%
Switzerland	21%
Cambodia[b]	16%

[a] Six states only.
[b] Phnom Penh and six provinces.

Source: L. Heise, M. Ellsberg, and M. Gottemoeller, *Population Reports* Series L, no. 11 (1999): table 1.

References

Lori Heise, Mary Ellsberg, and Megan Gottemoeller, "Ending Violence Against Women," *Population Reports* Series L, no. 11 (Baltimore: Johns Hopkins University, 1999); UNFPA, *The State of World Population 2000: Lives Together, Worlds Apart* (New York: UNFPA, 2000); and Yvette Collymore, ed., *Conveying Concerns: Women Report on Gender-Based Violence* (Washington, DC: Population Reference Bureau, 2000).

Women's Education

The World Bank calls women's education the "single most influential investment that can be made in the developing world." Many governments now support women's education not only to foster economic growth, but also to promote smaller families, increase modern contraceptive use, and improve child health. Educating women is an important end in itself, and it is also a long-term strategy for advancing women's reproductive health. The Cairo conference called for universal access to and completion of primary education, and for reducing the gender gap—

differences in boys' and girls' enrollment—in secondary education.

Worldwide, more men than women are literate (80 percent, compared with 64 percent). While nearly all boys and girls in more developed countries are enrolled in both primary and secondary school, women in less developed countries complete fewer years of education than men, on average, and are more likely to be illiterate.

According to the UN, primary and secondary school enrollments increased for both girls and boys during the 1990s in almost all world regions. The gender gap in school enrollments closed somewhat in recent years but remains pronounced at the secondary school level (see Table 2). Girls are more likely than boys to discontinue their schooling for a number of reasons: household duties; early marriage and childbearing; parents' perceptions that education is more beneficial for sons; worries about girls' safety as they travel to schools away from their villages; and limited job opportunities for women in sectors that require higher education. In some settings, gender bias among teachers and sexual harassment may lead to higher dropout rates among young women.[8]

Table 2

Secondary School Enrollment by Sex in World Regions, 1980 and 1990s

| | Percent enrolled* | | | |
| | 1980 | | 1990s | |
Region	Boys	Girls	Boys	Girls
More developed countries	88	89	99	102
Less developed countries	43	30	57	48
Northern Africa	47	29	63	57
Sub-Saharan Africa	19	10	29	23
Western Asia	49	31	63	48
South-Central Asia	38	20	55	37
Southeast Asia	40	35	53	49
East Asia	59	45	77	70
Central America	46	42	56	57
Caribbean	–	–	49	55
South America	38	42	–	–

–Not available.

* The percent enrolled is the ratio of the total number enrolled in secondary school (regardless of age) to the number of secondary-school-age children, or the gross enrollment ratio. Data from the 1990s are the latest available, generally between 1990 and 1996.

Source: A. Boyd, C. Haub, and D. Cornelius, *The World's Youth 2000* (Population Reference Bureau, 2000), based on national figures from UNESCO, *Statistical Yearbook 1999* (1999).

During the 1990s, the countries seeing the greatest gains in closing the gender gap were regions that had the lowest enrollments in the past: Northern Africa, sub-Saharan Africa, Southern Asia and Western Asia. Nevertheless, in the regions where almost a third of the world's women live (Southern Asia and sub-Saharan Africa), girls are much less likely than boys to attend secondary school. The populations of these two regions are among the world's fastest growing, which suggests that the number of illiterate women in these regions will continue to grow.[9]

Research over the last 20 years has shown that women with more education make a later transition to adulthood and have smaller, healthier families. Women with more education usually have their first sexual experience later, marry later, want smaller families, and are more likely to use contraception and other health care than their less educated peers (see Figure 1). In many less developed countries, women with no schooling have about twice as many children as do women with 10 or more years of school.[10] Expanding educational opportunities for women has been embraced as a means to lower national fertility rates and to slow population growth.

Figure 1

Women's Education and Childbearing, Selected Countries, 1995–1999

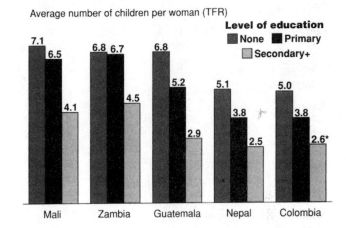

*Secondary level only.

Note: The TFR is the average total number of children born per woman given prevailing birth rates.

Source: Demographic and Health Surveys, Final Country Reports. Available online at: www.measuredhs.com.

Women's Work

Employment is another way that women can elevate their status: It enables them to earn income and have more control over resources. It can also increase women's involvement in the public sphere and help enhance decisionmaking skills. Women's participation in the labor force increased during the 1990s, but they are still at the lower end of the labor market in pay and authority.

Women make up an increasing share of the labor force in almost all regions of the world (see Table 3). Several factors explain this: Women's improved ability to limit and space pregnancies has enabled them to spend less time on child care and more on work outside the home; attitudes toward the employment of women have become more accepting; and new policies on family and child care ensure more flexibility and therefore favor working women. In addition, economic growth, and particularly the expansion of service industries (like finance, communications, and tourism), which tend to employ large numbers of women, has increased women's labor force opportunities. And finally, programs that have made credit available for small enterprises have benefited women.[10]

Table 3

Womens Share of the Labor Force, 1980 and 1997

Region	Women as percent of labor force	
	1980	1997
Africa		
Northern Africa	20	26
Sub-Saharan Africa	42	43
Latin America & Caribbean		
Caribbean	38	43
Central America	27	33
South America	27	38
Asia		
East Asia	40	43
Southeast Asia	41	43
Southern Asia	31	33
Central Asia	47	46
Western Asia	23	27
More developed regions		
Eastern Europe	45	45
Western Europe	36	42
Other more developed regions	39	44

Source: United Nations, *The World's Women 2000: Trends and Statistics* (2000): 110.

Figure 2

Women's Wages as a Percentage of Men's Wages, in Manufacturing, Selected Countries, 1992–1997

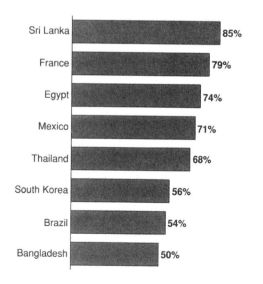

Sri Lanka	85%
France	79%
Egypt	74%
Mexico	71%
Thailand	68%
South Korea	56%
Brazil	54%
Bangladesh	50%

Source: United Nations, *The World's Women 2000: Trends and Statistics* (2000): chart 5.23.

These trends are positive, but equality in the work force is still a long way from reality. Women typically occupy lower-paid and lower-status jobs than men; women's unemployment rates are higher than men's; and far more women than men work in the "informal sector," occupations like street vending and market work, where wages are very low. Even when women work in the same sector as men, wages are typically lower (see Figure 2).

In addition, more women are remaining in the work force during their reproductive years, leading to a "dual burden": working outside the home while at the same time doing a larger share of work in the home than men—such as childrearing, cooking, and cleaning. The few studies that are available on how women's time is used (in more developed countries) show that women spend 50 percent to 70 percent as much time as men on paid work, but almost twice as much or more time as men on unpaid work.[11] A multicountry study on women's lives and family planning in less developed regions found that many women reported additional stress because of these dual responsibilities, and that many women in these situations would prefer not to work outside the home.[12]

The Cairo population and development conference and the Beijing women's conference called on governments to reduce disparities between men and women in the work force and to provide additional support to working women, such as maternity leave, child-care assistance, and other flexible arrangements. One area that has received much visibility and high-level political support is that of providing microcredit—small low-interest loans—that allow women to start their own small businesses. Experience has shown that women are good "credit risks" and repayment rates are high. These programs tend to be small, however, and they do not address the underlying social and cultural reasons for women's economic disadvantages.

Changing Family Dynamics

Women's strong attachment to family and responsibility for the household define much of their adult lives. Women usually marry at a younger age than do men, and the age gap between spouses tends to be wider in low-income countries. The age differences between spouses help perpetuate women's weaker authority. In societies where childbearing starts soon after marriage, women's opportunities to pursue careers or additional schooling, or to develop contacts beyond the family, are limited.

In many societies, laws pertaining to the family put women at a disadvantage and reinforce their dependence on men: Women may be unable to inherit land or other property, divorce their husbands, or get custody of their children if they can divorce. The gender gap in legal rights is gradually narrowing in some countries. Egypt revised its laws in 2000 to allow women similar divorce rights as men, for example, and Morocco revised its personal status laws in the 1990s to enhance women's rights in marriage—including polygamous marriages—and divorce.[13]

Women's Political Leadership

Women's right to equal participation in political life is guaranteed by a number of international conventions, most notably the

Convention on the Elimination of All Forms of Discrimination Against Women (CEDAW), which was adopted in 1979 but has not been ratified by all governments. In practice, in all countries, women are underrepresented at every level of government, especially in high-level executive positions and legislative bodies.

Figure 3
Legislative Seats Held by Women, Selected Countries, 1999

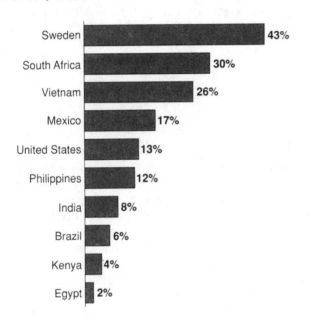

Sweden 43%
South Africa 30%
Vietnam 26%
Mexico 17%
United States 13%
Philippines 12%
India 8%
Brazil 6%
Kenya 4%
Egypt 2%

Source: United Nations, *The World's Women 2000: Trends and Statistics* (2000): table 6.A.

Worldwide, women held only 11 percent of seats in parliaments and congresses in 1999, up from 9 percent in 1987.[14] In the United States, sub-Saharan Africa, Latin America and the Caribbean, and East and Southeast Asia, these percentages ranged between 10 percent and 13 percent in 1999. In Northern Africa and South and West Asia, women held just 3 percent to 5 percent of seats in legislative bodies (see Figure 3). There are some exceptions: Women in Vietnam, South Africa, Mozambique, Cuba, and Argentina hold 25 percent to 30 percent of parliamentary seats.

In the 1990s, a number of countries, including India, Uganda, Brazil, and the Philippines, formally set aside a percentage of seats on national and local bodies to be held by women. In South Africa, the proportion of women in parliament rose from 1 percent to 30 percent following the establishment of a new constitution.

One drawback of mandated quotas is that women may be placed on election ballots and voted into office before they have developed the political and technical skills necessary to govern effectively. Until these reforms become more widespread and meaningful, women's lack of political power will limit their influence on laws and policies that affect their rights, as well as on priority-setting in health care and other public services.

Notes

1. Program for Appropriate Technology (PATH), *Reproductive Health Outlook*. Website accessed online at: www.rho.org, on Jan. 23, 2001.
2. AbouZahr, "Background Paper on Reproductive Health": 16
3. Ann Tinker, K. Finn, and Joanne E. Epp, *Improving Women's Health: Issues and Interventions* (Washington, DC: The World Bank, June 2000): 10.
4. Ibid.: 11.
5. David Dollar and Roberta Gotti, "Gender Inequality, Income and Growth: Are Good Times Good for Women?" *Policy Research Reports on Gender and Development*, Working Paper Series No. 1 (Washington, DC: World Bank, 1999).
6. Gita Sen and Srilatha Batliwala, "Empowering Women for Reproductive Rights," in *Women's Empowerment and Demographic Processes*, eds., Harriet B. Presser and Gita Sen (Oxford, England: Oxford University Press, 2000): 18.
7. Ibid.
8. Caroline Bledsoe, John Casterline, J. Johnson-Kuhn, and John Haaga, *Critical Perspectives on Schooling and Fertility in the Developing World* (Washington, DC: National Academy Press, 1999): 81–104.
9. United Nations, *The World's Women 2000: Trends and Statistics* (New York: United Nations, 2000): 85–86.
10. United Nations, *Linkages Between Population and Education* (New York: United Nations, 1997): 12.
11. United Nations, *The World's Women 2000: Trends and Statistics*: 110.
12. Ibid.: 126.
13. Barbara Barnett and Jane Stein, *Women's Voices, Women's Lives: The Impact of Family Planning* (Research Triangle Park, NC: Family Health International, 1998).
14. UNFPA, *State of World Population 2000*: 55; and Radouane Belouali and Najib Guédira, *Reproductive Health in Policy and Practice: Morocco* (Washington, DC: Population Reference Bureau, 1998): 8–10.

Reprinted with permission by *Population Bulletin*, by Lori S. Ashford, March 2001, Vol. 57, No. 1, pp. 21-29. © 2001 by "New Population Policies: Advancing Women's Health and Rights," Population Bulletin, (Washington, DC: Population Reference Bureau).

Article 41

Women & Development Aid

Ritu R. Sharma

Key Points

- Economic studies and program evaluations show that considering gender roles and targeting programs to women and girls dramatically enhances economic growth and project effectiveness.
- Women-headed households represent the majority of the poor worldwide. U.S. development programs, which aim to reduce poverty, should logically center on women.
- Despite economic evidence, evaluation results, and directives from Congress, U.S. development assistance programs have largely ignored gender integration.

Over the past 30 years, study after study by academics, development practitioners, and international agencies has demonstrated the seemingly self-evident fact that women are equal to men, and sometimes surpass men, in contributing to social and economic development.

Researchers have also documented the significant economic dividends of investing in women and girls. Studies conducted by the World Bank, United Nations, and various academics have shown that discrimination against women and girls in education, health care, financial services, and human rights dampens overall economic output, productivity,

and growth rates. One World Bank report found that gender inequality in education and employment suppresses Africa's annual per capita growth by 0.8%.

Beyond direct economic impacts, women's increased access to education, health care, and human rights brings a "virtuous" cycle of enhanced child health, improved food production, lower population growth rates, higher incomes, and, of course, better quality of life for women themselves.

In addition to undermining women's potential, discrimination and low status have relegated many women and their children to the ranks of the poor. Women-headed households make up a majority of the poorest of the poor both in developed and developing countries. More than 900 million women live on less than one dollar a day, and the number of rural women living in absolute poverty has risen by 50% over the past 20 years, as opposed to 30% for men.

Advocates, academics, and development practitioners have been working hard for more than thirty years to integrate gender roles—that is, the different roles males and females play in a society—into American aid policy and programming. Yet, despite the evidence that women are active in national development and that investing in women and girls yields a multitude of benefits, U.S. international assistance programs and policy have not caught up with the facts.

In 1970, the women-in-development movement was crystallized by Ester Boserup's groundbreaking book: Women's Roles in Economic Development. In her book, she debunked the myth that women are not economic actors, brought to light the extent to which the economies of poor countries are propelled by women, and asserted that programs that considered women's roles would lead to greater contributions to development.

In 1973, Congress passed an amendment that, for the first time, explicitly addressed women's roles in the development process. The Percy Amendment (after its sponsor, Senator Charles Percy) is still in effect. It requires U.S. bilateral assistance programs to enhance the integration of women into the national economies of developing countries, and it instructs the State Department to consider progress on women's issues when making decisions about funding international organizations (e.g., United Nations, World Bank). In 1974, the United States Agency for International Development (USAID) established the Office of Women in Development to assist USAID missions and regional bureaus in integrating women into their various projects in the field.

In 1993, the Government Accounting Office evaluated USAID'S progress in meeting the requirements of the Percy Amendment. The report found that USAID "has only recently begun to consider the role of women in its third-world development strategies, despite the fact that 20 years have passed since Congress directed that AID assistance programs focus on integrating women."

By 1995, First Lady Hillary Rodham Clinton's leadership as head of the U.S. delegation to the UN Conference on Women in Beijing created a flurry of activity within USAID. One outcome was the creation of the Gender Plan of Action (GPA) in 1996, a three-step plan for the total integration of gender dynamics into all USAID activities. This plan was significant in its willingness to use mechanisms that really matter—bids and contracting systems, performance evaluations and promotions, and USAID's annual "Results Review and Resource Request" process—to ensure real change on programming with a gender perspective.

Four years later, the Advisory Committee on Voluntary Foreign Assistance (ACVFA), an independent adviser to the USAID administrator, commissioned an in-depth analysis—including over 500 interviews—of the Gender Plan of Action. The summary report states: "Over 90% of those interviewed in USAID and the PVO/NGO community said that the GPA has not had any measurable impact on Agency operations." This was not due to faults in the plan; it was because the plan was never promoted or implemented by the agency's leadership.

Ritu Sharma is cofounder and executive director of Women's EDGE.

From *Foreign Policy In Focus Brief*, September 17, 2001, p. 1. Excerpted from Ritu Sharma's "Women and Development Aid" *Foreign Policy in Focus*, Vol. 6, No. 33, September 2001. © 2001 by Interhemispheric Resource Center. Reprinted by permission.

Women Waging Peace

You can't end wars simply by declaring peace. "Inclusive security" rests on the principle that fundamental social changes are necessary to prevent renewed hostilities. Women have proven time and again their unique ability to bridge seemingly insurmountable divides. So why aren't they at the negotiating table?

By Swanee Hunt and Cristina Posa

Allowing men who plan wars to plan peace is a bad habit. But international negotiators and policymakers can break that habit by including peace promoters, not just warriors, at the negotiating table. More often than not, those peace promoters are women. Certainly, some extraordinary men have changed the course of history with their peacemaking; likewise, a few belligerent women have made it to the top of the political ladder or, at the grass-roots level, have taken the roles of suicide bombers or soldiers. Exceptions aside, however, women are often the most powerful voices for moderation in times of conflict. While most men come to the negotiating table directly from the war room and battlefield, women usually arrive straight out of civil activism and—take a deep breath—family care.

Yet, traditional thinking about war and peace either ignores women or regards them as victims. This oversight costs the world dearly. The wars of the last decade have gripped the public conscience largely because civilians were not merely caught in the crossfire; they were targeted, deliberately and brutally, by military strategists. Just as warfare has become "inclusive"—with civilian deaths more common than soldiers'—so too must our approach toward ending conflict. Today, the goal is not simply the absence of war, but the creation of sustainable peace by fostering fundamental societal changes. In this respect, the United States and other countries could take a lesson from Canada, whose innovative "human security" initiative— by making human beings and their communities, rather than states, its point of reference—focuses on safety and protection, particularly of the most vulnerable segments of a population.

The concept of "inclusive security," a diverse, citizen-driven approach to global stability, emphasizes women's agency, not their vulnerability. Rather than motivated by gender fairness, this concept is driven by efficiency: Women are crucial to inclusive security since they are often at the center of nongovernmental organizations (NGOs), popular protests, electoral referendums, and other citizen-empowering movements whose influence has grown with the global spread of democracy. An inclusive security approach expands the array of tools available to police, military, and diplomatic structures by adding collaboration with local efforts to achieve peace. Every effort to bridge divides, even if unsuccessful, has value, both in lessons learned and links to be built on later. Local actors with crucial experience resolving conflicts, organizing political movements, managing relief efforts, or working with military forces bring that experience into ongoing peace processes.

International organizations are slowly recognizing the indispensable role that women play in preventing war and sustaining peace. On October 31, 2000, the United Nations Security Council issued Resolution 1325 urging the secretary-general to expand the role of women in U.N. field-based operations, especially among military observers, civilian police, human rights workers, and humanitarian personnel. The Organization for Security and Co-operation in Europe (OSCE) is working to move women off the gender sidelines and into the everyday activities of the organization—particularly in the Office for Democratic Institutions and Human Rights, which has been useful in monitoring elections and human rights throughout Europe and the former Soviet Union. Last November, the European Parliament passed a hard-hitting resolution calling on European Union members (and the European Commission and Council) to promote the equal participation of women in diplomatic conflict resolution; to ensure that women fill at least 40 percent of all reconciliation, peacekeeping, peace-enforcement, peace-building, and conflict-prevention posts; and to support the creation and strengthening of NGOs (including women's organizations) that focus on conflict prevention, peace building, and post-conflict reconstruction.

Ironically, women's status as second-class citizens is a source of empowerment, since it has made women adept at finding innovative ways to cope with problems.

But such strides by international organizations have done little to correct the deplorable extent to which local women have been relegated to the margins of police, military, and diplomatic efforts. Consider that Bosnian women were not invited to participate in the Dayton talks, which ended the war in Bosnia, even though during the conflict 40 women's associations remained organized and active across ethnic lines. Not surprisingly, this exclusion has subsequently characterized—and undermined—the implementation of the Dayton accord. During a 1997 trip to Bosnia, U.S. President Bill Clinton, Secretary of State Madeleine Albright, and National Security Advisor Samuel Berger had a miserable meeting with intransigent politicians elected under the ethnic-based requirements of Dayton. During the same period, First Lady Hillary Rodham Clinton engaged a dozen women from across the country who shared story after story of their courageous and remarkably effective work to restore their communities. At the end of the day, a grim Berger faced the press, offering no encouraging word from the meetings with the political dinosaurs. The first lady's meeting with the energetic women activists was never mentioned.

We can ignore women's work as peacemakers, or we can harness its full force across a wide range of activities relevant to the security sphere: bridging the divide between groups in conflict, influencing local security forces, collaborating with international organizations, and seeking political office.

BRIDGING THE DIVIDE

The idea of women as peacemakers is not political correctness run amok. Social science research supports the stereotype of women as generally more collaborative than men and thus more inclined toward consensus and compromise. Ironically, women's status as second-class citizens is a source of empowerment, since it has made women adept at finding innovative ways to cope with problems. Because women are not ensconced within the mainstream, those in power consider them less threatening, allowing women to work unimpeded and "below the radar screen." Since they usually have not been behind a rifle, women, in contrast to men, have less psychological distance to reach across a conflict line. (They are also more accepted on the "other side," because it is assumed that they did not do any of the actual killing.) Women often choose an identity, notably that of mothers, that cuts across international borders and ethnic enclaves. Given their roles as family nurturers,

women have a huge investment in the stability of their communities. And since women know their communities, they can predict the acceptance of peace initiatives, as well as broker agreements in their own neighborhoods.

As U.N. Secretary-General Kofi Annan remarked in October 2000 to the Security Council, "For generations, women have served as peace educators, both in their families and in their societies. They have proved instrumental in building bridges rather than walls." Women have been able to bridge the divide even in situations where leaders have deemed conflict resolution futile in the face of so-called intractable ethnic hatreds. Striking examples of women making the impossible possible come from Sudan, a country splintered by decades of civil war. In the south, women working together in the New Sudan Council of Churches conducted their own version of shuttle diplomacy—perhaps without the panache of jetting between capitals—and organized the Wunlit tribal summit in February 1999 to bring an end to bloody hostilities between the Dinka and Nuer peoples. As a result, the Wunlit Covenant guaranteed peace between the Dinka and the Nuer, who agreed to share rights to water, fishing, and grazing land, which had been key points of disagreement. The covenant also returned prisoners and guaranteed freedom of movement for members of both tribes.

On another continent, women have bridged the seemingly insurmountable differences between India and Pakistan by organizing huge rallies to unite citizens from both countries. Since 1994, the Pakistan-India People's Forum for Peace and Democracy has worked to overcome the hysterics of the nationalist media and jingoistic governing elites by holding annual conventions where Indians and Pakistanis can affirm their shared histories, forge networks, and act together on specific initiatives. In 1995, for instance, activists joined forces on behalf of fishers and their children who were languishing in each side's jails because they had strayed across maritime boundaries. As a result, the adversarial governments released the prisoners and their boats.

In addition to laying the foundation for broader accords by tackling the smaller, everyday problems that keep people apart, women have also taken the initiative in drafting principles for comprehensive settlements. The platform of Jerusalem Link, a federation of Palestinian and Israeli women's groups, served as a blueprint for negotiations over the final status of Jerusalem during the Oslo process. Former President Clinton, the week of the failed Camp David talks in July 2000, remarked simply, "If we'd had women at Camp David, we'd have an agreement."

Sometimes conflict resolution requires unshackling the media. Journalists can nourish a fair and tolerant vision of society or feed the public poisonous, one-sided, and untruthful accounts of the "news" that stimulate violent conflict. Supreme Allied Commander of Europe Wesley Clark understood as much when he ordered NATO to bomb transmitters in Kosovo to prevent the Milosevic media machine from spewing ever more inflammatory rhetoric. One of the founders of the independent Kosovo radio station RTV-21 realized that there were "many instances of male colleagues reporting with anger, which served to raise the tensions rather than lower them." As a result, RTV-

21 now runs workshops in radio, print, and TV journalism to cultivate a core of female journalists with a noninflammatory style. The OSCE and the BBC, which train promising local journalists in Kosovo and Bosnia, would do well to seek out women, who generally bring with them a reputation for moderation in unstable situations.

Nelson Mandela suggested at last summer's Arusha peace talks that if Burundian men began fighting again, their women should withhold "conjugal rights" (like cooking, he added).

INFLUENCING SECURITY FORCES

The influence of women on warriors dates back to the ancient Greek play *Lysistrata*. Borrowing from that play's story, former South African President Nelson Mandela suggested at last summer's Arusha peace talks on the conflict in Burundi that if Burundian men began fighting again, their women should withhold "conjugal rights" (like cooking, he added).

Women can also act as a valuable interface between their countries' security forces (police and military) and the public, especially in cases when rapid response is necessary to head off violence. Women in Northern Ireland, for example, have helped calm the often deadly "marching season" by facilitating mediations between Protestant unionists and Catholic nationalists. The women bring together key members of each community, many of whom are released prisoners, as mediators to calm tensions. This circle of mediators works with local police throughout the marching season, meeting quietly and maintaining contacts on a 24-hour basis. This intervention provides a powerful extension of the limited tools of the local police and security forces.

Likewise, an early goal of the Sudanese Women's Voice for Peace was to meet and talk with the military leaders of the various rebel armies. These contacts secured women's access to areas controlled by the revolutionary movements, a critical variable in the success or failure of humanitarian efforts in war zones. Women have also worked with the military to search for missing people, a common element in the cycle of violence. In Colombia, for example, women were so persistent in their demands for information regarding 150 people abducted from a church in 1999 that the army eventually gave them space on a military base for an information and strategy center. The military worked alongside the women and their families trying to track down the missing people. In short, through moral suasion, local women often have influence where outsiders, such as international human rights agencies, do not.

That influence may have allowed a female investigative reporter like Maria Cristina Caballero to go where a man could not go, venturing on horseback alone, eight hours into the jungle to tape a four-hour interview with the head of the paramilitary forces in Colombia. She also interviewed another guerilla leader and published an award-winning comparison of the transcripts, showing where the two mortal enemies shared the same vision. "This [was] bigger than a story," she later said, "this [was] hope for peace." Risking their lives to move back and forth across the divide, women like Caballero perform work that is just as important for regional stabilization as the grandest Plan Colombia.

INTERNATIONAL COLLABORATION

Given the nature of "inclusive" war, security forces are increasingly called upon to ensure the safe passage of humanitarian relief across conflict zones. Women serve as indispensable contacts between civilians, warring parties, and relief organizations. Without women's knowledge of the local scene, the mandate of the military to support NGOs would often be severely hindered, if not impossible.

In rebel-controlled areas of Sudan, women have worked closely with humanitarian organizations to prevent food from being diverted from those who need it most. According to Catherine Loria Duku Jeremano of Oxfam: "The normal pattern was to hand out relief to the men, who were then expected to take it home to be distributed to their family. However, many of the men did what they pleased with the food they received: either selling it directly, often in exchange for alcohol, or giving food to the wives they favored." Sudanese women worked closely with tribal chiefs and relief organizations to establish a system allowing women to pick up the food for their families, despite contrary cultural norms.

In Pristina, Kosovo, Vjosa Dobruna, a pediatric neurologist and human rights leader, is now the joint administrator for civil society for the U.N. Interim Administration Mission in Kosovo (UNMIK). In September 2000, at the request of NATO, she organized a multiethnic strategic planning session to integrate women throughout UNMIK. Before that gathering, women who had played very significant roles in their communities felt shunned by the international organizations that descended on Kosovo following the bombing campaign. Vjosa's conference pulled them back into the mainstream, bringing international players into the conference to hear from local women what stabilizing measures they were planning, rather than the other way around. There, as in Bosnia, the OSCE has created a quota system for elected office, mandating that women comprise one third of each party's candidate list; leaders like Vjosa helped turn that policy into reality.

In addition to helping aid organizations find better ways to distribute relief or helping the U.N. and OSCE implement their ambitious mandates, women also work closely with them to locate and exchange prisoners of war. As the peace processes in Northern Ireland, Bosnia, and the Middle East illustrate, a deadlock on the exchange and release of prisoners can be a major obstacle to achieving a final settlement. Women activists in Armenia and Azerbaijan have worked closely with the International Helsinki Citizens Assembly and the OSCE for the release

The Black and the Green

Grass-roots women's organizations in Israel come in two colors: black and green. The Women in Black, founded in 1988, and the Women in Green, founded in 1993, could not be further apart on the political spectrum, but both claim the mantle of "womanhood" and "motherhood" in the ongoing struggle to end the Israeli-Palestinian conflict.

One month after the Palestinian intifada broke out in December 1988, a small group of women decided to meet every Friday afternoon at a busy Jerusalem intersection wearing all black and holding hand-shaped signs that read: "Stop the Occupation." The weekly gatherings continued and soon spread across Israel to Europe, the United States, and then to Asia.

While the movement was originally dedicated to achieving peace in the Middle East, other groups soon protested against repression in the Balkans and India. For these activists, their status as women lends them a special authority when it comes to demanding peace. In the words of the

Asian Women's Human Rights Council: "We are the Women in Black... women, unmasking the many horrific faces of more public 'legitimate' forms of violence—state repression, communalism, ethnic cleansing, nationalism, and wars...."

Today, the Women in Black in Israel continue their nonviolent opposition to the occupation in cooperation with the umbrella group Coalition of Women for a Just Peace. They have been demonstrating against the closures of various Palestinian cities, arguing that the blockades prevent pregnant women from accessing healthcare services and keep students from attending school. The group also calls for the full participation of women in peace negotiations.

While the Women in Black stood in silent protest worldwide, a group of "grandmothers, mothers, wives, and daughters; housewives and professionals; secular and religious" formed the far-right Women in Green in 1993 out of "a shared love, devotion and concern for Israel." Known for the signature

green hats they wear at rallies, the Women in Green emerged as a protest to the Oslo accords on the grounds that Israel made too many concessions to Yasir Arafat's Palestinian Liberation Organization. The group opposes returning the Golan Heights to Syria, sharing sovereignty over Jerusalem with the Palestinians, and insists that "Israel remain a Jewish state."

The Women in Green boast some 15,000 members in Israel, and while they have not garnered the global support of the Women in Black, 15,000 Americans have joined their cause. An ardent supporter of Israeli Prime Minister Ariel Sharon, the group seeks to educate the Israeli electorate through weekly street theater and public demonstrations, as well as articles, posters, and newspaper advertisements.

While the groups' messages and methods diverge, their existence and influence demonstrate that women can mobilize support for political change—no matter what color they wear.

—*FP*

of hostages in the disputed region of Nagorno-Karabakh, where tens of thousands of people have been killed. In fact, these women's knowledge of the local players and the situation on the ground would make them indispensable in peace negotiations to end this 13-year-old conflict.

REACHING FOR POLITICAL OFFICE

In 1977, women organizers in Northern Ireland won the Nobel Peace Prize for their nonsectarian public demonstrations. Two decades later, Northern Irish women are showing how diligently women must still work not only to ensure a place at the negotiating table but also to sustain peace by reaching critical mass in political office. In 1996, peace activists Monica McWilliams (now a member of the Northern Ireland Assembly) and May Blood (now a member of the House of Lords) were told that only leaders of the top 10 political parties—all men—would be included in the peace talks. With only six weeks to organize, McWilliams and Blood gathered 10,000 signatures to create a new political party (the Northern Ireland Women's Coalition, or NIWC) and got themselves on the ballot. They were voted into the top 10 and earned a place at the table.

The grass-roots, get-out-the-vote work of Vox Femina convinced hesitant Yugoslav women to vote for change; those votes contributed to the margin that ousted President Slobodan Milosevic.

The NIWC's efforts paid off. The women drafted key clauses of the Good Friday Agreement regarding the importance of mixed housing, the particular difficulties of young people, and the need for resources to address these problems. The NIWC also lobbied for the early release and reintegration of political prisoners in order to combat social exclusion and pushed for a comprehensive review of the police service so that all members of society would accept it. Clearly, the women's prior work with individuals and families affected by "the Troubles" enabled them to formulate such salient contributions to the agreement. In the subsequent public referendum on the Good Friday Agreement, Mo Mowlam, then British secretary of state for Northern Ireland, attributed the overwhelming success of the YES Campaign to the NIWC's persistent canvassing and lobbying.

Women in the former Yugoslavia are also stepping forward to wrest the reins of political control from extremists (including women, such as ultranationalist Bosnian Serb President Biljana Plavsic) who destroyed their country. Last December, Zorica Trifunovic, founding member of the local Women in Black (an antiwar group formed in Belgrade in October 1991), led a meeting that united 90 women leaders of pro-democracy political campaigns across the former Yugoslavia. According to polling by the National Democratic Institute, the grass-roots, get-out-the-vote work of groups such as Vox Femina (a local NGO that participated in the December meeting) convinced hesitant women to vote for change; those votes contributed to the margin that ousted President Slobodan Milosevic.

International security forces and diplomats will find no better allies than these mobilized mothers, who are tackling the toughest, most hardened hostilities.

Argentina provides another example of women making the transition from protesters to politicians: Several leaders of the Madres de la Plaza de Mayo movement, formed in the 1970s to protest the "disappearances" of their children at the hands of the military regime, have now been elected to political office. And in Russia, the Committee of Soldiers' Mothers—a protest group founded in 1989 demanding their sons' rights amidst cruel conditions in the Russian military—has grown into a powerful organization with 300 chapters and official political status. In January, U.S. Ambassador to Moscow Jim Collins described the committee as a significant factor in countering the most aggressive voices promoting military force in Chechnya. Similar mothers' groups have sprung up across the former Soviet Union and beyond—including the Mothers of Tiananmen Square. International security forces and diplomats will find no better allies than these mobilized mothers, who are tackling the toughest, most hardened hostilities.

YOU'VE COME A LONG WAY, MAYBE

Common sense dictates that women should be central to peacemaking, where they can bring their experience in conflict resolution to bear. Yet, despite all of the instances where women have been able to play a role in peace negotiations, women remain relegated to the sidelines. Part of the problem is structural: Even though more and more women are legislators and soldiers, underrepresentation persists in the highest levels of political and military hierarchies. The presidents, prime ministers, party leaders, cabinet secretaries, and generals who typically negotiate peace settlements are overwhelmingly men. There is also a psychological barrier that precludes women from sitting in on negotiations: Waging war is still thought of as a "man's job," and as such, the task of stopping war often is delegated to men

(although if we could begin to think about the process not in terms of stopping war but promoting peace, women would emerge as the more logical choice). But the key reason behind women's marginalization may be that everyone recognizes just how good women are at forging peace. A U.N. official once stated that, in Africa, women are often excluded from negotiating teams because the war leaders "are afraid the women will compromise" and give away too much.

Some encouraging signs of change, however, are emerging. Rwandan President Paul Kagame, dismayed at his difficulty in attracting international aid to his genocide-ravaged country, recently distinguished Rwanda from the prevailing image of brutality in central Africa by appointing three women to his negotiating team for the conflict in the Democratic Republic of the Congo. In an unusually healthy tit for tat, the Ugandans responded by immediately appointing a woman to their team.

Will those women make a difference? Negotiators sometimes worry that having women participate in the discussion may change the tone of the meeting. They're right: A British participant in the Northern Ireland peace talks insightfully noted that when the parties became bogged down by abstract issues and past offenses, "the women would come and talk about their loved ones, their bereavement, their children and their hopes for the future." These deeply personal comments, rather than being a diversion, helped keep the talks focused. The women's experiences reminded the parties that security for all citizens was what really mattered.

The role of women as peacemakers can be expanded in many ways. Mediators can and should insist on gender balance among negotiators to ensure a peace plan that is workable at the community level. Cultural barriers can be overcome if high-level visitors require that a critical mass (usually one third) of the local interlocutors be women (and not simply present as wives). When drafting principles for negotiation, diplomats should determine whether women's groups have already agreed upon key conflict-bridging principles, and whether their approach can serve as a basis for general negotiations.

Moreover, to foster a larger pool of potential peacemakers, embassies in conflict areas should broaden their regular contact with local women leaders and sponsor women in training programs, both at home and abroad. Governments can also do their part by providing information technology and training to women activists through private and public partnerships. Internet communication allows women peace builders to network among themselves, as well as exchange tactics and strategies with their global counterparts.

"Women understood the cost of the war and were genuinely interested in peace," recalls retired Admiral Jonathan Howe, reflecting on his experience leading the U.N. mission in Somalia in the early 1990s. "They'd had it with their warrior husbands. They were a force willing to say enough is enough. The men were sitting around talking and chewing qat, while the women were working away. They were such a positive force.... You have to look at all elements in society and be ready to tap into those that will be constructive."

Want to Know More?

The Internet is invaluable in enabling the inclusive security approach advocated in this article. The Web offers not only a wealth of information but, just as important, relatively cheap and easy access for citizens worldwide. Most of the women's peace-building activities and strategies explored in this article can be found on the Web site of **Women Waging Peace**—a collaborative venture of Harvard University's John F. Kennedy School of Government and the nonprofit organization Hunt Alternatives, which recognize the essential role and contribution of women in preventing violent conflict, stopping war, reconstructing ravaged societies, and sustaining peace in fragile areas around the world. On the site, women active in conflict areas can communicate with each other without fear of retribution via a secure server. The women submit narratives detailing their strategies, which can then be read on the public Web site. The site also features a video archive of interviews with each of these women. You need a password to view these interviews, so contact Women Waging Peace online or call (617) 868-3910.

The Organization for Security and Co-operation in Europe (OSCE) is an outstanding resource for qualitative and quantitative studies of women's involvement in conflict prevention. Start with the final report of the *OSCE Supplementary Implementation Meeting: Gender Issues* (Vienna: UNIFEM, 1999), posted on the group's Web site. **The United Nations Development Fund for Women** (UNIFEM) also publishes reports on its colorful and easy-to-navigate site. The fund's informative book, *Women at the Peace Table: Making a Difference* (New York: UNIFEM, 2000), available online, features interviews with some of today's most prominent women peacemakers, including Hanan Ashrawi and Mo Mowlam.

For a look at how globalization is changing women's roles in governments, companies, and militaries, read Cynthia Enloe's *Bananas, Beaches and Bases: Making Feminist Sense of International Politics* (Berkeley: University of California Press, 2001). In *Maneuvers: The International Politics of Militarizing Women's Lives* (Berkeley: University of California Press, 2000), Enloe examines the military's effects on women, whether they are soldiers or soldiers' spouses. For a more general discussion of where feminism fits into academia and policymaking, see **"Searching for the Princess? Feminist Perspectives in International Relations"** (*The Harvard International Review*, Fall 1999) by J. Ann Tickner, associate professor of international relations at the University of Southern California.

The Fall 1997 issue of FOREIGN POLICY magazine features two articles that highlight how women worldwide are simultaneously gaining political clout but also bearing the brunt of poverty: **"Women in Power: From Tokenism to Critical Mass"** by Jane S. Jaquette and **"Women in Poverty: A New Global Underclass"** by Mayra Buvinic.

• For links to relevant Web sites, as well as a comprehensive index of related FOREIGN POLICY articles, access **www.foreign policy.com**.

Lasting peace must be homegrown. Inclusive security helps police forces, military leaders, and diplomats do their jobs more effectively by creating coalitions with the people most invested in stability and most adept at building peace. Women working on the ground are eager to join forces. Just let them in.

Swanee Hunt is director of the Women in Public Policy Program at Harvard University's John F. Kennedy School of Government. As the United States' ambassador to Austria (1993–97), she founded the "Vital Voices: Women in Democracy" initiative. Cristina Posa, a former judicial clerk at the United Nations International Criminal Tribunal for the former Yugoslavia, is an attorney at Cleary, Gottlieb, Steen & Hamilton in New York.

The True Clash of Civilizations

Samuel Huntington was only half right. The cultural fault line that divides the West and the Muslim world is not about democracy but sex. According to a new survey, Muslims and their Western counterparts want democracy, yet they are worlds apart when it comes to attitudes toward divorce, abortion, gender equality, and gay rights—which may not bode well for democracy's future in the Middle East.

By Ronald Inglehart and Pippa Norris

Democracy promotion in Islamic countries is now one of the Bush administration's most popular talking points. "We reject the condescending notion that freedom will not grow in the Middle East," Secretary of State Colin Powell declared last December as he unveiled the White House's new Middle East Partnership Initiative to encourage political and economic reform in Arab countries. Likewise, Condoleezza Rice, President George W. Bush's national security advisor, promised last September that the United States is committed to "the march of freedom in the Muslim world."

Republican Rep. Christopher Shays of Connecticut: "Why doesn't democracy grab hold in the Middle East? What is there about the culture and the people and so on where democracy just doesn't seem to be something they strive for and work for?"

But does the Muslim world march to the beat of a different drummer? Despite Bush's optimistic pronouncement that there is "no clash of civilizations" when it comes to "the common rights and needs of men and women," others are not so sure. Samuel Huntington's controversial 1993 thesis—that the cultural division between "Western Christianity" and "Orthodox Christianity and Islam" is the new fault line for conflict—resonates more loudly than ever since September 11. Echoing Huntington, columnist Polly Toynbee argued in the British *Guardian* last November, "What binds together a globalized force of some extremists from many continents is a united hatred of Western values that seems to them to spring from Judeo-Christianity." Meanwhile, on the other side of the Atlantic, Republican Rep. Christopher Shays of Connecticut, after sitting through hours of testimony on U.S.-Islamic relations on Capitol Hill last October, testily blurted, "Why doesn't democracy grab hold in the Middle East? What is there about the culture and the people and so on where democracy just doesn't seem to be something they strive for and work for?"

Huntington's response would be that the Muslim world lacks the core political values that gave birth to representative democracy in Western civilization: separation of religious and secular

The Cultural Divide

Approval of Political and Social Values in Western and Muslim Societies

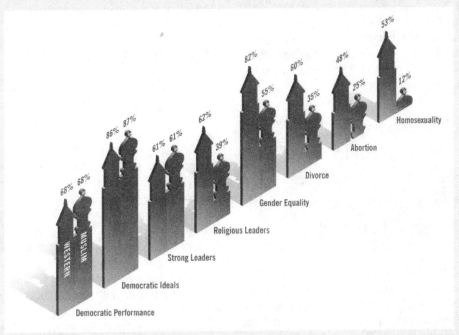

SOURCE: WORLD VALUES SERVEY, POOLED SAMPLE 1995-2001; CHARTS (3) BY JARED SCHNEIDMAN FOR FP

The chart above draws on responses to various political and social issues in the World Values Survey. The percentages indicate the extent to which respondents agree/disagree with or approved/disapproved of the following statements and questions:

DEMOCRATIC PERFORMANCE

- Democracies are indecisive and have too much quibbling. (Strongly disagree.)
- Democracies aren't good at maintaining order. (Strongly disagree.)

DEMOCRATIC IDEALS

- Democracy may have problems, but it's better than any other form of government. (Strongly agree.)
- Approve of having a democratic political system. (Strongly agree.)

STRONG LEADERS

- Approve of having experts, not government, make decisions according to what they think is best for the country. (Strongly disagree.)
- Approve of having a strong leader who does not have to bother with parliament and elections. (Strongly disagree.)

RELIGIOUS LEADERS

- Politicians who do not believe in God are unfit for public office. (Strongly disagree.)
- It would be better for [this country] if more people with strong religious beliefs held public office. (Strongly disagree.)

GENDER EQUALITY

- On the whole, men make better political leaders than women do. (Strongly disagree.)
- When jobs are scarce, men should have more right to a job than women. (Strongly disagree.)
- A university education is more important for a boy than for a girl. (Strongly disagree.)
- A woman has to have children in order to be fulfilled. (Strongly disagree.)
- If a woman wants to have a child as a single parent but she doesn't want to have a stable relationship with a man, do you approve or disapprove? (Strongly approve.)

DIVORCE

- Divorce can always be justified, never be justified, or something in between. (High level of tolerance for divorce.)

ABORTION

- Abortion can always be justified, never be justified, or something in between. (High level of tolerance for abortion.)

HOMOSEXUALITY

- Homosexuality can always be justified, never be justified, or something in between. (High level of tolerance for homosexuality.)

authority, rule of law and social pluralism, parliamentary institutions of representative government, and protection of individual rights and civil liberties as the buffer between citizens and the power of the state. This claim seems all too plausible given the failure of electoral democracy to take root throughout the Middle East and North Africa. According to the latest Freedom House rankings, almost two thirds of the 192 countries around the world are now electoral democracies. But among the 47 countries with a Muslim majority, only one fourth are electoral democracies—and none of the core Arabic-speaking societies falls into this category.

> ... the real fault line between the West and Islam... concerns gender equality and sexual liberation... the values separating the two cultures have much more to do with eros than demos.

Yet this circumstantial evidence does little to prove Huntington correct, since it reveals nothing about the underlying beliefs of Muslim publics. Indeed, there has been scant empirical evidence whether Western and Muslim societies exhibit deeply divergent values—that is, until now. The cumulative results of the two most recent waves of the World Values Survey (wvs), conducted in 1995–96 and 2000–2002, provide an extensive body of relevant evidence. Based on questionnaires that explore values and beliefs in more than 70 countries, the wvs is an investigation of sociocultural and political change that encompasses over 80 percent of the world's population.

A comparison of the data yielded by these surveys in Muslim and non-Muslim societies around the globe confirms the first claim in Huntington's thesis: Culture does matter—indeed, it matters a lot. Historical religious traditions have left an enduring imprint on contemporary values. However, Huntington is mistaken in assuming that the core clash between the West and Islam is over political values. At this point in history, societies throughout the world (Muslim and Judeo-Christian alike) see democracy as the best form of government. Instead, the real fault line between the West and Islam, which Huntington's theory completely overlooks, concerns gender equality and sexual liberalization. In other words, the values separating the two cultures have much more to do with eros than demos. As younger generations in the West have gradually become more liberal on these issues, Muslim nations have remained the most traditional societies in the world.

This gap in values mirrors the widening economic divide between the West and the Muslim world. Commenting on the disenfranchisement of women throughout the Middle East, the United Nations Development Programme observed last summer that "no society can achieve the desired state of well-being and human development, or compete in a globalizing world, if half its people remain marginalized and disempowered." But this "sexual clash of civilizations" taps into far deeper issues than

how Muslim countries treat women. A society's commitment to gender equality and sexual liberalization proves time and again to be the most reliable indicator of how strongly that society supports principles of tolerance and egalitarianism. Thus, the people of the Muslim world overwhelmingly want democracy, but democracy may not be sustainable in their societies.

TESTING HUNTINGTON

Huntington argues that "ideas of individualism, liberalism, constitutionalism, human rights, equality, liberty, the rule of law, democracy, free markets, [and] the separation of church and state" often have little resonance outside the West. Moreover, he holds that Western efforts to promote these ideas provoke a violent backlash against "human rights imperialism." To test these propositions, we categorized the countries included in the wvs according to the nine major contemporary civilizations, based largely on the historical religious legacy of each society. The survey includes 22 countries representing Western Christianity (a West European culture that also encompasses North America, Australia, and New Zealand), 10 Central European nations (sharing a Western Christian heritage, but which also lived under Communist rule), 11 societies with a Muslim majority (Albania, Algeria, Azerbaijan, Bangladesh, Egypt, Indonesia, Iran, Jordan, Morocco, Pakistan, and Turkey), 12 traditionally Orthodox societies (such as Russia and Greece), 11 predominately Catholic Latin American countries, 4 East Asian societies shaped by Sino-Confucian values, 5 sub-Saharan Africa countries, plus Japan and India.

Despite Huntington's claim of a clash of civilizations between the West and the rest, the wvs reveals that, at this point in history, democracy has an overwhelmingly positive image throughout the world. In country after country, a clear majority of the population describes "having a democratic political system" as either "good" or "very good." These results represent a dramatic change from the 1930s and 1940s, when fascist regimes won overwhelming mass approval in many societies; and for many decades, Communist regimes had widespread support. But in the last decade, democracy became virtually the only political model with global appeal, no matter what the culture. With the exception of Pakistan, most of the Muslim countries surveyed think highly of democracy: In Albania, Egypt, Bangladesh, Azerbaijan, Indonesia, Morocco, and Turkey, 92 to 99 percent of the public endorsed democratic institutions—a higher proportion than in the United States (89 percent).

Yet, as heartening as these results may be, paying lip service to democracy does not necessarily prove that people genuinely support basic democratic norms—or that their leaders will allow them to have democratic institutions. Although constitutions of authoritarian states such as China profess to embrace democratic ideals such as freedom of religion, the rulers deny it in practice. In Iran's 2000 elections, reformist candidates captured nearly three quarters of the seats in parliament, but a theocratic elite still holds the reins of power. Certainly, it's a step in the right direction if most people in a country endorse the idea of democracy. But this sentiment needs to be complemented by

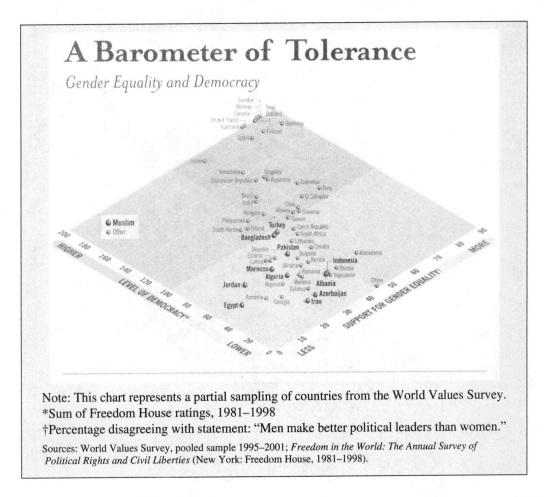

A Barometer of Tolerance

Gender Equality and Democracy

Note: This chart represents a partial sampling of countries from the World Values Survey.
*Sum of Freedom House ratings, 1981–1998
†Percentage disagreeing with statement: "Men make better political leaders than women."

Sources: World Values Survey, pooled sample 1995–2001; *Freedom in the World: The Annual Survey of Political Rights and Civil Liberties* (New York: Freedom House, 1981–1998).

deeper underlying attitudes such as interpersonal trust and tolerance of unpopular groups—and these values must ultimately be accepted by those who control the army and secret police.

The wvs reveals that, even after taking into account differences in economic and political development, support for democratic institutions is just as strong among those living in Muslim societies as in Western (or other) societies [see chart, The Cultural Divide]. For instance, a solid majority of people living in Western and Muslim countries gives democracy high marks as the most efficient form of government, with 68 percent disagreeing with assertions that "democracies are indecisive" and "democracies aren't good at maintaining order." (All other cultural regions and countries, except East Asia and Japan, are far more critical.) And an equal number of respondents on both sides of the civilizational divide (61 percent) firmly reject authoritarian governance, expressing disapproval of "strong leaders" who do not "bother with parliament and elections." Muslim societies display greater support for religious authorities playing an active societal role than do Western societies. Yet this preference for religious authorities is less a cultural division between the West and Islam than it is a gap between the West and many other less secular societies around the globe, especially in sub-Saharan Africa and Latin America. For instance, citizens in some Muslim societies agree overwhelmingly with the statement that "politicians who do not believe in God are unfit for public office" (88 percent in Egypt, 83 percent in Iran, and 71 percent in Bangladesh), but this statement also garners

strong support in the Philippines (71 percent), Uganda (60 percent), and Venezuela (52 percent). Even in the United States, about two fifths of the public believes that atheists are unfit for public office.

Today, relatively few people express overt hostility toward other classes, races, or religions, but rejection of homosexuals is widespread. About half of the world's populations say that homosexuality is "never" justifiable.

However, when it comes to attitudes toward gender equality and sexual liberalization, the cultural gap between Islam and the West widens into a chasm. On the matter of equal rights and opportunities for women—measured by such questions as whether men make better political leaders than women or whether university education is more important for boys than for girls—Western and Muslim countries score 82 percent and 55 percent, respectively. Muslim societies are also distinctively less permissive toward homosexuality, abortion, and divorce.

These issues are part of a broader syndrome of tolerance, trust, political activism, and emphasis on individual autonomy

211

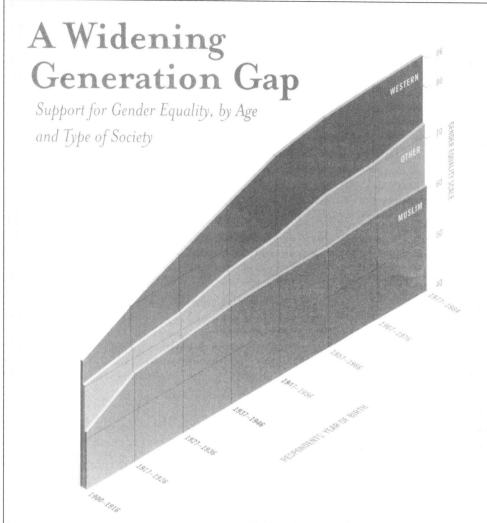

A Widening Generation Gap

Support for Gender Equality, by Age and Type of Society

* The 100-point Gender Equality Scale is based on responses to the following five statements and questions: "If a woman wants to have a child as a single parent but she doesn't want to have a stable relationship with a man, do you approve or disapprove?"; "When jobs are scarce, men should have more right to a job than women"; "A university education is more important for a boy than a girl"; "Do you think that a woman has to have children in order to be fulfilled or is this not necessary?"; and "On the whole, men make better political leaders than women do." The scale was constructed so that if all respondents show high scores on all five items (representing strong support for gender equality), it produces a score of 100, while low scores on all five items produce a score of 0.

Source: World Values Surveys, pooled 1995–2001

that constitutes "self-expression values." The extent to which a society emphasizes these self-expression values has a surprisingly strong bearing on the emergence and survival of democratic institutions. Among all the countries included in the wvs, support for gender equality—a key indicator of tolerance and personal freedom—is closely linked with a society's level of democracy [see chart, A Barometer of Tolerance].

In every stable democracy, a majority of the public disagrees with the statement that "men make better political leaders than women." None of the societies in which less than 30 percent of the public rejects this statement (such as Jordan, Nigeria, and Belarus) is a true democracy. In China, one of the world's least democratic countries, a majority of the public agrees that men make better political leaders than women, despite a party

line that has long emphasized gender equality (Mao Zedong once declared, "women hold up half the sky"). In practice, Chinese women occupy few positions of real power and face widespread discrimination in the workplace. India is a border-line case. The country is a long-standing parliamentary democracy with an independent judiciary and civilian control of the armed forces, yet it is also marred by a weak rule of law, arbitrary arrests, and extra-judicial killings. The status of Indian women reflects this duality. Women's rights are guaranteed in the constitution, and Indira Gandhi led the nation for 15 years. Yet domestic violence and forced prostitution remain prevalent throughout the country, and, according to the wvs, almost 50 percent of the Indian populace believes only men should run the government.

Want to Know More?

Samuel Huntington expanded his controversial 1993 article into a book, *The Clash of Civilizations and the Remaking of World Order* (New York: Simon and Schuster, 1996). Among the authors who have disputed Huntington's claim that Islam is incompatible with democratic values are Edward Said, who decries the clash of civilizations thesis as an attempt to revive the "good vs. evil" world dichotomy prevalent during the Cold War (**"A Clash of Ignorance,"** *The Nation*, October 22, 2001); John Voll and John Esposito, who argue that "The Muslim heritage… contains concepts that provide a foundation for contemporary Muslims to develop authentically Muslim programs of democracy" (**"Islam's Democratic Essence,"** *Middle East Quarterly*, September 1994); and Ray Takeyh, who recounts the efforts of contemporary Muslim scholars to legitimize democratic concepts through the reinterpretation of Muslim texts and traditions (**"Faith-Based Initiatives,"** FOREIGN POLICY, November/December 2001).

An overview of the Bush administration's **Middle East Partnership Initiative**, including the complete transcript of Secretary of State Colin Powell's speech on political and economic reform in the Arab world, can be found on the Web site of the U.S. Department of State. Marina Ottaway, Thomas Carothers, Amy Hawthorne, and Daniel Brumberg offer a stinging critique of those who believe that toppling the Iraqi regime could unleash a democratic tsunami in the Arab world in **"Democratic Mirage in the Middle East"** (Washington: Carnegie Endowment for International Peace, 2002).

In a poll of nearly 4,000 Arabs, James Zogby found that the issue of "civil and personal rights" earned the overall highest score when people were asked to rank their personal priorities (***What Arabs Think: Values, Beliefs and Concerns***, Washington: Zogby International, 2002). A poll available on the Web site of the Pew Research Center for the People and the Press (**"Among Wealthy Nations …U.S. Stands Alone in Its Embrace of Religion,"** December 19, 2002) reveals that Americans' views on religion and faith are closer to those living in developing nations than in developed countries.

The Web site of the **World Values Survey** (WVS) provides considerable information on the survey, including background on methodology, key findings, and the text of the questionnaires. The second iteration of the A.T. Kearney/FOREIGN POLICY Magazine Globalization Index (**"Globalization's Last Hurrah?"** FOREIGN POLICY, January/February 2002) found a strong correlation between the WVS measure of "subjective well-being" and a society's level of global integration.

For links to relevant Web sites, access to the FP Archive, and a comprehensive index of related FOREIGN POLICY articles, go to **www.foreignpolicy.com.**

> Muslim societies are neither uniquely nor monolithically low on tolerance toward sexual orientation and gender equality.… However, on the whole, Muslim countries not only lag behind the West but behind all other societies as well.

The way a society views homosexuality constitutes another good litmus test of its commitment to equality. Tolerance of well-liked groups is never a problem. But if someone wants to gauge how tolerant a nation really is, find out which group is the most disliked, and then ask whether members of that group should be allowed to hold public meetings, teach in schools, and work in government. Today, relatively few people express overt hostility toward other classes, races, or religions, but rejection of homosexuals is widespread. In response to a WVS question about whether homosexuality is justifiable, about half of the world's population say "never." But, as is the case with gender equality, this attitude is directly proportional to a country's level of democracy. Among authoritarian and quasi-democratic states, rejection of homosexuality is deeply entrenched: 99 per-

cent in both Egypt and Bangladesh, 94 percent in Iran, 92 percent in China, and 71 percent in India. By contrast, these figures are much lower among respondents in stable democracies: 32 percent in the United States, 26 percent in Canada, 25 percent in Britain, and 19 percent in Germany.

Muslim societies are neither uniquely nor monolithically low on tolerance toward sexual orientation and gender equality. Many of the Soviet successor states rank as low as most Muslim societies. However, on the whole, Muslim countries not only lag behind the West but behind all other societies as well [see chart, A Widening Generation Gap]. Perhaps more significant, the figures reveal the gap between the West and Islam is even wider among younger age groups. This pattern suggests that the younger generations in Western societies have become progressively more egalitarian than their elders, but the younger generations in Muslim societies have remained almost as traditional as their parents and grandparents, producing an expanding cultural gap.

CLASH OF CONCLUSIONS

"The peoples of the Islamic nations want and deserve the same freedoms and opportunities as people in every nation," President Bush declared in a commencement speech at West Point last summer. He's right. Any claim of a "clash of civilizations" based on fundamentally different political goals held by

Western and Muslim societies represents an oversimplification of the evidence. Support for the goal of democracy is surprisingly widespread among Muslim publics, even among those living in authoritarian societies. Yet Huntington is correct when he argues that cultural differences have taken on a new importance, forming the fault lines for future conflict. Although nearly the entire world pays lip service to democracy, there is still no global consensus on the self-expression values—such as social tolerance, gender equality, freedom of speech, and interpersonal trust—that are crucial to democracy. Today, these divergent values constitute the real clash between Muslim societies and the West.

But economic development generates changed attitudes in virtually any society. In particular, modernization compels systematic, predictable changes in gender roles: Industrialization brings women into the paid work force and dramatically reduces fertility rates. Women become literate and begin to participate in representative government but still have far less power than men. Then, the postindustrial phase brings a shift toward greater gender equality as women move into higher-status economic roles in management and gain political influence within elected and appointed bodies. Thus, relatively industrialized Muslim societies such as Turkey share the same views on gender equality and sexual liberalization as other new democracies.

Even in established democracies, changes in cultural attitudes—and eventually, attitudes toward democracy—seem to be closely linked with modernization. Women did not attain the right to vote in most historically Protestant societies until about 1920, and in much of Roman Catholic Europe until after World War II. In 1945, only 3 percent of the members of parliaments around the world were women. In 1965, the figure rose to 8 percent, in 1985 to 12 percent, and in 2002 to 15 percent.

The United States cannot expect to foster democracy in the Muslim world simply by getting countries to adopt the trappings of democratic governance, such as holding elections and having a parliament. Nor is it realistic to expect that nascent democracies in the Middle East will inspire a wave of reforms reminiscent of the velvet revolutions that swept Eastern Europe in the final days of the Cold War. A real commitment to democratic reform will be measured by the willingness to commit the resources necessary to foster human development in the Muslim world. Culture has a lasting impact on how societies evolve. But culture does not have to be destiny.

Out of Madness, A Matriarchy

They survived machetes and mass rapes.
Now, Rwanda's women—nearly two-thirds of the population—are learning
how to lead their country out of the darkness.

Kimberlee Acquaro & Peter Landesman

ON APRIL 7, 1994, when the genocide was in its second day, Joseline Mujawamariya, then 17, huddled with her twin sister and younger brother in the tall grass on the outskirts of her village, Butamwe, in central Rwanda. They hid for three days as Hutu men and boys they had grown up with, armed with machetes, began a rampage of butchery and rape, burning homes and hunting down their Tutsi friends and neighbors. Then the Hutus set fire to the fields.

> Joseline Mujawamariya was a teenager when her parents were killed in the massacres of 1994. Today she is her town's top development official—one of thousands of women who are taking charge of rebuilding Rwanda.

Joseline waited until nightfall, and as black smoke blotted the last light from the sky, she and the children fled. They joined a group of refugees that moved under the cover of night, living for more than two months on food scavenged from corpse-littered gardens and rainwater collected in cupped hands. Worse than the starvation and fatigue was the terror, Joseline says. "The Hutus would stop you to look at your fingers, your nose and ears to see if you were Tutsi." In late June, the Tutsi-led Rwandan Patriotic Front, invading from neighboring Uganda and Burundi, began to sweep through the country, driving out the Hutus. Ragged groups of refugees began to trek home, Joseline among them. When Joseline reached Butamwe, she found a ghost town. "There was no one left," she says.

One morning last fall, Joseline stood on a hilltop overlooking Butamwe, her five-month-old son tied to her back. Before her stretched an undulating sea of banana groves and valleys creeping with hanging lakes of mist. Columns of smoke from cooking fires and controlled burns seemed to dangle groundward from the sky. A hundred Tutsi survivors were building a road over the mountainside to the capital, Kigali. Machetes arched through the high grass. Women with infants sleeping on their backs chopped through the rocky ground with hoes.

Joseline was their leader. Now 25 and a mother of three, with only a primary school education, she was elected in 1999 as the area's head of development, after twice campaigning for other positions. "I didn't know what I was supposed to do when I was first elected," she said. "I thought I was supposed to buy a cow for the village." What she now does is supervise the reconstruction of Butamwe's shattered infrastructure, as well as its health services and systems of justice, education, culture, and economy. She is rebuilding her neighbors' lives while struggling to rebuild her own. From the hill, Joseline pointed to the ruins of a mud structure crumbling in an overgrown field. "That was my home," she said quietly, "where my parents and my brothers and sisters were all killed."

THE 1994 GENOCIDE, one of the worst mass slaughters in recorded history, was triggered by the assassination of Rwanda's Hutu president, after a lengthy civil war between the Hutu-led government and the Tutsi-dominated Rwandan Patriotic Front. It was a deliberate effort to eliminate the country's Tutsi "problem"; books about Hitler and the Holocaust, and lists of potential victims, were later discovered in the offices of top government officials. In all, at least 1 million Tutsis and moderate Hutus died.

But it isn't just the numbers that set the genocide apart from other horrors of the late 20th century. The ferocity and concentration of the killing were unprecedented, as was its intimate methodology. The murderers were neighbors, relatives, teachers, doctors, even nuns and priests, and they killed not with machine guns or gas chambers, but with machetes, clubs, knives, and their bare hands. So many men were killed that Rwanda was left overwhelmingly female and became a nation of traumatized widows, orphans, and mothers of murdered children. Even today, the population remains 60 percent female.

"Men think this is a revolution," says Angelique Kanyange, a student leader at Rwanda's national university. "It's not a revolution—it's a development strategy."

Among the most nefarious tools of the genocide was a planned mass sexual assault on Tutsi women, with Hutu officials encouraging HIV-positive soldiers to take part in gang rapes. The United Nations has estimated that at least 250,000 women were raped, most of them repeatedly and over the course of weeks or months. (Some women we met remember being raped up to five times every day for 10 weeks.) Most of those women were killed afterward, but others were purposely kept alive to give birth to a population of fatherless "un-Tutsi." According to one study by AVEGA, an association of genocide widows, 70 percent of women who survived the rapes—and many of their children—now have AIDS.

The genocide lasted three months, from April to June of 1994. It left Rwanda in physical ruins, completely destroying the country's political, economic, and social structures. In a culture that historically prohibited its female population from performing the most rudimentary chores—from climbing on roofs to milking cows—women were now forced to take on tasks that had previously been out of reach. The result has been an unplanned—if not inadvertent—movement of female empowerment driven by national necessity. Traveling in Rwanda for three months last year, we found women heading households and businesses, serving as mayors and assuming cabinet positions. Rwanda's Parliament is now 25 percent female, by far the highest proportion of women in national leadership in the world outside Scandinavia and nearly double the share of women in the U.S. Congress. "Men think this is a revolution," says Angelique Kanyange, a student leader at Rwanda's national university. "It's not a revolution; it's a development strategy."

AFTER THE GENOCIDE ENDED, about 800,000 Tutsi refugees returned to Rwanda, swelling a rebounding flood of 2 million Hutus who had fled in fear of Tutsi retribution. Waiting for them was a population of "living dead,"

dazed and grief-stricken survivors. In a country of only 7 million, national reconstruction required more than replacing murdered leaders and rebuilding razed homes. It was "a process of deconstruction as much as reconstruction—deconstructing ethnic perceptions, deconstructing gender stereotypes," says Jack Hjelt, the former chief of the United States' post-genocide humanitarian and development program in Rwanda. "From the destruction comes a unique opportunity to rebuild from the ground up."

Before the genocide, women in positions of power had been rare, though not unheard of. (Rwanda's prime minister at the time, a moderate Hutu who opposed the mass killing of Tutsis, was a woman; she was murdered and sexually mutilated in the first hours of the slaughter.) Now 18 percent of all top government jobs are held by women. Four women hold cabinet posts, including Angelina Muganza, who has been appointed to head the newly created Ministry for Women in Development.

"There is a history of division and misinformation particularly for illiterate, undereducated women in our country," Muganza told us. Before the genocide, 55 percent of Rwanda's women were illiterate (48 percent of the men couldn't read), and many became easy prey to a relentless propaganda campaign dominated by newspaper cartoons portraying Tutsi women as salacious harlots set on seducing Hutu men and subverting Hutu families. With their wives' misinformed consent, Hutu men were encouraged to rape and kill Tutsi women for the sake of, Hutu "unity."

Akimana's mother was one of an estimated 2,500 Rwandan women who became pregnant after being gang-raped. The girl's name means "child of God."

Joseph Nzabirinda, a benevolent-looking 48-year-old former school bus driver, told us that he'd killed so many Tutsis he'd lost count. Joseph wore pink shorts and matching pink shirt, the universal prisoner's uniform in Rwanda. He was awaiting sentencing by the local courts. Sitting straight-backed on his chair, his palms resting on his knees, he told us about his youngest victim, a one-year-old girl. "I had a boy the same age," he said, struggling to compose himself. "I stood at the edge of a latrine and I remember dropping the baby down the hole and throwing stones on top of her. She was alive. When I threw her, I could hear she was still screaming.

"My wife knew I was killing Tutsis," he said. "She told me, 'If you don't go, they'll come here and kill me.' She told me to go kill. Go so they don't come."

In some cases, women more directly participated in, even orchestrated, the slaughter. Last year, two Rwandan nuns, Gertrude Mukangango and Maria Kisito Mukabutera, were tried in Belgium and convicted of murder for their roles in the massacre of 7,000 Tutsis who had taken shelter in their convent. Rwanda's minister for family and women's affairs at the time the genocide began, Pauline Nyiramasuhuko, is currently on trial at the International Criminal Tribunal for Rwanda in Arusha, Tanzania. She is alleged to have personally directed squads of Hutu men to torture and butcher Tutsi men, and to rape and mutilate Tutsi women.

> Odette Mukakabera, one of 200 women in Rwanda's new national police force, is also raising four children, going to law school, and speaking publicly about living with HIV.

Today, Rwanda's women and girls have one of the highest literacy rates in Africa—61 percent. Boys and girls attend school at about the same rate, nearly 70 percent; before the genocide, boys outnumbered girls 9 to 1. Nearly half of university graduates are women, compared to 6 percent just 10 years ago. And for the first time, women have the right to own property; in the past, they could not keep their homes, or even their children, when their husbands or male relatives died.

Delphine Umutesi was 10 when she watched her father butchered by Hutus in her home. She was the eldest of five children, the youngest less than a year old. After the genocide, the children were housed in an orphanage. When it closed, Delphine, barely 14, found herself one of Rwanda's 65,000 underage heads of household, a child taking care of other children. "I asked myself, 'How will we live?'"

She moved her siblings back into their family's mud home in rural Kigali. Technically she was a squatter there; then, in 1999, Parliament passed a law allowing women to inherit land, and she now owns the property. She has no time to go to school herself, but each morning she sends the others off to class and walks several miles to work. She feeds her family by making greeting cards from banana leaves—a micro-business that looks modest on the surface, but would have been impossible before the genocide, when women were not allowed to earn money independently. "I try to help them with their homework," she says. "But soon they'll know more than I learned in school."

AFTER JOSEPHINA MUKAHKUSI'S HUSBAND, five daughters, and two sons were slaughtered with machetes before her eyes, their Hutu killers beat her, then dumped her with her girls into a pit latrine, leaving her for dead. Eventually, she was rescued and lifted to the surface drenched in her daughters' blood. Now 45, Josephina can't recall if she lay there for hours or days. What she does remember is the desolation of being left alive, childless and alone. "The Hutu women looted after their husbands killed," she told us. "Many of those women were my friends. We were godmothers to each other's children."

After the genocide, afraid to return home alone, Josephina moved into a shelter where she met and adopted a young Hutu orphan named Jane. Josephina and Jane have since returned to her village, where they live among people who took part in the killings. "We are raising their children," she told us. "Jane's mother was Hutu, but Jane is innocent." Her hope, she said, is that Jane will grow up thinking of herself as Rwandan rather than Hutu or Tutsi.

Severa Mukakinani was forced to watch her seven children butchered, then was gang-raped repeatedly by their murderers. "The raping went on for a long time—I don't know how long," Severa, now 43, told us. "When they tired of me, they cut and beat me and threw me in the river." She was left for dead, but survived to find that she was pregnant from the rapes. "I wanted to remove the baby. I decided to keep it, because I believe the child is innocent." She named her daughter Akimana, which means "child of God."

"Rwanda is like no other situation in the world," says Barbara Ferris, founder and president of the International Women's Democracy Center, which trains women to participate in politics in emerging democracies. Rwandan women have come this far, she says, in part by legally redefining women's roles. "It's logical that women move into formal leadership roles," Ferris says. "Informally, women already are leaders. They want what every other woman wants for her children: a better life than they had, access to education, food on the table, and a roof over their head. If they have to enter politics to do it, that's what they will do."

> For the first time, Rwandan women have the right to own property; in the past, they could not keep their homes, or even their children, when their husbands died.

At the State House in Kigali last June, Rwanda's President Paul Kagame became animated when we asked him about the role of women. "Now we want to encourage women's responsibility, not to just look to men," he said. "We know in our conscience that there are women who can go to high levels of leadership, whereas before, women didn't even know they should be playing leadership roles. So their consciousness is developing." He believes, he said, that women's empowerment is a national imperative. Rwanda's new constitution, now being drafted, will include passages that specifically address women's rights.

Of Rwanda's 106 mayors, five are now women. (There were none before the genocide.) One of them, Specioze Mukandutiye, spoke to us in her office in Save, in southern Rwanda. She bears a particular responsibility to rebuild the country, she said soberly, because she is a Hutu. Her Tutsi husband was murdered, though her four children survived. "Many times the Hutus tried to kill us," she said, "because I was a moderate and any Hutu who didn't follow them was killed."

After the genocide, Specioze started a support group for genocide widows; she was the only Hutu member. She was appointed mayor by the government in 1995, then elected in 2001 by a mostly Tutsi constituency. In her years in office, Specioze told us, her ethnicity has been less of a barrier than her gender—though once exposed to women in leadership positions, she added, "the men began to understand that women are as able as men."

Still, she noted with a mixture of sadness and irony, she would have gone nowhere politically had her husband survived. "I would have had to follow his instruction."' It's difficult to balance her work and family, she told us, but she was determined to do it because "now is the moment for Rwandan women to demonstrate they can do the same jobs as men, and have the same value."

Odette Mukakabera is one of the 200 women in Rwanda's new 5,000-member national police force (the second-highest police official, the assistant commissioner, is also a woman). A teacher for 10 years before the civil war, Odette answered a recruitment call for female officers in 2000. Since joining the force, she has also gone public about the fact that she has AIDS, a rare step for women in Africa. Her husband died of AIDS shortly after the genocide, leaving her to raise four children alone while attending law school full time. "I am writing my dissertation on the rights of people living with HIV," she told us in her office in police headquarters in Kigali. "I speak publicly about my condition to raise awareness. It is diffi-

cult to do all this—to be a mother and policewoman and student. But I am trying as long as I am still strong."

Two years ago, Specioze Mukandutiye was elected mayor of her town, an office she couldn't have sought if her husband were still alive.

Odette has not experienced discrimination within the force, she said, but out on the beat people often seem surprised to see a female officer. To most Rwandans, Deputy Police Commissioner Dennis Karera told us, women in uniform are still a novelty: "It's a transformation process that everyone is going through." But female officers are often more effective than men at solving crimes, he added. Rape and domestic violence are now reported more often than ever. "At first, women asked us, 'How will we do this?' We answered, 'You don't need muscles to do investigations.'"

Whatever their position, says Minister for Women Muganza, women throughout Rwanda are learning to use their new authority to improve their, and their children's, lives. "Rwanda's women work together despite their differences," she says. "Their attitude is, 'What has happened has happened, but let's make the future different.'"

Severa Mukakinani told us that she knows her daughter will have a better life, even though the present is a struggle. "I had seven children before, but I had a husband to feed them," she said. "Now I only have one, but I can barely put breakfast on the table." Many of the women we spoke to echoed the same sentiment. History, they said, has handed them both an extraordinary burden and an unprecedented opportunity. "Rwanda's future," Severa said, "is on our backs."

From *Mother Jones,* January/February 2003, pp. 59-63. © 2003 by Foundation for National Progress. Reprinted by permission.

Index

Index

Test Your Knowledge Form

We encourage you to photocopy and use this page as a tool to assess how the articles in *Annual Editions* expand on the information in your textbook. By reflecting on the articles you will gain enhanced text information. You can also access this useful form on a product's book support Web site at *http://www.dushkin.com/online/*.

NAME:

DATE:

TITLE AND NUMBER OF ARTICLE:

BRIEFLY STATE THE MAIN IDEA OF THIS ARTICLE:

LIST THREE IMPORTANT FACTS THAT THE AUTHOR USES TO SUPPORT THE MAIN IDEA:

WHAT INFORMATION OR IDEAS DISCUSSED IN THIS ARTICLE ARE ALSO DISCUSSED IN YOUR TEXTBOOK OR OTHER READINGS THAT YOU HAVE DONE? LIST THE TEXTBOOK CHAPTERS AND PAGE NUMBERS:

LIST ANY EXAMPLES OF BIAS OR FAULTY REASONING THAT YOU FOUND IN THE ARTICLE:

LIST ANY NEW TERMS/CONCEPTS THAT WERE DISCUSSED IN THE ARTICLE, AND WRITE A SHORT DEFINITION:

We Want Your Advice

ANNUAL EDITIONS revisions depend on two major opinion sources: one is our Advisory Board, listed in the front of this volume, which works with us in scanning the thousands of articles published in the public press each year; the other is you—the person actually using the book. Please help us and the users of the next edition by completing the prepaid article rating form on this page and returning it to us. Thank you for your help!

ANNUAL EDITIONS: Developing World 04/05

ARTICLE RATING FORM

Here is an opportunity for you to have direct input into the next revision of this volume.
We would like you to rate each of the articles listed below, using the following scale:

1. **Excellent: should definitely be retained**
2. **Above average: should probably be retained**
3. **Below average: should probably be deleted**
4. **Poor: should definitely be deleted**

Your ratings will play a vital part in the next revision.
Please mail this prepaid form to us as soon as possible.
Thanks for your help!

RATING	ARTICLE	RATING	ARTICLE
_____	1. The Great Divide in the Global Village	_____	35. The Population Implosion
_____	2. The Poor Speak Up	_____	36. Local Difficulties
_____	3. Institutions Matter, but Not for Everything	_____	37. A Dirty Dilemma: The Hazardous Waste Trade
_____	4. Development as Poison: Rethinking the Western Model of Modernity	_____	38. 'Undoing the Damage We Have Caused'
_____	5. Why People Still Starve	_____	39. Withholding the Cure
_____	6. Putting a Human Face on Development	_____	40. Empowering Women
_____	7. The Free-Trade Fix	_____	41. Women & Development Aid
_____	8. Trading for Development: The Poor's Best Hope	_____	42. Women Waging Peace
_____	9. Rich Nations' Tariffs and Poor Nations' Growth	_____	43. The True Clash of Civilizations
_____	10. Unelected Government	_____	44. Out of Madness, a Matriarchy
_____	11. The IMF Strikes Back		
_____	12. Ranking the Rich		
_____	13. Eyes Wide Open: On the Targeted Use of Foreign Aid		
_____	14. The Cartel of Good Intentions		
_____	15. The WTO Under Fire		
_____	16. Playing Dirty at the WTO		
_____	17. Market for Civil War		
_____	18. Engaging Failing States		
_____	19. Progressing to a Bloody Dead End		
_____	20. An Indian 'War on Terrorism' Against Pakistan?		
_____	21. The Terror War's Next Offensive		
_____	22. Mugabe's End-Game		
_____	23. Tashkent Dispatch: Steppe Back		
_____	24. North Korea: The Sequel		
_____	25. Blaming the Victim: Refugees and Global Security		
_____	26. Democracies: Emerging or Submerging?		
_____	27. Two Theories		
_____	28. Not a Dress Rehearsal		
_____	29. One Country, Two Plans		
_____	30. Iran's Crumbling Revolution		
_____	31. New Hope for Brazil?		
_____	32. Latin America's New Political Leaders: Walking on a Wire		
_____	33. The Many Faces of Africa: Democracy Across a Varied Continent		
_____	34. NGOs and the New Democracy: The False Saviors of International Development		

(Continued on next page)

BUSINESS REPLY MAIL
FIRST CLASS MAIL PERMIT NO. 551 DUBUQUE IA

POSTAGE WILL BE PAID BY ADDRESEE

McGraw-Hill/Dushkin
2460 KERPER BLVD
DUBUQUE, IA 52001-9902

ABOUT YOU

Name _____ Date _____

Are you a teacher? ☐   A student? ☐
Your school's name
_____

Department
_____

Address _____ City _____ State _____ Zip _____

School telephone # _____

YOUR COMMENTS ARE IMPORTANT TO US!

Please fill in the following information:
For which course did you use this book?
_____

Did you use a text with this ANNUAL EDITION?   ☐ yes ☐ no
What was the title of the text?
_____

What are your general reactions to the *Annual Editions* concept?
_____

Have you read any pertinent articles recently that you think should be included in the next edition? Explain.
_____

Are there any articles that you feel should be replaced in the next edition? Why?
_____

Are there any World Wide Web sites that you feel should be included in the next edition? Please annotate.
_____

May we contact you for editorial input?   ☐ yes ☐ no
May we quote your comments?   ☐ yes ☐ no